Physics of Baseball & Softball

Rod Cross

Physics of Baseball & Softball

Rod Cross
School of Physics
University of Sydney
Sydney, New South Wales
Australia
cross@physics.usyd.edu.au

ISBN 978-1-4419-8112-7 e-ISBN 978-1-4419-8113-4
DOI 10.1007/978-1-4419-8113-4
Springer New York Dordrecht Heidelberg London

Library of Congress Control Number: 2011921934

Printed on acid-free paper

Springer is part of Springer Science+Business Media (www.springer.com)

Preface

The number of baseball and softball fans in the world is probably around 100 million. The number of people who are interested in physics might also be about 100 million. In theory, therefore, this book should appeal to somewhere between 100 million and 200 million people. However, the number of people who are seriously interested in both physics and baseball or softball is somewhat less than this. Only a handful of physicists in the world have actually conducted serious studies of the physics of baseball and softball. Not because the subject does not interest them, but because they are usually too busy doing other serious physics work. If they were caught out doing fun baseball experiments on the side, it might give the false impression that they were not being properly employed. Partly because of the nature of the subject, there have been many more engineers and biomechanists and even historians and economists who have engaged in academic studies of baseball and softball.

While baseball is known as the national pastime in the USA and softball is even more popular in terms of the total number of players, sport is not a high priority area when it comes to government or even private funding of physics research. Nevertheless, physics laboratories are usually sufficiently well equipped for anyone who is so inclined to sneak in some sports research on the side. That is how I first managed to get involved, in 1995. I found it absolutely fascinating and I still do. Part of the fascination is in discovering things that were not previously known. The physics of sport is not a rich field for "new" physics, but it is fun using "old" physics to provide new insights into some of nature's mysteries. The bounce of a ball is just one of those mysteries. Very little was known about the subject when I started in 1995, apart from some early work done by Sir Isaac Newton around 1670 and a few additional studies during the next 300 years. Much more was known about the flight of balls through the air. My background before getting sidetracked into the physics of sport was 30 years experience in high temperature plasma physics research. It had no particular relevance to baseball or softball, apart from the fact that it helps to teach and study physics for 30 years or more to get on top of the subject.

In 1990, Professor Robert Adair at Yale University wrote a very popular book called "The physics of baseball." It is currently in its third edition and provides an easy-to-read and entertaining account of the subject. During the last 15 years there have been many advances in our understanding of the physics and engineering of baseball and softball, and there is now room for a second book on the subject. Given that baseball and softball are both very similar sports, the physics of one is essentially the same as the physics of the other. In fact, the physics of sport is essentially the same as the physics of many other topics and likely to be a lot more interesting to anyone with an active interest in sport. Basic mechanics features prominently in this book, almost all of the examples being taken from baseball and softball. I would have achieved a useful objective if some of the material finds its way into classrooms.

The wide range of interests and abilities of the 100 million baseball and softball fans in the world presented me with a problem. Even if I assumed that only 10 million of those fans were interested in physics, only a tiny fraction of that number would have studied physics at University level. I have, therefore, written the book assuming that most readers will have a basic understanding of high school physics, but are not necessarily proficient at that level. One of my objectives is to try to boost that proficiency by emphasizing the physics issues in baseball and softball in more detail than in Adair's book. Professor Adair achieved an excellent result in explaining baseball in terms of the known laws of physics. A difference between his book and mine is that I have placed greater emphasis on explaining the physics in terms of the known behavior of bats and balls. I have tried to discuss the physics in a conversational manner, using simple equations where necessary, but I have also included more advanced material in the Appendices at the end of each chapter. That way, the reader can skip the hard parts or can refer to them later, depending on his or her prior knowledge of, and interest in, physics and mathematics.

I am especially grateful to Professor Alan Nathan at the University of Illinois for his assistance in helping me prepare this book. Alan and I have collaborated on many physics of sports projects over the years, despite living 9,240 miles apart. Both of us maintain web sites that contain additional material on the physics of baseball and softball, including some interesting video film of various topics described in this book. The sites are www.physics.usyd.edu.au/~cross and http://go.illinois.edu/physicsofbaseball. Professor Lloyd Smith also provided valuable assistance, especially on the physics of softball, and has a very nice site at www.mme.wsu.edu/~ssl, as does Professor Dan Russell at http://paws.kettering.edu/~drussell/bats.html. A web search on "physics of baseball" will reveal thousands of other sites, indicating that there are indeed many thousands of people actively interested in the topic.

Sydney University
August 2010

Rod Cross

Contents

Chapter 1
Basic Physics

Most of the physics in this book can be understood in terms of basic high school mechanics and slightly beyond. This chapter is provided for those who might need some extra guidance or a kick-start. The author has taught a course in sports mechanics for physical education teachers for many years. Many of the students arrive at University without ever having studied physics and only a small amount of mathematics. Some arrive after being in the work force for a few years and have forgotten everything they knew from high school. Those students take a few weeks to get the hang of it and usually do quite well after extra help from their tutors. This chapter contains material that is taught to such students, and also at high school, but it comes with a warning. In most countries around the world, including the USA, physics is taught using the MKS or SI system of units, the basic units being metres, kilograms, and seconds. In baseball and softball, length is more commonly measured in inches, feet, and miles, while mass or weight is commonly measured in ounces or pounds. Many of the physics equations in this book are described using MKS units and the answers are then translated into English units. The advantage of the MKS system is that the mathematics is easier, since 10 mm = 1 cm, 100 cm = 1 m, and 1,000 m = 1 km. Similarly, 1,000 g = 1 kg. In the English system, 12 in. = 1 ft, 5,280 ft = 1 mile, and 16 oz = 1 lb, which makes the math a little more complicated. Things get even more complicated when slugs and poundals are introduced to get around the problem that a pound is commonly used both as a unit of mass and a unit of weight or a force. In physics, mass and force are very different things. A list of conversion factors is given at the end of the book.

1.1 Linear Motion

Linear motion normally refers to motion along a straight line. A batter running from one base to the next is an example. A baseball or a softball does not normally move in a straight line, but the force of gravity on the ball is vertical at all times and is zero in the horizontal direction. The effect of gravity on the ball can be treated by regarding the motion of a ball as a combination of linear motion at constant speed

R. Cross, *Physics of Baseball & Softball*, DOI 10.1007/978-1-4419-8113-4_1,

in the horizontal direction, plus linear, accelerated motion in the vertical direction. In the real world, air resistance also acts on the ball but the effect is relatively small at low ball speeds.

Speed

The average speed of an object is given by the formula speed = distance/time. For example, if a baseball takes 4 s to travel 400 ft then the average speed of the ball is $400/4 = 100\,\mathrm{ft\,s^{-1}}$ (Fig. 1.1). To convert that result to other units, we can use the following conversion factors: $1\,\mathrm{ft\,s^{-1}} = 0.6818\,\mathrm{mph} = 0.3048\,\mathrm{m\,s^{-1}}$. For example, if we multiply by 100, then $100\,\mathrm{ft\,s^{-1}} = 68.18\,\mathrm{mph} = 30.48\,\mathrm{m\,s^{-1}}$.

We use the phrase "average speed" here because the ball slows down through the air. The ball might be struck at $110\,\mathrm{ft\,s^{-1}}$ and slow down to $90\,\mathrm{ft\,s^{-1}}$ by the time it lands. Using a video camera we could record the position of the ball as it travels through the air and measure the actual decrease in speed due to air resistance. Suppose that near the end of its flight, the ball moves 3 ft from one frame to the next. Video film is recorded at 30 frames $\mathrm{s^{-1}}$, so the time between frames is $1/30 = 0.033$ s. The average speed over this time is $3/0.033 = 90.9\,\mathrm{ft\,s^{-1}}$. The formula for the speed in this case is $v = \mathrm{d}x/\mathrm{d}t$ where $\mathrm{d}x$ is the change in position = 3 ft and $\mathrm{d}t$ is the change in time = 0.033 s.

Mathematically, the expression $\mathrm{d}x/\mathrm{d}t$ is called the derivative of x with respect to t. If you have studied calculus in math you will recognize this. If you haven't studied calculus then at least you now know what a derivative is. It is just a small increase in one quantity divided by a small increase in another quantity. In physics, "small" here means something that is small enough for practical purposes. In mathematics, "small" is much smaller. For example, if the ball travels 1 in. in 0.001 s, then that would definitely be small enough for a physicist, but it would still be too large for mathematicians. They would prefer that $\mathrm{d}x$ and $\mathrm{d}t$ are both infinitesimally small, but then it would be impossible to measure such tiny changes in x and t.

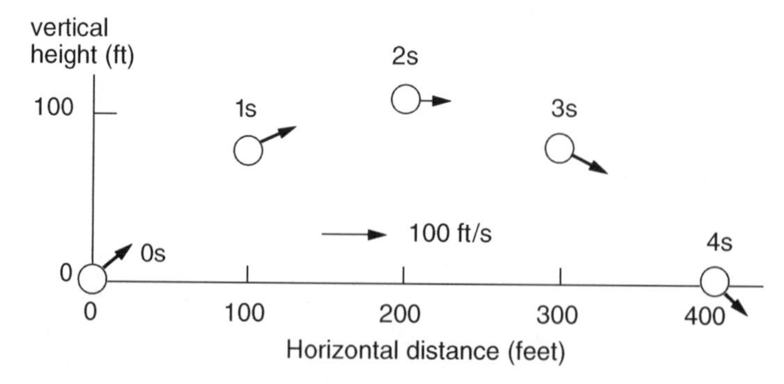

Fig. 1.1 A baseball will take about 4 s to travel 400 ft horizontally, at $100\,\mathrm{ft\,s^{-1}}$ in the horizontal direction. The position of the ball is shown at 1 s intervals

When a baseball travels 400 ft horizontally through the air, it rises to a height of about 110 ft in about 2 s and drops back to the ground during the next 2 s. The average horizontal speed is 100 ft s^{-1}. The average vertical speed on the way up or on the way down is $110/2 = 55$ ft s^{-1}. When the ball gets to its maximum height it comes to a stop in the vertical direction. At that point the vertical speed is zero (since it is not actually moving up or down at that instant) but the horizontal speed is still 100 ft s^{-1}. Given that the average vertical speed is 55 ft s^{-1}, the batter actually struck the ball with a vertical speed of 110 ft s^{-1} (since the average of 0 and 110 is 55).

Batter Decision Time

In baseball, the ball can be pitched at speeds up to 90 mph or 132 ft s^{-1}. The front edge of the pitcher's plate is 60.5 ft from the rear point of home base, and 10 in. above it, but the pitcher leans forward and releases the ball about 5 ft from the pitcher's plate. Similarly, the batter strikes the ball about 2 ft in front of the rear point of home base so the ball travels about 53 ft from the pitcher's hand to the bat. The batter starts the early part of the swing even before the pitcher releases the ball, but the final decision as to where to swing the bat must wait until the pitcher releases the ball. The decision time, while the ball is in the air, can be calculated from the ball speed. If the average ball speed was 90 mph, then the time is $53/132 = 0.40$ s. But the ball slows down through the air by about 10% and arrives at home plate at about 81 mph. The average ball speed is then 85.5 mph or 125 ft s^{-1} so the ball travel time is actually about $53/125 = 0.42$ s.

Acceleration

The acceleration due to gravity is $g = 9.8$ m s^{-2} (or 32 ft s^{-2}). The value of g gives the increase in speed each second when an object falls to the ground. After 1 s, the speed of a ball dropped from a large height will be 9.8 m s^{-1} (or 32 ft s^{-1}). After 2 s the speed will be $2 \times 9.8 = 19.6$ m s^{-1}. If a ball is dropped from a small height, say 2 m, then it will take only 0.64 s to hit the ground. It starts off with zero speed. After 0.1 s, it has accelerated to 0.98 m s^{-1}. After 0.2 s, the ball speed is $0.2 \times 9.8 = 1.96$ m s^{-1}, as shown in Fig. 1.2. After 0.6 s, the ball speed is $0.6 \times 9.8 = 5.88$ m s^{-1}. The formula for the ball speed here is $v = 9.8\,t$, and the fall height is $y = 4.9\,t^2$, assuming that the ball starts at $t = 0$ with $y = 0$ and $v = 0$.

When a batter swings a bat, the tip of the bat increases in speed from about 5 m s^{-1} to about 30 m s^{-1} over the last 0.2 s of the swing. The tip moves in an approximately circular path. The average acceleration, a, of the tip along that path is given by the formula a = increase in speed/increase in time. The increase in speed is 25 m s^{-1}. The increase in time is 0.2 s. So $a = 25/0.2 = 125$ m s^{-2}, which is 12.7

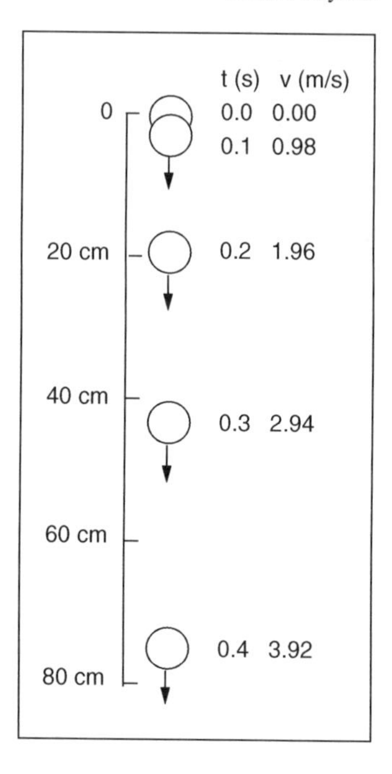

Fig. 1.2 A falling ball accelerates as it falls. The speed and position of the ball are shown here at intervals of 0.1 s after release

times larger than g. The average acceleration is therefore 12.7 g. The acceleration is not constant during the whole swing. Over any small interval of time, $\mathrm{d}t$, the speed increases by $\mathrm{d}v$ and $a = \mathrm{d}v/\mathrm{d}t$.

Momentum

The momentum of an object of mass m moving at speed v is defined to be m multiplied by v. If an object has a large amount of momentum then it has a large mass or a large speed or both. If one object (a bat) is about to collide with another (a ball), then the total momentum of the two objects is just the sum of the two separate values, taking the sign of the momentum into account. For example, the momentum of a bat traveling left to right might be $+20\,\mathrm{kg\,m\,s^{-1}}$ and it might collide with a ball traveling in the opposite direction with momentum $-2\,\mathrm{kg\,m\,s^{-1}}$. The total momentum just before the collision is $+18\,\mathrm{kg\,m\,s^{-1}}$ (Fig. 1.3).

Momentum is an important quantity when describing collisions since the total momentum doesn't change during a collision. The bat will slow down and the ball will turn around and head off in the same direction as the bat after the collision, but the total momentum will still be $+18\,\mathrm{kg\,m\,s^{-1}}$ after the collision. Momentum lost by the bat is given to the ball. We describe this effect as "conservation of momentum."

For a baseball or softball bat, the tip and the handle of the bat usually move at different speeds. In that case, we define the momentum of the bat as the mass of

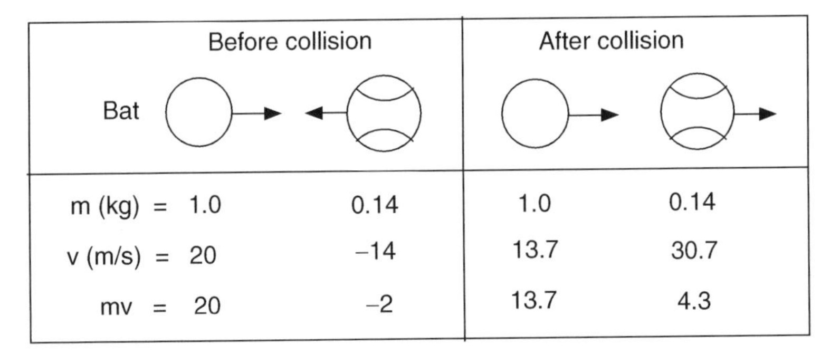

Fig. 1.3 A head-on collision between a bat and ball. The total momentum (mv) after the collision (18 kg m s^{-1} here) is the same as that before the collision. The total mass remains the same, but the total v does not remain the same, nor does the relative velocity remain the same

the bat multiplied by the speed of its center of mass (CM). The CM of a bat can be found by balancing the bat across a rod or tube. The CM is directly above the balance point, on the long axis of the bat.

Force

In the absence of any force acting on an object, an object at rest will remain at rest, and an object that is moving will continue to move at a constant speed. If Newton had lived in the USA, he might have described this result as his First Law of Baseball. If there is no net force on an object then there is no acceleration. If you stand still on the ground then the force of gravity pulls you down but the ground exerts an equal and opposite force upward. The total force is zero so you remain at rest.

In order for an object to accelerate there must be a net force acting on the object. The force is given by $F = ma$ where m is the mass of the object. This is Newton's Second Law of Baseball. If there are two or more forces acting on the object, then F is the total force on the object. The object accelerates in the same direction as the total force acting on the object.

If a baseball is flying through the air then it follows a curved path since gravity pulls it back down to earth. The acceleration of the ball in the vertical direction is $g = 9.8\ \mathrm{m\,s^{-2}}$ and the vertical force on the ball is $F = mg$. If $m = 0.142$ kg then $F = 0.142 \times 9.8 = 1.39$ N.

The units here are important, at least when describing the physics of the situation. The words are also important. The mass of the baseball is 0.142 kg. The force of gravity acting on the ball is called its weight. The weight of the baseball is 1.39 N.

In everyday use, we say that a baseball weighs about 5 oz or 142 g. That is actually the mass of the ball. In everyday use, mass and weight are treated as the same thing. We might say that a bat weighs 30 oz or the batter pushes on the bat with a force of 20 lb. We know what people mean when they say this, and there is no real confusion in conversational terms, but it is technically incorrect and leads to problems when doing physics calculations. If a 160 lb astronaut weighed himself in the weightless environment of a space vehicle then he would weigh nothing on a set of scales. His mass would still be 160 lb but his weight would be zero since he would exert no force on the scales.

Force on a Baseball

The force exerted on a baseball can be estimated from its change in speed. In a big hit, the ball will be pitched at about 90 mph and will struck at about 110 mph. The collision time is only about 0.001 s. The change in velocity is 90 + 110 = 200 mph (89.4 m s^{-1}) so the average acceleration $a = 89{,}400$ m s^{-2}. The mass of a baseball is 0.145 kg so the average force on the baseball is $ma = 12{,}963$ N or 2,914 lb (since 1 N = 0.2248 lb). During the collision the force increases from zero to a maximum value and drops back to zero. The maximum force is about twice the average value and is therefore about 5,800 lb. If the collision time is as short as 0.6 ms then the maximum force could be as large as 9,700 lb.

Work and Energy

When a force is exerted on an object then the object accelerates in the direction of the force. Suppose you throw a baseball of mass m by pushing the ball forward through a distance d with a force F (Fig. 1.4). The work, W, done on the ball is defined to be $W = Fd$ and the answer is expressed in Joules. Starting from rest, the ball will accelerate up to a speed v given by $W = \frac{1}{2}mv^2$. On release, the ball will continue to travel through the air at speed v. The quantity $\frac{1}{2}mv^2$ is called the kinetic energy (KE) of the ball and it is also measured in Joules. In this case, the kinetic energy acquired by the ball is equal to the work done when you throw it. For example, if $m = 0.145$ kg (5.1 oz) and $v = 40$ m s^{-1} (90 mph) then $W = KE = 0.5 \times 0.145 \times 40^2 = 116$ J. The force needed to throw the ball at this speed can be estimated in terms of the throw distance d. If $d = 2$ m, then $F \times 2 = 116$ so $F = 58$ N (13 lb).

Power

Power is a term in physics that refers to the amount of energy that is delivered or consumed in 1 s. It is measured in Watts. For example, a 100 W light bulb is one that uses 100 J of electrical energy every second. A small 2 kW engine delivers 2,000 J of energy every second.

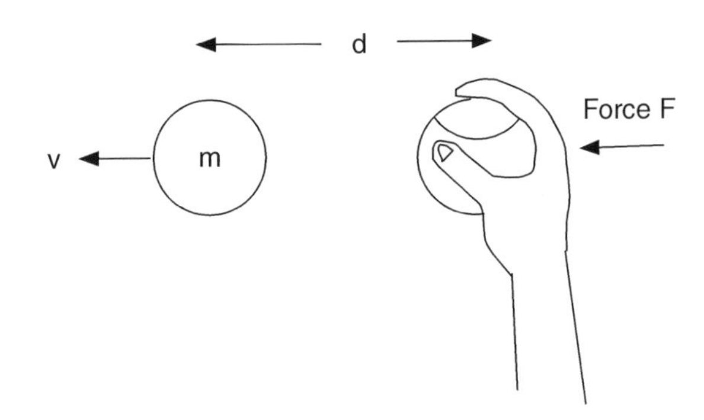

Fig. 1.4 A ball is thrown by exerting a force F on the ball and by maintaining that force over a distance d. The work done is Fd. The speed of the ball on release is given by $\frac{1}{2}mv^2 = Fd$ where $\frac{1}{2}mv^2$ is the kinetic energy of the ball

In common usage, power can mean lots of things. The USA is a powerful country in terms of its military might, and there are a lot of powerful people in the USA. An atomic bomb is a very powerful weapon. People refer to mind power or spiritual power or the power of positive thinking. A boxer can pack a powerful punch. Heavy bats are more powerful than light bats. And so on. Not all of these power terms mean the same thing as "power" in physics or engineering terms.

For example, consider the "power" of a baseball bat. All the energy gained by the bat is supplied by the batter. The bat is just an instrument that helps send the ball on its way. If it does its job well, then we usually say that the bat is powerful. In physics terms we should really describe the bat in terms of its efficiency. An efficient bat would be one that allows the batter to transfer the energy in his arms to the ball without too much loss of energy in the process. In fact, all bats are very inefficient in the sense that only a small fraction of the energy in the arms is given to the ball. Most of that energy is retained in the bat and in the arms as a result of the "follow through" after the bat strikes the ball. It is almost a case of using a sledgehammer to crack a nut, but that is what is needed to hit a ball at high speed.

The reason that bats are not very efficient is that the ball is much lighter than the bat, so the bat follows through after the collision. If the ball was as heavy as the bat then the bat would be much more efficient in terms of the transfer of energy to the ball but the ball speed would be very low. The combination of a heavy bat and a light ball is used in baseball and softball to make sure the ball comes off the bat at high speed. In common usage, the power of a bat refers to the outgoing ball speed rather than the energy transfer.

Reference Frames

If you sit in a train at rest next to another train at rest then you are not moving and neither is the other train. If you see the other train start moving, it is often difficult to tell whether your train started to move or whether the other train started moving. That is, you can't tell whether you are moving forward or the other train is moving backward or whether both trains started to move at the same time. Your train is your

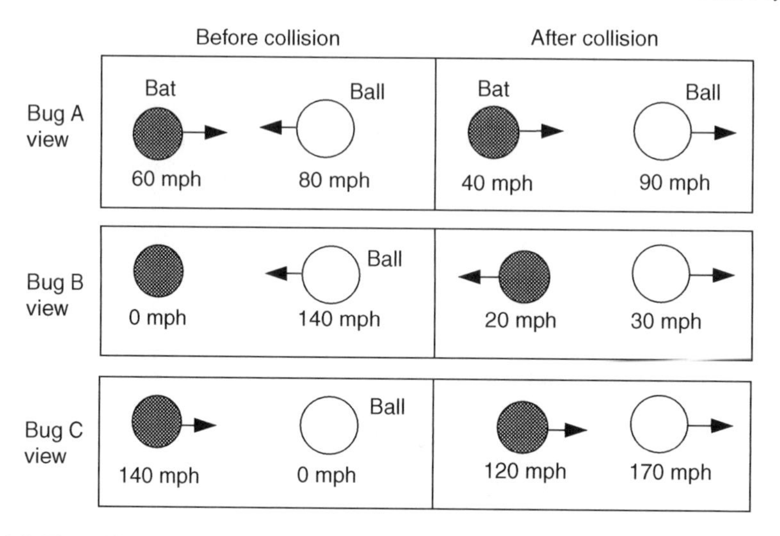

Fig. 1.5 The collision between a bat and ball viewed in three different reference frames

reference frame and you sit in your seat at rest. If your train moves forward you will probably think that the other train is moving backward. If the other train moves forward you see it move forward. But a person in the other train sees things the other way around.

Suppose that a ball moves toward a batter at 80 mph and the batter swings the bat at 60 mph. Relative to the bat, the ball approaches the bat at 140 mph. Relative to the ball, the bat approaches the ball at 140 mph. After the collision, the bat might slow down to 40 mph and the ball might head off at 90 mph, as shown in Fig. 1.5. After the collision, the ball moves away at a speed of 50 mph relative to the bat.

We can describe the motion of the bat and the ball in three different reference frames, in the same way that we described the motion of the two trains. One reference frame is the playing field. Bug A sitting on the ground sees the ball approach at 80 mph and then fly off at 90 mph in the other direction. Another is the reference frame of bug B flying along beside the bat at 60 mph. Bug B thinks the bat is at rest. It sees the ball approach at 140 mph and then head off at 30 mph in the other direction. The third reference frame is that of bug C flying beside the ball at 80 mph. This bug thinks the ball is at rest. It sees the bat approach at 140 mph and then sees the ball suddenly head off at 170 mph.

It is easy to change from the playing field reference frame to bug B's reference frame just by subtracting 60 mph from all the playing field speeds, as shown in Fig. 1.5. Similarly, the speeds with respect to bug C can be found by adding 80 mph to all the playing field speeds. The collision itself is the same collision in each case, and the force on the ball and the bat is the same, but the speeds before and after the collision depend on the reference frame. To measure the effect of a bat striking a ball in the laboratory, it is easier to fire a ball at a bat at rest rather than to measure what happens when the bat and the ball are both moving toward each other. That is how most bats are tested.

1.2 Trajectory of a Ball Through the Air

When a pitcher throws a ball or when a batter strikes a ball, the ball follows a curved path through the air until it reaches its destination. An obvious physics question we can ask is "What keeps the ball going?" If you hold a ball in your hand and then let it go, the ball will drop straight down to the ground. That's because the earth pulls it toward the center of the earth with a certain force known as the force of gravity. The question we can ask about a ball flying through the air is, "How come the ball doesn't fall straight down out of the air?" A struck ball eventually falls to the ground, but it might travel a long way toward the fence before it does.

A common answer is that the air holds the ball up and stops it falling straight down, in the same way that a bird or an airplane is supported by the air. The problem with that answer is that a ball doesn't have wings and it doesn't have an onboard engine or propeller. Another common answer is that when a ball is thrown or struck, the force on the ball that got it started keeps it going. In other words, the force on the ball is carried away from the bat or the pitcher and is retained by the ball until it lands. Both answers are good guesses but they are wrong. The correct answer is that nothing is needed to keep a ball moving. If gravity didn't pull the ball back to earth, and if air resistance didn't slow it down, a ball moving through the air would keep moving in a straight line forever, or at least until it collided with a star or planet or something else in outer space.

Common experience tells us that an object that is moving at constant speed will eventually slow down and come to a stop unless there is something to keep it moving at constant speed. An example is a vehicle or a bicycle on a road. If the driver takes his or her foot off the accelerator, or if the cyclist stops pedaling, then the vehicle or the bicycle will slow down and come to a stop. The problem here is that there is a friction force between the wheels and the road, acting backwards, and it eventually acts to bring the bicycle to a stop if the cyclist stops pedaling. If the friction force could be reduced then the cyclist would not have to pedal so hard. And if friction and air resistance could be eliminated completely then the cyclist could coast along at constant speed, at least on a level surface, without having to pedal at all.

Newton's First Law of Baseball

Galileo and Sir Isaac Newton were the first people to recognize that an object can move at constant speed along a straight line without needing any force to keep it going. Before their time, philosophers believed that the natural state of an object was when it came to rest. Galileo argued that if the friction force on a moving object was reduced then it would take longer for the object to come to rest, and if the friction force could be eliminated completely then the object would never come to rest. Newton incorporated this argument into his first law, which can be rephrased for baseball fans as follows: a baseball at rest will remain at rest and a moving baseball will continue to move in a straight line at constant speed unless some force acts on the ball, such as that due to (a) gravity or (b) air resistance or (c) a batter or (d) a pitcher or (d) a fielder or (e) the ground.

Effects of Gravity

The effect of gravity on a baseball can be determined by various experiments such as the one described in Project 1 at the end of this book. Regardless of how a baseball is launched or how fast it is moving, the force of gravity acts vertically downward on a baseball, pulling it toward the center of the earth. The force causes the ball to accelerate downward in the vertical direction. Thrown or hit in an approximately horizontal direction, the ball will follow a curved path, curving down towards the ground.

If a ball, or any other object is released from rest then it falls in a vertical direction, gathering speed as it falls. The speed, v, is given by $v = 9.8\,t$ when v is measured in metres per second and where t is the time. After 1 s, $v = 9.8\,\mathrm{m\,s^{-1}}$ (22 mph). After 2 s, $v = 2 \times 9.8 = 19.8\,\mathrm{m\,s^{-1}}$ (44 mph). The speed increases by $9.8\,\mathrm{m\,s^{-1}}$ every second, so the acceleration is $9.8\,\mathrm{m\,s^{-2}}$.

If a ball is thrown vertically upward then it slows down because the earth pulls it back to the earth. The ball will come to a complete stop for an instant when it reaches its maximum height and then it falls back to earth as if it was dropped from its maximum height. The faster the ball is thrown, the higher it will go before it comes to a stop. If the ball is thrown twice as fast it doesn't rise to twice the height. It rises four times higher. That's because it takes twice as long to come to a stop and it travels twice as fast on average so it travels four times higher.

If a ball is thrown vertically upward at $9.8\,\mathrm{m\,s^{-1}}$ then it will take 1 s to rise to its maximum height because it decelerates at a rate of $9.8\,\mathrm{m\,s^{-1}}$ every second. It rises to a height of 4.9 m during that 1 s, at an average speed of $4.9\,\mathrm{m\,s^{-1}}$. If the ball is thrown upwards at $19.6\,\mathrm{m\,s^{-1}}$ then it takes 2 s to reach its maximum height, and it reaches a height of 19.6 m. The time taken to reach maximum height provides a simple estimate of how fast you can throw or hit the ball. If it takes 2 s to reach its maximum height it will take another 2 s to fall back to the starting point. It is difficult to measure the height of the ball but you can measure the time in the air to estimate the vertical launch speed.

In baseball or softball, it is rare that a ball is launched in a vertical direction. Usually, the ball is launched at some angle to the horizontal, and the ball then travels along a curved path, as shown in Fig. 1.6. The trajectory in Fig. 1.6 was calculated for a ball launched at a speed of $36\,\mathrm{m\,s^{-1}}$ (80 mph) at an angle of 33.7° above the horizontal. The ball is shown at intervals of 1 s, and it took just over 2 s to reach maximum height and just over 4 s to fall back down to its original launch height. During that time, it reached a maximum height of 20 m and it traveled a horizontal distance of 120 m (394 ft).

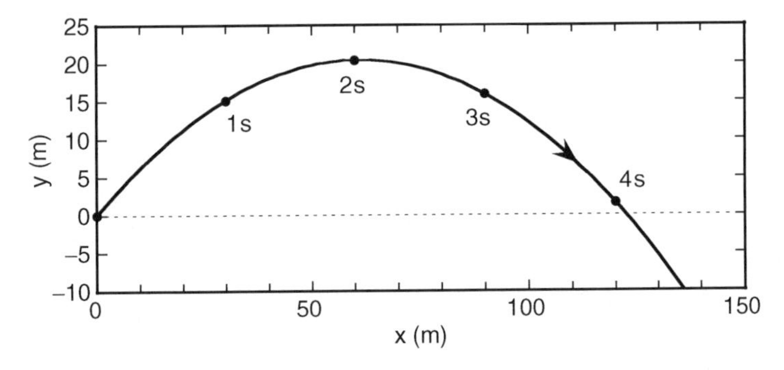

Fig. 1.6 The trajectory of a ball launched at a speed of $36\,\mathrm{m\,s^{-1}}$ at 33.7° to the horizontal, ignoring air resistance. Each *dot* shows the ball position at 1 s intervals. The ball can drop below $y = 0$ if $y = 0$ is a point above ground level

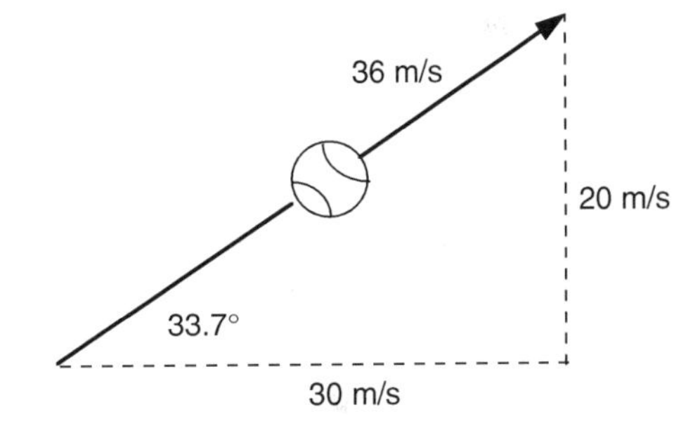

Fig. 1.7 A ball launched at a speed of $36\,\mathrm{m\,s^{-1}}$ at 33.7° to the horizontal has a horizontal speed of $30\,\mathrm{m\,s^{-1}}$ and a vertical speed of $20\,\mathrm{m\,s^{-1}}$. The lengths of the three sides of the *triangle* are proportional to these three speeds

A launch angle of 33.7° might seem a bit strange, but it was chosen to simplify the discussion. If a ball is launched at a horizontal speed of $30\,\mathrm{m\,s^{-1}}$ and with a vertical speed of $20\,\mathrm{m\,s^{-1}}$, then we can use a right angle triangle to work out the actual launch speed and angle, as shown in Fig. 1.7. If we draw a horizontal line say 3 in. long and a vertical line say 2 in. long, then the length of the two lines is proportional to the horizontal and vertical speeds, respectively. The length of the hypotenuse of the triangle is then 3.6 in., and the angle between the horizontal and the hypotenuse is 33.7°, indicating that the ball was launched at a speed of $36\,\mathrm{m\,s^{-1}}$ and at an angle of 33.7° above the horizontal. If the sun was directly overhead, the shadow of the ball would travel along the ground at $30\,\mathrm{m\,s^{-1}}$.

To simplify the discussion even further, air resistance was ignored and it was assumed that the ball traveled in the horizontal direction at a constant speed of $30\,\mathrm{m\,s^{-1}}$. It traveled at constant speed because there is no force and no acceleration in the horizontal direction when we ignore air resistance. After 1 s, the ball reached a point 30 m from where it started, and after 2 s it was 60 m from the starting point. The motion in the vertical direction was calculated by assuming the acceleration in the vertical direction was $9.8\,\mathrm{m\,s^{-2}}$ and that the ball was launched with a vertical speed of $20\,\mathrm{m\,s^{-1}}$. At any given time, the vertical height of the ball is the same as that for a ball thrown at $20\,\mathrm{m\,s^{-1}}$ in the vertical direction.

Suppose that the ball in Fig. 1.6 is launched at $36\,\mathrm{m\,s^{-1}}$ but at an angle greater than 33.7°. The ball will then spend a longer time in the air before it comes to a stop in the vertical direction and falls back to earth. Does that mean it will travel a greater horizontal distance? Not necessarily. For example, if the ball is thrown vertically up, it will rise to a considerable height before it stops, and will fall straight back down after having traveled zero distance horizontally. To throw the ball as far as possible, we would like the horizontal speed and the vertical speed to both be as large as possible. But if we are already throwing the ball as fast as possible, then any increase in the launch angle will act to increase the vertical speed and simultaneously reduce the horizontal speed. It turns out that the best compromise is to throw (or hit) the ball at an angle of 45° to throw it as far as possible, at least when we ignore air resistance. At a launch angle of 45°, the horizontal launch speed is exactly the same as the vertical launch speed.

When launched at an angle of 45°, and at speed v, the horizontal distance traveled by the ball is $v^2/9.8$, at least when we ignore air resistance. A ball struck or thrown at a speed of 100 mph ($44.7\,\mathrm{m\,s^{-1}}$) would then travel a horizontal distance of $44.7^2/9.8 = 203.9\,\mathrm{m} = 669\,\mathrm{ft}$.

In practice, a greater throw distance results when the launch angle is less than 45°. One of the reasons is that a person can throw a ball faster in the horizontal direction than in the vertical direction since the muscles involved in a horizontal throw are stronger and since the rest of the body is utilized more effectively. In the shot put, the athlete throws at a relatively low angle since the launch commences not at ground level but higher up. When throwing something off a tall building or a cliff, the flight time is determined mainly by the height of the building or the cliff, in which case a greater throw distance will be achieved by throwing almost horizontally. When air resistance is considered, the optimum launch angle is reduced even further, especially if the ball has backspin.

1.3 Circular Motion

In linear motion, the force on a moving object acts in the same direction at all times, even though the object itself might move in a parabolic arc. Circular motion is different since the force on an object moving in a circular path keeps changing direction. The force on an object moving in a circular path at constant speed always acts toward the center of the circle.

Angular Speed

When a bat is swung in a circular arc, the tip travels faster than the knob. The "speed" of the bat varies from end to the other. But all parts of the bat rotate through the same angle at the same rate. Suppose the bat rotates through 90° in 0.1 s.

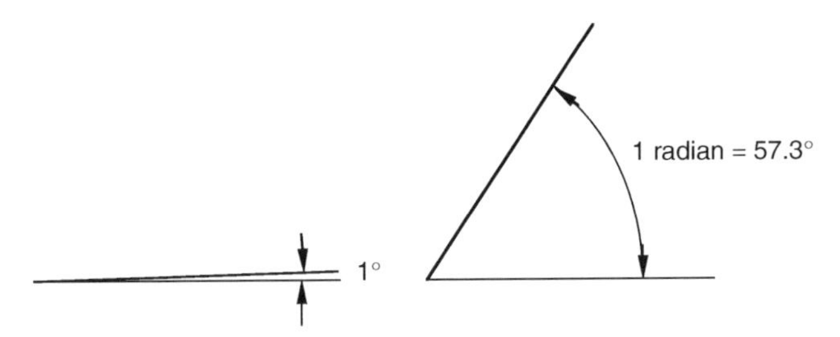

Fig. 1.8 One radian is about 57.3°. It is a more convenient angle to use when describing circular motion

We define the angular speed as the change in rotation angle divided by the time interval. In this case, the angular speed is $90/0.1 = 900° \, s^{-1}$. The tip rotates at the same angular speed as the knob.

The rotation speed of a ball or a wheel or an electric drill or any other rapidly rotating object is usually quoted in revolutions per minute or rpm. If something is rotating at say 3,000 rpm, then it rotates 3,000 times in 1 min or 3,000/60 = 50 times in 1 s. Since one revolution corresponds to a rotation through an angle of 360°, the rotation speed can also be expressed as $50 \times 360 = 18{,}000° \, s^{-1}$.

In physics, we use yet another conversion factor to describe rotation speed. Angles can be measured in degrees, or they can be measured in radians. One radian is 57.3°, and it is defined this way so that there are $2\pi = 6.28$ radians in a circle (Fig. 1.8). This might sound like a really strange thing to do, but there is a good reason. Suppose that the tip of a bat is swung in a complete circle of radius R and it takes a time T to travel once around the circle. Since the circumference of the circle is $2\pi R$, the actual speed of the tip is $v = 2\pi R/T$. The whole bat rotates through 2π radians in time T so the angular velocity of the bat is given by $\omega = 2\pi/T$ radians per second, where ω is the Greek letter omega. The relation between v and ω is therefore $v = R\omega$. This is a nice, simple formula since it does not contain the 2π factor of the other expressions. That's why physicists like to use radians rather than degrees. It's basically because we like simple formulas rather than ones with nasty π's in them.

For example, if $R = 1.2$ m and $T = 0.5$ s then $v = 2 \times 3.14 \times 1.2/0.5 = 15 \, m \, s^{-1}$. The angular speed is given by $\omega = 360/0.5 = 720° \, s^{-1}$ or $720/57.3 = 12.6 \, rad \, s^{-1}$. In our bat tip example, the whole bat is rotating at $\omega = 12.6 \, rad \, s^{-1}$ and the tip is at $R = 1.2$ m so the tip speed is $v = R\omega = 1.2 \times 12.6 = 15 \, m \, s^{-1}$ (Fig. 1.9).

Angular Acceleration

When a batter swings a bat, the bat does not rotate at a constant angular speed. Just after the start of the swing the bat will rotate slowly and just before impact with the ball it will be rotating rapidly. Suppose that the angular speed increases from

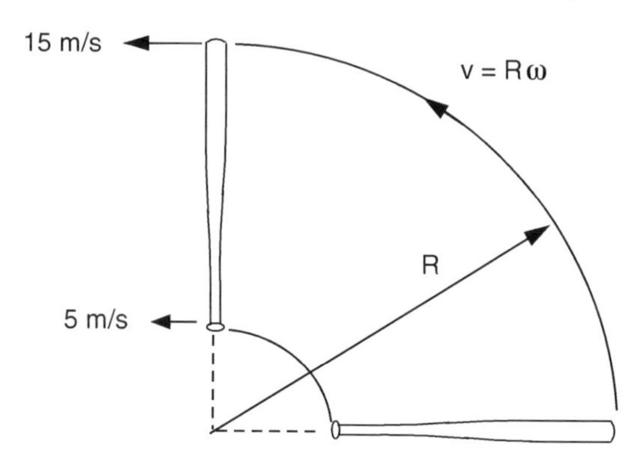

Fig. 1.9 When a bat rotates about an axis near the knob, the tip travels further and faster than the knob but the tip and the knob rotate at the same angular speed since both take the same time to rotate through 90°

$5\,\text{rad}\,\text{s}^{-1}$ to $25\,\text{rad}\,\text{s}^{-1}$ over a period of 0.2 s. The angular acceleration, α, is given by $\alpha = \text{d}\omega/\text{d}t$ where $\text{d}\omega$ is the increase in ω occurring over the time interval $\text{d}t$. In the present case, $\text{d}\omega = 20\,\text{rad}\,\text{s}^{-1}$ and $\text{d}t = 0.2$ s, so $\alpha = 20/0.2 = 100\,\text{rad}\,\text{s}^{-2}$.

Centripetal Force

The tip of a bat travels in a roughly circular path. It might be headed south just after the start of the swing, at a speed of $5\,\text{m}\,\text{s}^{-1}$. A short time later it is headed east, and when it collides with the ball it could be headed north at $30\,\text{m}\,\text{s}^{-1}$. The speed in each direction keeps changing so the acceleration along the east-west line changes, and so does the acceleration along the north-south line. Even if the tip rotated at constant speed there would be an acceleration of the tip due to its change in direction. That acceleration is always in a direction pointing to the center of the circle. It is called a centripetal acceleration, meaning that it points to the center. If the radius of the circle is R and the speed of the tip at any instant is v, then the centripetal acceleration is given by $a = v^2/R$. For example, if $v = 30\,\text{m}\,\text{s}^{-1}$ and $R = 1.0$ m then $a = 900\,\text{m}\,\text{s}^{-2}$. If the tip is headed north at this time and is veering around to the west, then it is accelerating in the west direction at $900\,\text{m}\,\text{s}^{-2}$ (Fig. 1.10).

Torque

Suppose that you open a door by pushing with a force F on the door knob as shown in Fig. 1.11. If the distance from the knob to the door hinge is d then the quantity $\tau = F \times d$ is defined as the torque acting on the door. If you push at a point closer to the hinge then the door will be more difficult to open since the torque will be smaller, assuming you push with the same force F but at a smaller distance d. The difference between the two situations can be explained in terms of the work required

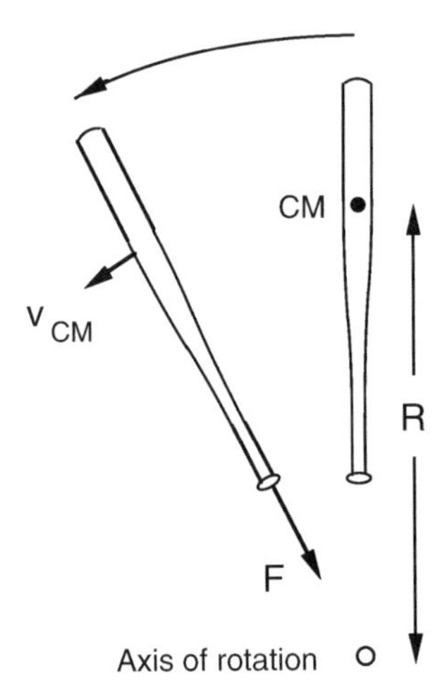

Fig. 1.10 A bat moving at speed v will continue to move in a straight line at speed v unless there is a force on the bat. To rotate a bat so its CM rotates in an arc of radius R at speed v_{CM}, there needs to be a centripetal force $F = mv_{CM}^2/R$ acting along the bat toward the axis of rotation. That force is generated by the batter pulling on the handle and is the largest single force on the bat, exceeding the weight of the batter by the end of the swing

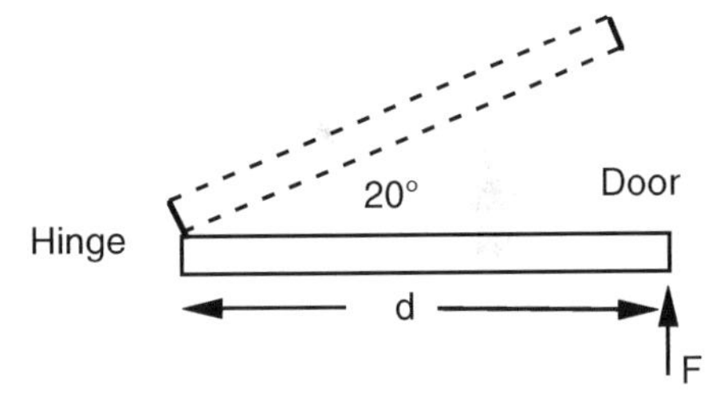

Fig. 1.11 The torque applied to the door is given by $\tau = Fd$. If F is applied near the hinge then the door is more difficult to open

to open the door. Suppose the door rotates by 20°. If the handle moves 12 in. then a point half way out will move 6 in. The work done to rotate the door is given by F times the distance moved. By pushing on the handle, the work done is $F \times 12$. By pushing half way between the hinge and the handle, the work done is $F \times 6$. In the latter case, you need to push twice as hard to do the same amount of work, and then the torque will be the same in the two cases.

If you keep pushing the door it will rotate faster and faster assuming that there is no friction in the hinges. The angular acceleration, α, of the door is given by $\tau = I \times \alpha$ where I is called the moment of inertia of the door. This relation is similar to the relation $F = ma$ that applies when an object moves in a straight line.

In this book, we are concerned primarily with bats and balls rather than doors, but the physics of the rotation process is essentially the same. Bats and balls don't have hinges. Nevertheless, a bat or a ball will rotate about a certain axis if a torque is applied by the batter or the pitcher. The torque is given by the applied force multiplied by the distance to the axis. Each will rotate faster and faster while the torque is applied, according to the relation $\tau = I\alpha$. The quantity I represents the resistance to rotation. Large, heavy objects are harder to rotate than small, light objects. Some relevant formulas are given in the following section.

Moment of Inertia

If an object is pushed along a straight horizontal line then the resulting acceleration of the object depends on its mass. Obviously, a light object will accelerate faster than a heavy object assuming that the push force is the same. The resistance to motion in a straight line is described as inertia. Heavy objects accelerate just as fast as light objects when they are dropped from a height and fall vertically to the ground. That's because the force of gravity on a heavy object, given by $F = mg$, is larger than it is on a light object.

All objects (including baseballs, bats, people and their arms and legs) also have a certain resistance to rotation. The rotational inertia of an object depends not only on its mass but also on how that mass is distributed in relation to the rotation axis. If most of the mass is close to the axis then the object will be easy to rotate, Bats with a small rotational inertia (or small moment of inertia) can be described as being "quick." Coaches often describe a batter as having a "quick bat" as if the bat speed depends only on the swing technique of the player, but the swing speed also depends on the moment of inertia of the bat itself. If a bat has a heavy barrel then it will have a large moment of inertia and it will be difficult to swing. It is easier to swing a bat by holding onto the barrel end rather than the handle end since most of the mass is then closer to the axis. Similarly, it is easier to swing a bat by holding the bat one foot from the knob end rather than at the knob end.

The formula for the moment of inertia of a mass M located at a long distance R from the axis is $I = MR^2$. If the axis of rotation is close to the object or passes through the object, then each part of the object is at a different distance from the axis. In that case, we can imagine that the whole object consists of say 100 different small parts, each of mass $m = M/100$ and each at some distance r from the axis. The moment of inertia of each small part is mr^2. The moment of inertia of the whole object is just the total value of all the separate 100 parts. This calculation is best done using calculus and the results for some cases relevant to bats and balls are shown in Fig. 1.12.

The moment of inertia of a bat is difficult to calculate because of its complicated shape. Nevertheless, it is approximately the same as that for a uniform cylindrical rod of the same mass M and the same length L. However, the moment of inertia of a bat also depends on the axis about which it rotates. It is normally measured about an axis in the handle 6 in. from the knob, in which case the moment of inertia is commonly called the swing weight of the bat, and is denoted in this book by the symbol I_6. The moment of inertia about a different axis, parallel to one through the CM of a bat, is given by the parallel axis theorem,

$$I = I_{\text{CM}} + Mh^2,$$

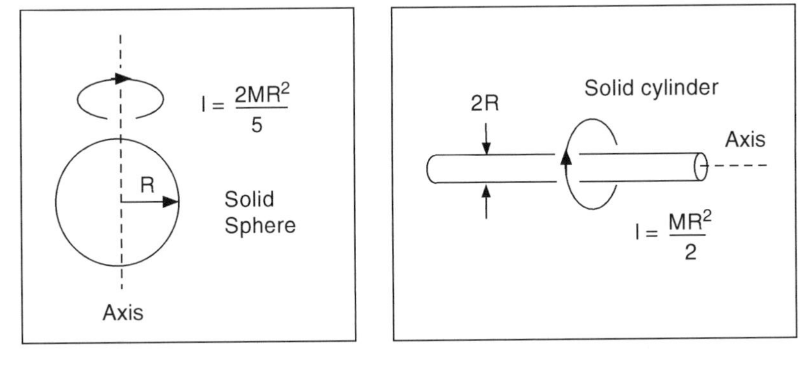

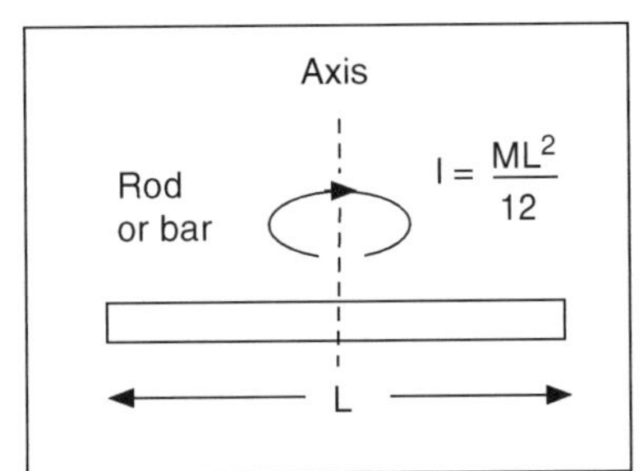

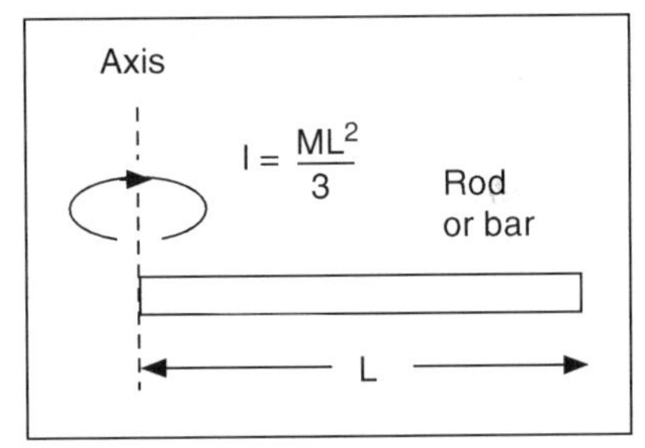

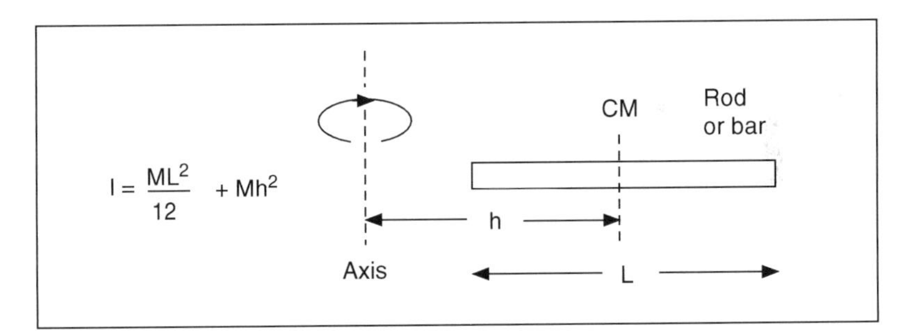

Fig. 1.12 The moment of inertia of an object depends on it mass, length, shape, and the chosen axis

where I_{CM} is the moment of inertia about an axis through the CM, M is the bat mass and h is the distance between the two axes, as shown in Fig. 1.12.

If extra mass m is added (or removed) at a distance r from the rotation axis of any object, then the moment of inertia about that axis increases (or decreases) by an amount mr^2. End loading a bat in this way makes it more difficult to swing the bat but it increases the "inbuilt power" of the bat.

Chapter 2
Bats and Balls

The ball is a sphere weighing not less than 5 nor more than 5 1/4 ounces avoirdupois and measuring not less than 9 inches nor more than 9 1/2 in. in circumference. It shall be formed by yarn wound around a small core of rubber, cork or combination of both and covered by two pieces of white horsehide or cowhide tightly stitched together. The coefficient of restitution (COR) of a baseball cannot exceed 0.555.

Wood bat. The bat must be a smooth, rounded stick not more than 2 3/4 in. in diameter at its thickest part nor more than 42 in. in length. Nonwood bat: The maximum length is 36 in. and the maximum diameter is 2 5/8 in.

– NCAA Baseball Rule 1, Sections 11 and 12, 2010

2.1 Introduction

A good place to start our discussion of the physics of baseball and softball will be to examine the basic equipment used, namely bats and balls. We will have a lot more to say about bats and balls in other chapters, but first it will be useful to look at the basic properties of bats and balls, mention briefly some of the history of the subject, and then consider how bats are designed to appreciate the main differences between solid wood, and hollow aluminum or composite bats.

From a practical point of view, the technical question of greatest interest to baseball and softball players is likely to be one that concerns the performance of various bats. The same questions are asked by golfers and tennis players. If they were allowed one technical question, it would probably be "Which club, or which racquet will work best for me?" Many such questions can be posed about bat performance. Is maple better than ash? Is a heavy bat better than a light bat? Is an aluminum bat better than a wood bat? Are composite bats better than aluminum bats? What is the best shape for a bat? Should extra weight be added at the tip end, or is it better to add weight at the handle end? Does it help to cork a bat? And so on. We will try to answer all of these questions in this book, and more, at least from the point of view of the physics of the problem.

When a physicist looks at the performance of a bat, he or she does so with a biassed point of view. It is not necessarily the best point of view, since physicists tend to ignore many of the practical issues that batters themselves regard as being important. For example, when we measure the performance of any given bat, we are not interested in the durability of the bat, or how many ball impacts it takes to dent or crack the bat, we would not be concerned with how it feels or sounds or smells, we would not be interested in the color or the fancy decals stuck on the bat, and we would not be interested in the price. All of these things might be important to the player, but they have no direct effect on the performance. They might have a

R. Cross, *Physics of Baseball & Softball*, DOI 10.1007/978-1-4419-8113-4_2,

psychological effect, in that a player with a fancy, new, brightly colored, expensive bat might feel more confident with such a bat, concentrate harder on hitting the ball, and end up performing better because he or she is in a better frame of mind to perform well.

The sound of a bat, like the sound of a club or a racquet, also has a strong psychological influence on a player. A ball impacting a wood bat makes a nice "crack" sound when hit properly, while a hollow bat "pings" like a bell. Some players prefer to play in wood bat leagues simply because the crack of the bat sounds more like the real thing, or more like the bat used by a Major League player. Batters are not alone. Professional tennis players are very fussy about the sound of the shot. About half of them use string dampeners to change the ping to a thud, and the other half don't, depending on what they are accustomed to hearing. It makes no difference to the performance of the racquet, but if the shot sounds "wrong" then it won't "feel" right. However, bats that ping generally perform better than bats that crack, for reasons that we will explore later.

From a physics point of view, the performance of a bat depends only on its physical properties, the five most important being the length, diameter, weight, weight distribution, and stiffness. The actual materials used to construct the bat are important only insofar as they might affect those five properties. In that respect, there is very little difference between maple and ash since they are both good quality wood materials. Maple is slighter harder and its surface is not so easily damaged, but that is a durability issue rather than a physics issue. Nevertheless, bats can develop tiny cracks with repeated use, and if that affects the stiffness and hence the performance of the bat then it can become a physics as well as an engineering issue.

There is one other very important issue that we won't dwell on in this book, and that is the performance of the player. The best bat in the world, if such a thing could be found, will not help to dramatically improve a batter's performance. Good batters need, above all else, good eyesight, good athletic ability, lots of practice, strength, good coaching, and a natural ability to swing the bat and connect with the ball. A good bat also helps, but it ranks below most of the other factors. The reason that batters are so interested in bat performance is that an extra few mph in batted ball speed can make a significant difference in the batting average and in the number of home runs scored in a season. In that respect, every little bit helps a batter to perform better.

Baseball

Baseball had its origins in the 1850s when players started making their own wood bats. In those days, they were free to experiment with bats of any size, shape and weight they liked. Players discovered that they could hit the ball farther with round bats rather than using bats with a flat surface, mainly because round objects are stiffer and stronger than flat objects. In general, the greatest strength to weight ratio is achieved when a material is constructed as a hollow tube, but hollow wood bats are not heavy enough or strong enough to be of any practical use. A rule was introduced in 1859 to limit the maximum bat diameter to 2 1/2 in. Another rule was introduced in 1869 to limit the maximum length to 42 in. The maximum allowable bat diameter was increased to 2 3/4 in. in 1895. These dimensions have been maintained in the rules of baseball for the last 115 years. The pitching mound was first moved to its present position, 60 ft 6 in. from the home plate, in 1893.

Most of the early bats were made from hardwoods such as hickory or elm. About half of today's wood bats are made from white ash, while other wood bats are made from maple and other hardwoods. Hickory is harder, denser, and stronger than maple, while maple is harder, denser, and stronger than ash. Hickory has a density of about 0.46 oz in.$^{-3}$, maple has a density of about 0.40 oz in.$^{-3}$, and white ash has a density of about 0.37 oz in.$^{-3}$. The actual density depends on the moisture content and on the species of wood. For example, sugar maple is generally preferred for bats since it is denser than red or black maple. Water has a density of 0.578 oz in.$^{-3}$, while softwoods like cedar, pine, or fir have densities of around 0.17–0.28 oz in.$^{-3}$. Softwoods have never been used to make bats since they are not heavy enough or strong enough. Light maple bats with thin handles have a reputation for splintering and sending sharp spears through the air when they break. From a physics point of view, it is interesting to note that a broken bat often signals a well struck ball rather than a weak hit. The point here is that takes a short time for the bend in the barrel to propagate down to the handle, by which time the ball is on its way to its destination. However, broken bats are a safety issue and continue to be closely monitored and researched by most baseball and softball organizations.

If two bats have the same length and shape, so that they have the same volume, then a hickory bat will be heavier and stiffer than a maple bat, and a maple bat will be heavier and stiffer than an ash bat. Many batters still prefer the feel of the softer ash bats, possibly because it is what they are used to, while others prefer to use maple. Personal preferences can be a more significant issue than small differences in bat performance.

The construction of a wood bat is an art that has developed over the last 150 years, and involves careful selection of the right grades of wood, control over moisture content and drying times, and even the number of growth rings [1]. About eight growth rings per inch in a white ash bat seems to be favored by most players, although some bats are made with 15 growth rings per inch since a large number of growth rings indicates an increase in the density and strength of the wood. The direction of the wood grain is also important, since bats bend and break more easily in one particular direction, which is why players are taught to swing a bat with the manufacturer's trademark facing upward. Viewed end-on, a wood bat appears to be made from many thin slices of wood glued together. That is just the visual effect of the grain, but the end result is the same. The bat is strongest when the ball impacts each slice edge-on rather than at right angles. The effect is similar to bending a ruler, or a stack of rulers that can slide on each other. It is much harder to bend or break a ruler if it is bent edge-on rather than across the wide face of the ruler. At least, that is the case for ash bats. Recent research has shown that maple bats are stronger in the opposite direction, so maple bats are now rotated 90° before the trademark is attached.

Softball

Softball was invented in Chicago, Illinois, in 1887 as a winter version of baseball that could be played indoors, and was originally called "indoor baseball." The name changed to softball in 1926 since by that time a relatively large, soft ball was being used which was 16 in. in circumference and could be fielded safely with bare hands. These days, slow pitch and fast pitch versions of the game are played outdoors, usually with a much harder ball that is 10–12 in. in circumference depending on the particular league. Eleven inch balls are commonly used in youth softball and 12-in. balls are used in most adult versions of the game. There are many different versions of softball rules, designed to be played by men or women or by mixed teams of all ages. As a result, softball is now one of the most popular outdoor sports in the USA, and it is a game that is played throughout the world by more than 40 million players. Women's fast pitch softball was introduced into the Olympic Games in 1996 but softball and baseball were both dropped as Olympic sports for the 2012 Summer Olympics.

There are many organizations throughout the world that govern the sport. In the USA, three of the largest are the Amateur Softball Association (ASA), established in 1933, the United States Speciality Sports Association (USSSA) founded in 1968 and the National Collegiate Athletic Association (NCAA). The International Softball Federation has over 100 member countries.

One of the differences between softball and baseball is that softball is played with a bigger ball and a thinner bat. A softball bat can be no more than 34 in. (86.4 cm) long, 2 1/4 in. (57 mm) in diameter or 38 oz (1.08 kg) in weight. The bat can be made of wood, aluminum or composite materials, although wood bats are not commonly used in softball.

Aluminum Bats

Hollow aluminum bats were first introduced in the 1970s, their main advantage being that they did not break as often as wood bats. When manufacturers improved the design to outperform wood bats, they quickly became the preferred choice of bat type in most amateur leagues. The improvements involved the addition of small amounts of copper or scandium for increased strength, thereby allowing for thinner walls, lighter bats, and greater durability. Another innovation was to construct bats with two thin walls close together rather than one thick wall. Today, many more aluminum bats are sold than wood bats.

One of the advantages of aluminum, from a design and performance point of view, is that the weight distribution can be altered by varying the shape and wall thickness. Solid wood bats have most of their weight in the barrel end. A hollow bat of the same outside dimensions, and the same overall weight, has its center of mass closer to the handle end. As a result, a hollow bat of any given weight is easier to swing than a solid wood bat of the same weight, making it easier for the batter to direct the bat onto the ball, and allowing the batter to swing the bat faster.

The disadvantage is that weight is moved away from the region where most balls are struck, closer to the barrel end. However, the wall of a hollow bat is flexible, allowing for a trampoline effect that makes up for the lower mass of the barrel. The physics of the trampoline effect is examined in Chap. 13. Mainly as a result of the enhanced ball speed arising from the trampoline effect and the higher swing speed, all baseball and softball associations, since about 2000, have placed strict limits on the performance of aluminum and other hollow bats made from composite materials such as fibreglass, graphite, and kevlar. The measures adopted by the associations are described and explained in Chap. 11.

Composite Bats

Carbon or graphite composite materials are very light and very strong and are used in the construction of the frames of tennis racquets, many other sporting goods and in the aircraft industry and elsewhere. Composite bats were originally introduced around 2000 by bonding a braided graphite sleeve with an epoxy resin onto a wood or aluminum bat to increase the strength of the bat. Such bats are still manufactured, although a modern trend is to construct 100% composite bats that are made completely from composite materials that include graphite and fibreglass and other materials such as kevlar. Another modern trend is to construct bats with a composite handle and an aluminum barrel, in which case the bat is classed as a hybrid bat.

Composite bats tend to be relatively stiff when they are new and then soften up with repeated impacts. The softer the bat, the bigger the trampoline effect. Impacts create small, almost invisible cracks, which soften the bat. Baseball and softball associations are well aware of this effect and now insist that composite bats be tested after an initial softening up period. Softening up can be accelerated by rolling a bat between rollers, a practice that is relatively widespread (see, for example, BatRolling4u.com). It is not illegal but it gives those with a softer bat a competitive edge.

2.2 Typical Properties of Bats and Balls

Bats are about six times heavier than balls and about six times lighter than a batter's two arms. This is no mere coincidence. Tennis racquets are also about six times heavier than tennis balls. The factor of 6 is about the best ratio to ensure that energy in the batter's arms is well coupled to the bat, and that energy in the bat is well coupled to the ball. If a bat was a lot lighter than a batter's arms, then most of the effort of the batter would be used up in swinging his arms, and only a small part of the total energy available would end up in the bat. Furthermore, a very light bat would tend to bounce off the ball rather than transferring its energy to the ball.

The factor of six can be partly understood in terms of an analogous problem in mechanics. If a heavy ball makes a head-on collision with a light ball at rest then the heavy ball will transfer some of its energy to the light ball, but the heavy ball will continue to move forward and will retain most of its original energy. The transfer of energy from a heavy ball to a light ball is not very efficient. The efficiency can be improved if the heavy ball first collides with a medium weight ball and then the medium weight ball collides with the light ball. It turns out that the efficiency is a maximum if the mass ratios (heavy/medium and medium/light) are the same. If the heavy ball happens to be 36 times heavier than the light ball, then the energy transfer can be maximized by using an intermediate weight ball that is six times lighter than the heavy ball and six times heavier than the light ball. In this respect, a bat functions as a device that helps to improve the transfer of energy from a batter's arms to the ball. The arms of a batter don't actually collide with the bat, but they do act to transfer energy from the batter to the bat.

Some typical properties of bats are shown in the following table (1 oz = 28.35 g, 1 in. = 25.4 mm). These are popular values, players being free to choose bats of some other length, weight or diameter if they want to, within certain allowed limits.

Property	Little league	Softball	Baseball
Length	30 in.	34 in.	33 in.
Weight	20 oz	28 oz	30 oz
Diameter	2-1/4 in.	2-1/4 in.	2-5/8 in.

In Little League, there is no difference between a baseball and a softball bat. At the adult level, softball bats are thinner and usually lighter than baseball bats, and the shape of the barrel is also different. Most adult baseball bats have a barrel that tapers over most of its length, being fattest near the far end. Softball bats have a cylindrical barrel that is constant in width over the whole length of the barrel. Fast pitch softball bats usually have a longer barrel than slow pitch bats and have a shorter taper region connecting the barrel to the handle. The longer barrel allows players to make better contact with the ball for inside pitches. Fast pitch bats are also lighter since the player needs to swing the bat into position more quickly.

Typical ball properties are shown in the following table, although the coefficient of restitution (COR) and stiffness values vary with ball speed and can't be taken too literally. The values in the table are those commonly listed in various rule books and are measured under conditions that don't necessarily represent playing conditions. The COR is a measure of how well the ball bounces, and is normally measured by firing a ball at a speed of 60 mph at a heavy wood block. The COR is the ratio of the rebound speed to the incident speed.

Property	Little league	11 in. softball	12 in. softball	9 in. baseball
Weight	5.25 oz	6 oz	6.75 oz	5.25 oz
Diameter	2.9 in.	3.5 in.	3.8 in.	2.9 in.
COR	0.55	0.44 or 0.47	0.44 or 0.47	0.55
Stiffness	1,500 lb in.$^{-1}$	1,400 lb in.$^{-1}$	1,400 lb in.$^{-1}$	1,500 lb in.$^{-1}$

In most leagues, softballs are not soft. They are just as hard as baseballs. Baseballs are constructed by winding wool yarn around a central cork and rubber pill. The yarn is sourced from Australia since it is stronger than most and can be wound more tightly. Most softballs are now made with a solid polyurethane core. The stiffness of a ball is officially defined by the force in lbs needed to squash a ball by 1/4 in. That value is multiplied by 4 in the above table, assuming that it takes four times the force to compress the ball by 1 in. The stiffness written on a ball (e.g., 350) is the force in lbs needed to compress the ball slowly by 1/4 in. If a ball was compressed slowly by 1 in. then it would probably be destroyed or at least permanently deformed in the process. Nevertheless, the ball can easily squash by 1/2 in. or more in a solid hit but it expands back to its original diameter very quickly before any permanent damage is done. The ball stiffness increases the more it is compressed, with the result that the stiffness can actually be as large as 10,000 lb in.$^{-1}$ when the ball squashes rapidly by 1/2 in. or more. The rules refer only to a slow compression of 1/4 in. in a materials testing machine.

The rules of the game allow for variations in all these quantities, partly since manufacturers cannot guarantee that all balls will be absolutely identical. In fact, there is a wide variation in ball stiffness between different manufacturers. Hendee et al. [2] selected 11 different baseballs in 1998 and found that their mass varied by 4% (from 140.1 to 145.8 g), their COR at 60 mph varied by 6% (from 0.546 to 0.577), their static (slow compression) stiffness varied by 70% (from 1939 to 3307 N cm^{-1}) and the maximum force on the ball in a 90 mph impact on a force plate varied by 48% (from 21 to 31 kN). The impact force is a measure of the dynamic (very rapid compression) stiffness of a ball.

In 2004, Lloyd Smith and Joseph Duris at Washington State University tested 150 different softballs by compressing each ball by 1/4 in. in a materials testing machine [3]. The balls varied in static stiffness from 1,000 to 2,000 lb in.$^{-1}$. They then fired each ball at 95 mph onto a fixed, solid cylinder and found that the dynamic stiffness varied from 5,000 to 10,000 lb in.$^{-1}$. The stiffness of a softball can therefore vary by 100% from one manufacturer to another and by much more than 100% with increasing ball compression or ball speed.

Such wide variations in ball stiffness have a strong effect not only on batted ball speed with non-wood bats, but also on the impact force on a player if he or she is struck by a ball. These issues are discussed in detail in Chaps. 10 and 12.

2.3 Bat and Ball Rules

There are many different rules concerning bats and balls, developed by the various national organizations that govern each sport. Most of the rules developed by each organization are similar, but there are often significant differences. The interested reader will probably be familiar with bat and ball rules issued by their own favorite organization. The specific parameters are important for the success of each sport, but are not essential in terms of the physics involved. One of the nice aspects of physics is that the equations we use do not depend on the specific parameters of the problem. For example, if we denote the mass of a ball by m, and the force on the ball by F, then the acceleration of the ball will be F/m regardless of the actual mass or the actual force. For that reason, it is not necessary in this book to list the mass and diameter of every bat and every ball used in every version of the sport. The following examples are chosen simply to quote some typical values.

Major League Baseball

In Major League baseball (MLB), bats can be up to 42 in. long, up to 2 3/4 in. in diameter, can be fitted with a grip extending up to 18 in. along the handle, and are allowed to have a small cupped section in the far end of the barrel up to 1 in. deep and between 1 and 2 in. in diameter. In other words, they can be about 1 oz lighter at the far end than an equivalent uncupped bat. Major League bats must be made from a smooth, round piece of solid wood. In practice, most bats used in Major League are 32–34 in. long and 32–34 oz in weight.

Major League balls must be between 5 and 5.25 oz avoirdupois in weight, must have a circumference between 9 and 9.25 in., must be made from yarn wound tightly around a small core of cork, rubber or similar material, and covered with two strips of white horsehide or cowhide tightly stitched together. The word avoirdupois here refers to the system of units where 16 oz = 1 lb. There are other ounce measures such as the troy ounce used to measure gold and silver, and the fluid ounce which is a measure of volume and is approximately equal to the volume of one avoirdupois ounce of water.

NCAA Baseball

The rules for wood bats are the same as those for Major League. Non-wood bats can be up to 36 in. long, up to 2 5/8 in. in diameter and the weight of the bat in oz must be greater than the length (in inches) −3. For example, a 34-in. bat must weigh 31 oz or more. The ball is essentially the same as a major league ball but its circumference can be between 9 and 9.5 in., and its COR can be no larger than 0.555.

NCAA Softball

Softball bats used by the NCAA can be no more than 34 in. long, no more than 38 oz in weight and no more than 2.25 in. in diameter. In practice, most bats used in adult softball are 33 or 34 in. long. The bats used in adult slow pitch softball are typically about 28 oz in weight, while in adult fast pitch softball bats tend to be lighter, around

23 oz since the batter needs to react faster. In fact, many bats in fast pitch softball have a bat drop of −10 or −11. The bat drop is the weight in oz minus the length in inch. The bat drop in baseball is typically −3, while a 33 in., 23 oz bat used in fast pitch softball has a drop of −10.

The ball must be optic yellow, with a circumference between 11 7/8 and 12 1/8 in., a weight between 6.25 and 7.0 oz, and a COR no larger than 0.47. The compression force, to compress the ball by 1/4 in., must be between 300 and 400 lb.

2.4 Bat Performance

In theory, a bat can be constructed to have almost any mass and length and barrel diameter that the designer or the player wants, but in practice there are now strict rules in all baseball and softball leagues, both amateur and professional, that govern the allowed properties of bats and balls. The rules are determined in the USA by MLB, the ASA, the NCAA, National Federation of State High School Associations (NFHS), Little League, USSSA and others. Each organization has a different set of rules, but they are all designed with the common objectives of having a good balance between offense and defense and maintaining the sport's safety. The safety issue is related primarily to increases in batted-ball speed with improvements in bat technology.

A slightly confusing aspect of the various tests is that different organizations use different tests, and each organization varies the test method from time to time as they each gain more experience in the practical aspects of designing and performing the various tests. At first sight, it might seem like a relatively simple task to test a bat to see how well it performs. One could simply swing each bat at a speed of say 60 mph at a ball pitched at say 70 mph and then measure the exit speed of the ball coming off the bat. If the ball exits at more than say 100 mph then the bat might be declared illegal. That is indeed the basic method now used to test all bats, but there are many subtle features that require careful consideration to interpret the results in a valid manner.

Some of the problems in testing bats this way are the following:

What allowance should be made for the fact that some balls bounce better than others, and that the bounce of a ball depends on its temperature and moisture content or on the humidity?

What allowance should be made for the fact that light bats can generally be swung faster than heavy bats?

What allowance should be made for the fact that the ball bounces best at a point near the sweet spot?

What allowance should be made for the fact that the best bounce point is not actually a fixed point on the bat but varies according to the actual bat speed and the actual ball speed?

Anyone reading the details of these tests for the first time will probably have trouble understanding all the various technical terms that are used. Some of those terms include the BBS = Batted Ball Speed, the BESR = Ball Exit Speed Ratio, the BPF = Bat Performance Factor, COR = Coefficient of Restitution, COP = Center of Percussion, and two new terms introduced in 2011 called the BBCOR = Ball–Bat COR and the BWCOR = Ball–Wall COR. The idea behind the BWCOR is that the bat and the ball need to be tested separately so that any differences in ball properties don't interfere with the primary objective of testing the bat. A discussion of the various test methods will be given in Chap. 11.

2.5 Real Bats and Toy Bats

The design of modern baseball bats has evolved over many years, partly by trial and error, partly by engineering calculations and innovations [4–8] and partly driven by marketing requirements to produce new, improved models almost every year. Cosmetic features can be added to bats so that they appear to have certain advantages, and can be advertised as being better, but the actual difference in performance might be negligible. The main physics and engineering principles behind bat design can be understood by considering a few hypothetical bats with simple geometric shapes to keep the calculations simple. We will shortly describe a few such "toy" bats to see how they compare with real bats. But first we look at two real bats for clues to see how they were designed.

Two Real Bats

Figure 2.1 shows the profile of two real bats. One is a Louisville Slugger R161 wood bat of length 33 in., weight 31 oz and barrel diameter 2-5/8 in. The other is an Easton BK7 aluminum bat of length 33 in., weight 30 oz and barrel diameter 2-5/8 in. Both bats look very similar, but the balance point of the wood bat is 22.2 in. from the

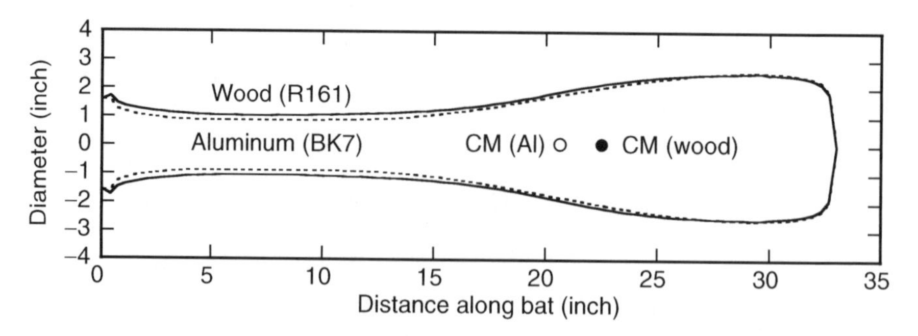

Fig. 2.1 Profiles of two real bats, one wood (with the larger diameter handle) and one aluminum. The vertical scale is not the same as the horizontal scale, hence the bats appear stubby

knob, and the balance point of the aluminum bat is 20.7 in. in from the knob. The balance point, or the center of mass, is the point where the bat can be balanced by supporting it on one finger.

The measured swing weights were 10,600 oz in.^{2} for the wood bat, and 9,530 oz in.^{2} for the aluminum bat, both measured about an axis 6 in. from the knob. The aluminum bat is therefore easier to swing. The term "swing weight" is explained in Chap. 1 (Basic Physics) and in Project 7. It is a number that describes how the weight is distributed along the bat and it determines how easy it is to swing the bat.

The total volume of the wood bat is 81.92 in.^{3} so the wood density is 0.38 oz in.$^{-3}$, indicating that it is made from ash. Small holes drilled in the aluminum bat indicated that the wall thickness was 3.5 mm (0.138 in.) along its whole length, which is consistent with a total aluminum volume of 19.2 in.^{3}, given that the density of aluminum is 1.56 oz in.$^{-3}$.

Each bat can be regarded as consisting of a handle of length 11 in., a barrel of length 11 in. and a middle section of length 11 in. From the bat profiles, it was found that the handle of the wood bat weighed 4.4 oz, while the handle of the aluminum bat weighed 5.6 oz. The middle 11 in. of the wood bat weighed 7.2 oz, and the barrel weighed 19.4 oz. The middle section of the aluminum bat weighed 7.5 oz and the barrel weighed 16.9 oz. The aluminum bat therefore had a lighter barrel and a heavier handle, which resulted in a balance point closer to the knob and a smaller swing weight.

Toy Bats with One or Two Sections

Three simple toy bats are shown in Fig. 2.2, all of the same weight (31 oz) and length (33 in.). Bat A is the simplest possible bat design, being a straight hickory cylinder of diameter 1.61 in. Such a bat could be used for training purposes. Bat B is an improved design, consisting of an ash handle of length 23 in. and diameter 1.28 in., with a cylindrical ash barrel of length 10 in. and diameter 2-5/8 in. (the maximum allowed diameter). Bat C is similar in shape to bat B but it is made from two lengths of aluminum tube welded together, each tube having a wall thickness of 0.136 in. Bat C has a handle of length 22 in. and diameter 1.0 in., plus a barrel of length 11 in. and diameter 2.62 in. Real bats are tapered and have a knob, but the same basic design principles apply to both real bats and our simplified bats. The question is, how will the three bats differ from each other and from real bats? Will they all be fairly similar or will there be some significant differences?

The average baseball or softball player looking at the toy bats might be excused for thinking that the bats are so badly designed that they couldn't possibly work and that they wouldn't use one even if they cost only $1 each. That might indeed be the case, but if we can pinpoint exactly why they are so bad, then we will be in a much better position to understand what it is that does make a good bat, and why. Simply saying that our bats don't "look" right is not going to tell us anything useful.

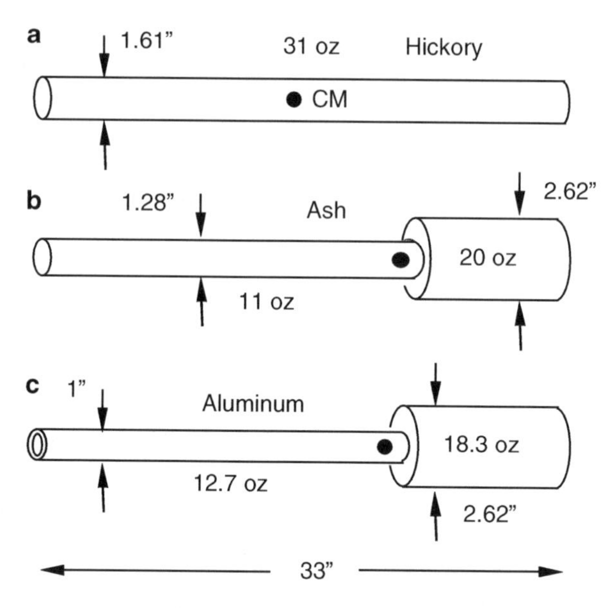

Fig. 2.2 Three 31 oz, 33 in. toy bats. The *black dots* denoted the balance point (center of mass)

The weight of each bat is determined by the density and by the volume of the material used. Hickory has a density of about 0.46 oz in.$^{-3}$ and ash has a density of about 0.37 oz in.$^{-3}$. Hickory was more popular in the past when players liked using heavy bats, and it is a suitable wood for Bat A given that the handle and the barrel have the same diameter. Obviously, a thinner handle and a thicker barrel would be better, but the question we first need to ask is whether Bats B and C are dramatically different or just slightly different from Bat A in terms of their balance point and swing weight.

Bat A has a volume of 67.2 in.^{3} so its mass is $67.2 \times 0.46 = 31$ oz. Bat B has a total volume of 83.8 in.^{3} so its mass is $83.8 \times 0.37 = 31$ oz. Bat C has a total volume of only 19.9 in.^{3} of aluminum but it also has a mass of 31 oz since aluminum is about four times denser than wood. Aluminum bats therefore need to be hollow if the barrel is to be more than about 1 in. in diameter. We could make a solid aluminum bat if we wanted to but it would need to be a solid rod similar to Bat A and it would be only about 0.9 in. in diameter.

For a given mass or volume of material, the dimensions of Bats B and C could be almost anything. Bat B was chosen so that it had the same length, mass and balance point as the Louisville Slugger R161 wood bat. The balance point is just another term for the center of mass. An older term no longer used is the "center of gravity." The balance point of Bat B was 22.2 in. from the end of the handle, and its swing weight was 10,664 oz in.$^{-2}$. Bat B therefore has physical properties that are almost identical to the real wood bat and should therefore perform in a very similar manner. However, Bat B would not be as strong and would probably break at a point near where the handle joins to the barrel.

It was difficult to design Bat C so that it had the same balance point as Bat B without ending up with an impractically small wall thickness in the handle or an impractically small diameter and very flexible handle. Bat C was, therefore, designed with a wall of uniform thickness in both the handle and the barrel, but the balance point then ended up being 20.7 in. from the end of the handle, 1.4 in. closer to the handle than Bat B. The weight and balance point of Bat C ended up being exactly the same as the real aluminum bat (the Easton BK7) and it also had an almost identical wall thickness (0.136 in. vs. 0.138 in.). The swing weight of Bat C was 9,473 oz in.^{2}, only 0.6% less than the Easton aluminum bat. Bat C should therefore behave in a very similar manner to the "real" bat.

The aluminum bat, like the real one, was just as heavy as the wood bats but it was not possible to design an aluminum bat with as much weight in the barrel as in a wood bat. The essence of the problem is that the area and volume of any section of a solid bat is proportional to its diameter squared. For example, if the barrel is twice the diameter of the handle then any given length of the barrel will be four times heavier than the same length of the handle. For a hollow cylinder with a thin wall, the volume of material in any given length is proportional to the diameter, not the diameter squared. So, if the barrel is twice the diameter of the handle, and if it has the same wall thickness, then any given length of the barrel will be only twice as heavy as the same length of the handle. A hollow bat will therefore have a relatively light barrel and a relatively heavy handle compared with a solid bat. The balance point of a hollow bat will therefore be closer to the handle.

Bat A ended up with a swing weight of only 6,231 oz in.^{2} so it would be much easier to swing than the other bats. However, it would not work as well as the other bats since the barrel end would be too light, even allowing for the fact that Bat A can be swung faster. That is why all real bats have a skinny handle and a fat barrel. Bats are tapered so that the stress on the bat is spread out over a reasonable length of the bat and is not concentrated at a small transition region.

2.6 Stiffness of Bats and Balls

One of the properties of a bat that determines how well it performs is its stiffness. We use the word "stiffness" here in its usual sense, to describe how easily the bat bends. However, we will also use the word "stiffness" to describe how easily a bat or a ball can be compressed. For example, a baseball is lot stiffer than a tennis ball since it is much easier to squash a tennis ball than a baseball. Whenever we use the word stiffness in this book it will be clear from the context whether we are referring to bending or compression, but the reader should remain alert to the fact that bending and compression are two different, but related things. If you bend a long wood or metal rod or bar, one side lengthens or stretches and the opposite side shortens or compresses. Consequently, a material that is easy to stretch or compress will also be easy to bend.

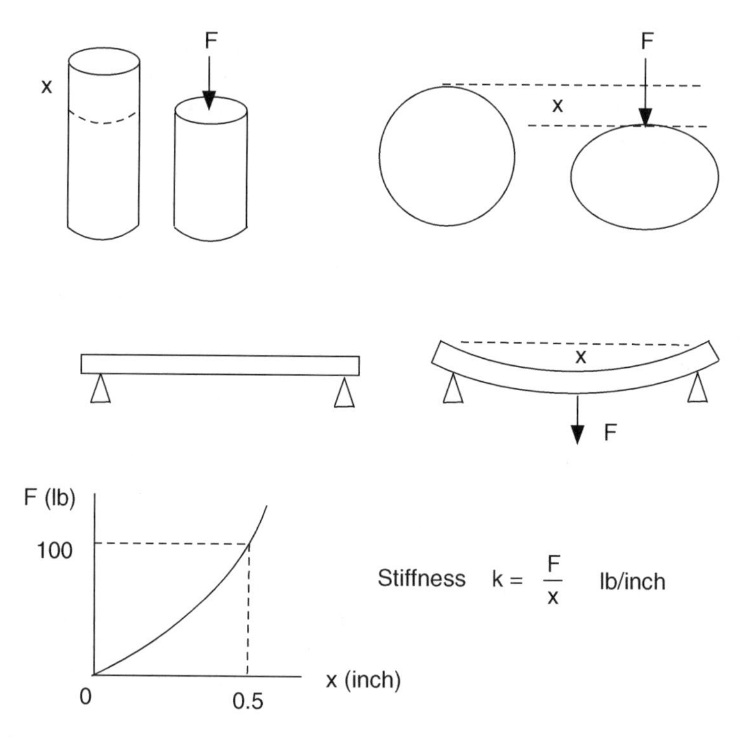

Fig. 2.3 The stiffness of a cylinder, a ball or a length of wood is defined by stiffness $k = F/x$. A graph of F vs. x is usually not a straight line. If the line curves upwards, then the stiffness, F/x, increases as x increases

A bat can bend in different ways. For example, if you were to place each end of a bat on a brick and stand on the bat in the middle, then the bat would bend in the middle, as shown in Fig. 2.3. If you put the barrel in a vice and tighten the vice, the bat would bend out of shape and squash across its diameter. In either of these circumstances, a stiff bat or ball will not bend or squash as much as a flexible bat or ball.

A bat is not necessarily stiff along its whole length. In fact, the handle end is always more flexible than the barrel end since the handle is thinner. Stiffness depends on several factors. One is the nature of the material. For example, rubber is a very flexible material, wood is stiffer, and steel is stiffer than wood. But stiffness also depends on the shape and thickness of the material in the bending direction. It would be easy to bend a thin steel wire, and much harder to bend a thick plank of wood. A ruler is very stiff if you try to bend it edge-on, and quite flexible if you bend it across the flat face.

Stiffness is measured in lb in.$^{-1}$ (or in N m^{-1} in SI units). If a force F is applied to a ball and if it compresses by an amount x, then we define the stiffness, k, of the ball by $k = F/x$. For example, if it takes a force $F = 375$ lb to compress a ball by $x = 0.25$ in., then $k = 375/0.25 = 1,500$ lb in.$^{-1}$.

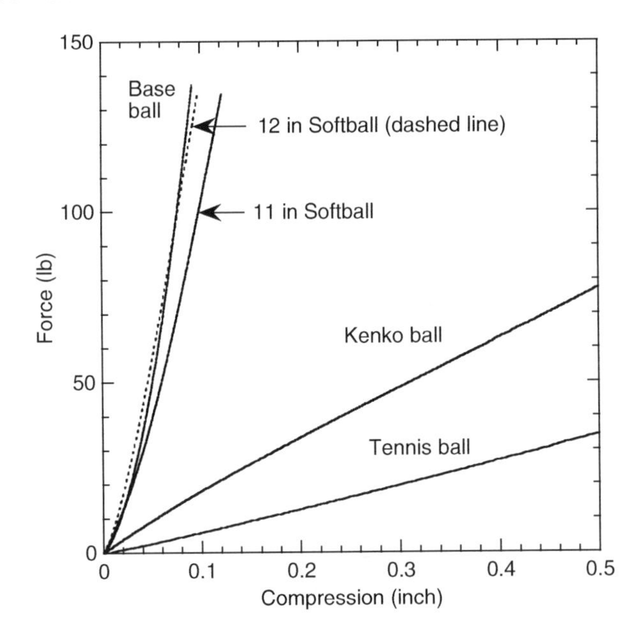

Fig. 2.4 Force vs. compression for several different balls, showing that baseballs and softballs are typically about 30 times stiffer than a tennis ball. The Kenko ball is a hollow rubber youth baseball used in Japan and elsewhere

F vs. x graphs for several different balls are shown in Fig. 2.4, obtained by compressing each ball in a materials testing machine at a rate of 20 mm per minute. The stiffness of a baseball or a softball increases the more it is compressed, being about 1,500 lb in.$^{-1}$ at a compression of $1/4$ in. and about 8,000 lb in.$^{-1}$ at a compression of 1 in., since it takes a force of about 8,000 lb to squash a ball by 1 in. The adult baseballs and 12 in. softballs tested were almost equal in stiffness, while the 11 in. softball was softer. The Kenko ball (a hollow rubber ball) is softer still, and is used in many countries around the world as a youth baseball for safety reasons. For comparison, a tennis ball was also tested, indicating that baseballs and softballs are typically about 30 times stiffer than a tennis ball.

Results of a simple experiment to compare the stiffness of bats and balls are shown in Fig. 2.5. Using a materials testing machine, the author squashed a baseball using a force of 120 lb and found that it compressed by 0.08 in. The ball stiffness was therefore $120/0.08 = 1{,}500$ lb in.$^{-1}$. An attempt was made to compress the barrel of some bats in the machine, but the barrels were tapered and could not be compressed evenly between the two parallel plates. So a ball was placed on top of the bat and compressed together, as if the bat and ball were colliding in the usual way. That way, the same force was applied to both the bat and the ball. The results showed that an aluminum bat compressed slightly more than a white ash wood bat, but neither compressed as much as the ball. If they had, then the total compression of the bat and the ball, at a force of 120 lb, would have been $2 \times 0.08 = 0.16$ in.

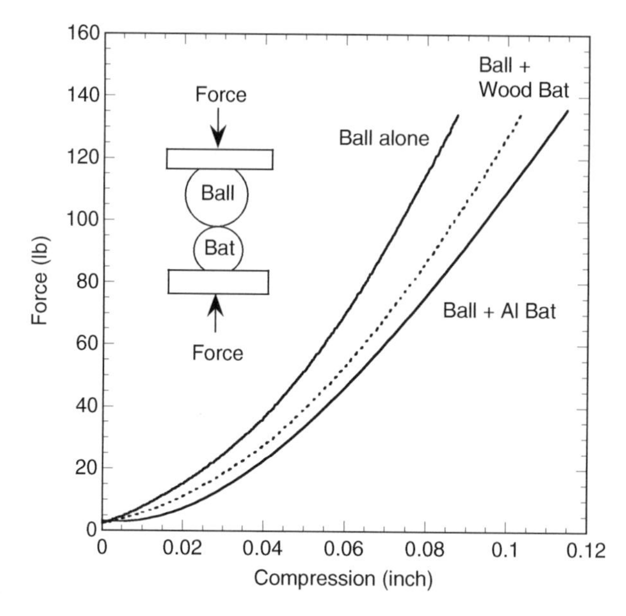

Fig. 2.5 Force vs. compression for a baseball by itself and for a baseball plus a wood or aluminum bat squashed together as shown in the inset

When a bat collides with a ball, the ball squashes, and the bat also squashes across its diameter. The bat doesn't squash as much as the ball, so the bat is stiffer than the ball. Nevertheless, hollow bats squash more than solid wood bats, leading to a bigger trampoline effect. We will examine the trampoline effect in Chap. 13. For the moment, we simply want to point out that bat stiffness is one of the important properties that determine how the bat will perform.

It was surprising that the wood bat compressed almost as much as the hollow aluminum bat, given that there is a much stronger trampoline effect with aluminum bats than with wood bats. The usual explanation is that wood bats are much stiffer than the ball, so there is very little compression of the wood and almost no elastic energy is stored in the wood. What I found was that the wood bat was about five times stiffer than the ball, while the aluminum bat was about three times stiffer than the ball. The ball or the bats were not compressed very far, and different results can be expected at large compressions, but the results should not be drastically different. Nevertheless, the relative softness of a wood bat at low compressions is not all that surprising. It is easy to dent the surface of a wood bat by striking it on a hard surface, since only a small amount of material on the surface is squashed in the process. In a similar way, only a small amount of surface material is squashed when a bat is squeezed gently in a vice, so the required force is relatively small. It would be far more difficult to squash a wood bat in half by squeezing it in a vice, partly because more of the material is being squashed, so the bat would then be much stiffer.

The difference between wood and aluminum was not as large as expected so a drastic experiment was devised to investigate whether wood is elastic. The end of the wood bat was cut off and machined into a white ash wood ball. The wood ball did not bounce very well at all. Dropped onto a concrete slab, the wood ball bounced to a slightly smaller height than a baseball. So, part of the reason that wood bats show no trampoline effect is that wood is not very elastic. An advantage of aluminum is that it is more elastic, behaving more like a spring. When a ball collides with a hollow aluminum bat, the bat compresses like a spring and then ejects the ball as it springs back to its original shape.

References

1. P.R. Blankenhorn, B.D. Blankenhorn, A.G. Norton, Selected quality characteristics of white ash used in professional baseball bats. Forest Prod. J. **53**(3), 43–46 (2003)
2. S.P. Hendee, R.M. Greenwald, J.J. Crisco, Static and dynamic properties of various baseballs. J. Appl. Biomech. **14**(4), 390–400 (1998)
3. J. Duris, L.V. Smith, Evaluating test methods used to characterize softballs. 5th International Conference on the Engineering of Sport, vol. 2 (Davis, CA, 2004), pp. 80–86. www.mme.wsu.edu/~ssl/pubs/pubs.html
4. A.T. Bahill, W.J. Karnavas, The ideal baseball bat. New Scientist **130**, 26–31 (1991)
5. L.P. Fallon, J.A. Sherwood, A study of the barrel constructions of baseball bats. 4th International Conference on the Engineering of Sport, Kyoto, Japan, 2002
6. L. Noble, Inertial and vibrational characteristics of softball and baseball bats: research and design implications, in Proceedings of the 1998 Conference on International Society of Biomechanics in Sports, 1998
7. P.J. Drane, J.A. Sherwood, J.W. Jones, T.P. Connelly, The effect of baseball construction on the game of baseball, in Proceedings 7th ISEA Conference, Biarritz, Paper 275, 2008
8. R.K. Adair, The Physics of Baseball. Phys. Today **48**, 26–31 (1995)

Chapter 3
Ball Trajectories

To set the cylinder in rotation, a fine silk thread was wrapped round the same. A sharp pull at this string gave the cylinder a rotation, which continued for two or three minutes. If the centrifugal fan was made to follow the cylinder, the latter moved laterally. When made to rotate in an opposite direction, it moved towards the opposite side.

– G. Magnus, 1852

3.1 Introduction

The sight of a baseball or softball flying through the air evokes different responses in different people. If the ball is headed toward the batter then the batter will be frantically trying to figure out what the ball is doing and where it is going. If the ball is headed toward one of the bases then the player running to that base will be wondering whether he will make it in time. If the ball is headed toward the fence then some of the spectators will be cheering and some will be moaning. These are the sorts of emotional reactions that make the game what it is. It is the combined effect of the importance and the uncertainty of the outcome that keeps players on their toes and spectators on the edge of their seats.

From a scientific point of view, the outcome of any given ball launch situation is predetermined at the moment the ball commences its flight, either from the throwing arm of the pitcher or a fielder or as it comes off the bat. The flight of a ball is determined by the laws of physics. That is not of any immediate help to players, but an understanding of the situation can provide players and coaches with useful insights. If nothing else, it can also help to settle arguments about the way the ball curves and swerves in flight, and by how much. In the 1940s, there was considerable debate as to whether spin has any effect at all on the way a ball curves [1, 2]. The argument was settled in 1959 by a retired physicist, Lyman Briggs, who measured the effect in a wind tunnel [3]. Players may not be able to explain the effects that are involved, and press on regardless. However, they will know that the effects exist, from practical experience. Pitchers know *how* to hold and spin the ball in many different ways and they know how hard to throw it to achieve a desired result. People who are curious about the game want to know *why* the ball does what it does. In this chapter we consider both the effects of gravity on ball flight and the aerodynamics involved. There is not enough room in this chapter to cover everything that is known on the subject. The aerodynamics of a baseball has been studied in much greater detail than any other aspect of the physics of baseball [1–29], despite the fact that

R. Cross, *Physics of Baseball & Softball*, DOI 10.1007/978-1-4419-8113-4_3,

there have been very few wind tunnel measurements of the relevant lift and drag coefficients. The coefficients are still not known as well as they need to be known.

A brief introduction to the effects of gravity on ball trajectories is given in Chap. 1 and in Appendix 3.1 for those readers who are not familiar with the basic physics. In this chapter we consider the combined effects of gravity and air resistance on the flight of batted balls. In Chap. 4 we will describe trajectories of pitched balls.

3.2 Typical Ball Trajectories

The trajectories of several different balls launched from a height of 3 ft above the ground at 100 mph and at an angle of 40° above the horizontal are shown in Fig. 3.1. In the absence of air resistance, all balls would follow the same path and land 660 ft from the launch point. Air resistance causes a ball to land short, not by a small amount but by a surprisingly large amount. The amount depends on the mass of the ball and its diameter. At any given launch speed and launch angle, the distance traveled by a ball approximately doubles when the mass increases four times, and it approximately halves when the diameter is doubled. That information is of no use to a baseball or softball fan, but it is useful if we want to compare the flight of a baseball with say a softball or a golf ball and it summarizes the effect of air resistance in a simple and practical manner.

The distance traveled by a ball also depends on the spin of the ball, but it was assumed in Fig. 3.1 that the ball was launched without spin and that the drag coefficient had a value of 0.5 for each ball. The force on a ball arising from air resistance

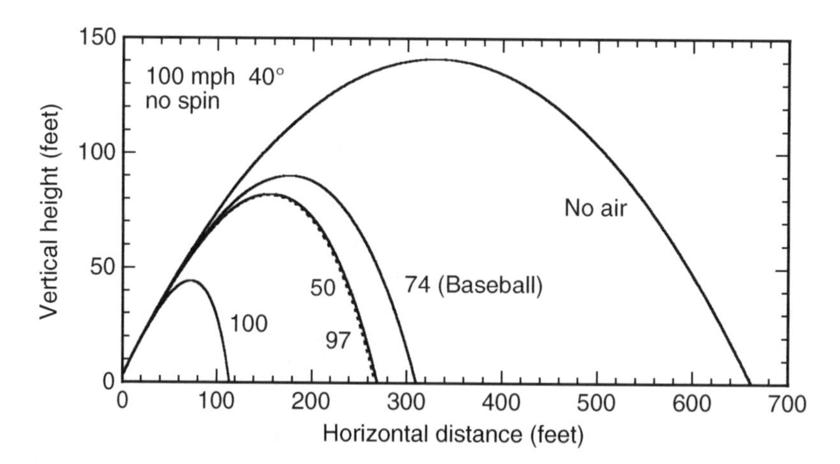

Fig. 3.1 The trajectory of various balls launched at a speed of 100 mph at 40° to the horizontal, without spin and with a drag coefficient $C_D = 0.5$. The ball diameters are indicated on each trajectory. All balls travel the same distance when there is no air. The mass and diameter of each ball are: (50 g, 50 mm), (50 g, 100 mm), (145 g, 74 mm baseball), and (185 g, 97 mm softball, *dashed curve*)

is directly proportional to the drag coefficient. The drag coefficient depends on the shape of the object and varies from about 0.1 for a streamlined object such as the wing of an aeroplane to 1.0 for a blunt object such as a flat disk. For a sphere, the drag coefficient is about 0.5 at low speeds, and decreases to about 0.3 at high speeds due to a change in the level of air turbulence around the sphere. For a baseball or a softball, the average value of the drag coefficient over the whole flight distance is typically about 0.4 for a ball launched at a speed above 70 mph, as is the usual situation when a batter strikes the ball.

As shown in Fig. 3.1, the effect of air resistance is quite dramatic, especially for light balls of large diameter. Conversely, the effect of air resistance is smallest for heavy balls with small diameter. The drag force due to air resistance does not depend on the mass of the ball. It depends primarily on the diameter of the ball and its speed, but the smoothness or roughness of the ball surface is also important, which is why a golf ball has dimples. A dimpled ball traveling at high speed has a drag coefficient of only about 0.2. If two balls of the same diameter and surface roughness are launched at the same speed, the drag force on each ball will be the same but the effect of that force will depend on the mass of the ball. A light ball will decelerate rapidly, while a heavy ball will decelerate at a smaller rate.

For two balls to travel the same distance, the ratio of their cross-sectional area to their mass must be the same. That is why, in Fig. 3.1, a ball of mass 50 g and diameter 50 mm travels essentially the same distance as a softball of mass 185 g and diameter 97 mm. The softball is larger in diameter by a factor of $97/50 = 1.94$, larger in area by a factor of $1.94^2 = 3.76$ and larger in mass by a factor of $185/50 = 3.70$.

According to the rules of baseball, the ball must have a mass between 5.0 oz and 5.25 oz (141.7–148.8 g) and a circumference between 9.0 and 9.25 in. or a diameter between 72.8 mm and 74.8 mm. If a ball has the minimum allowed diameter and the maximum allowed mass it will travel about 15 ft further when struck at 100 mph than a ball with the minimum allowed mass and the maximum allowed diameter. Assuming that both teams use the same ball, then there is no advantage to either team. But if one team uses a heavier or smaller ball than the other team, then there could be a distinct advantage to one team, at least for the team using the smaller diameter ball. There is no real advantage in using a light or a heavy ball. Light balls can be struck faster but air resistance slows light balls faster. As a result, the distance traveled by a light baseball is almost identical to that traveled by a heavy ball. The only real advantage in selecting a particular ball would be to use the smallest diameter ball allowed, in which case it will travel about 8 ft farther than the largest diameter ball allowed.

3.3 Soft vs. Hard Balls

The results in Fig. 3.1 raise an interesting question concerning the hardness of baseballs and softballs. The question is, why use a very hard ball when a soft ball might do the same job and would be safer? For example, a tennis ball of mass 57 g and diameter 65 mm struck at 100 mph will travel about the same distance as a softball, and it will do less damage to a player if the player is struck in the head or chest by the ball. The force on the player will be about 12 times smaller. Similarly, a rubber ball similar to a tennis ball but somewhat larger and heavier, and relatively soft if it is hollow, could be used as a safer version of a conventional baseball or softball. That idea is not new. In fact, such a ball has been used for many years as the standard youth baseball in many countries around the world, and is known as a Kenko ball. In the USA, it is used mainly as a training ball.

Despite the name "softball," balls now used in softball are about 30 times stiffer and about three times heavier than a tennis ball. At any given ball speed, the impact force of a ball striking an object or a player increases if the mass of the ball increases, and it increases if the stiffness of the ball increases. There is no obvious or logical answer to this soft vs. hard ball question apart from the fact that the balls used in softball and baseball are specified by the rules and those rules have developed over many years of trial and error. The balls used in softball used to be relatively soft but they tended to deform with use. The harder balls used today maintain their spherical shape much longer, but so do hollow tennis and Kenko balls. The safety of a high speed ball has not been a major issue in determining the main rules of each game, apart from the fact that protective head gear and gloves are worn by all players in recognition of the fact that hard, high speed balls are potentially dangerous. Bat specifications have changed over recent years to reflect concerns regarding batted ball speeds but ball specifications have not changed significantly. Statistical data on baseball and softball injuries (for example the NEISS (National Electronic Injury Surveillance System) web page or the CPSC (US Consumer Product Safety Commission) web page shows that baseball and softball are relatively safe sports compared with other sports but there are still around 274,000 injuries each year in baseball and softball that require hospital treatment. By comparison, the 2008 figures for other sports were basketball (487,000), football (478,000) and bicycles (516,000). Accidents in the home are even more common, with 594,000 cases in 2008 involving beds, 515,000 involving chairs, and 1,213,000 involving steps and stairs.

3.4 Air Resistance

Air seems to be so light that it hard to believe that it could slow down anything as heavy as a baseball or softball, except perhaps by a tiny fraction. Common experience says that air does not slow you down when you walk or run through it or if you wave your hand through it. It is not like walking or running through water or pulling your arm through water. Air is 826 times less dense than water. Nevertheless, air is surprisingly heavy. One cubic foot of air weighs 1.21 oz, and 1 m^3 of air at room temperature weighs 1.21 kg. An average size room full of air, with a volume of 100 m^3, contains 121 kg or 267 lb of air.

When thinking about the force exerted on a ball by the air, it helps to imagine that the ball is at rest and the air is rushing past the ball. If a ball is moving at 90 mph through still air, the force on the ball is the same as that on a stationary ball suspended in mid air when the air is rushing past the ball at 90 mph. That is why measurements are made in a wind tunnel using a large fan to blow air past objects placed at rest in the wind tunnel. If you were to sit in a wind tunnel in a 90 mph gale you would notice that air can exert a very large force when it is moving rapidly. In a light, 9 mph breeze, you feel almost no effect at all. At 90 mph the force of the air on your body is not ten times larger. It is one hundred times larger. An average size person facing a 90 mph gale will feel a backward force of about 100 lb.

Consider what happens when a 145 g baseball is thrown 60 ft from the pitcher to the batter. Over that distance, the ball has to plough its way through 2.74 cubic feet of air weighing 94 g. If all that 94 g of air was right in front of the batter, with nothing in between, then the ball would sail straight through to the batter without slowing down at all and suddenly collide with that 94 g mass of air. If the air behaved like a rubber ball then the baseball would slow down from say 90 mph to about 20 mph when it struck the air. In fact, air does not behave like a solid rubber ball at all. Air behaves like a liquid and flows around the ball, from the front to the back, as the ball ploughs its way through the air. By sneaking around the back like that, only about 1/16 of the air in front of the ball gets pushed forward by the baseball. If a 145-g baseball collided with a $94/16 = 6$ g rubber ball then the baseball would slow down by about 10%. This is indeed what happens as the ball ploughs through the air on its way to the batter. The ball slows down by about 10% regardless of the initial speed of the ball. The loss in momentum of the ball results in an equal and opposite gain in momentum of the air around it, just as if the baseball collided with a 6 g rubber ball.

All objects experience a drag force as they pass through air. The same effect occurs in water, but the effect is 826 times bigger at the same speed. If a baseball is dropped vertically into a swimming pool then the ball will slow down rapidly, come to a stop and then float to the top. A ball traveling through the air decelerates much more slowly. In fact, a baseball dropped vertically to the ground speeds up as it falls since the force of gravity is normally much larger than the drag force. We will see in a moment that the drag force on a baseball is equal to the force of gravity when the ball speed is about 90 mph.

The drag force depends on both the cross-sectional area of the object and its aerodynamic shape. Long, pointy objects experience only a small drag force, partly because of their streamline shape but also because of their small cross-sectional area. The drag force on a flat, circular disk of area A is given by $F = \frac{1}{2}\rho v^2 A$ where ρ is the density of the air (1.21 kg m^{-3} at room temperature) and v is the speed of the object. For objects of any other shape, the drag force is given by

$$F = \frac{1}{2} C_{\mathrm{D}} \rho v^2 A, \tag{3.1}$$

where C_D is called the drag coefficient and A is the cross-sectional area of the object (not its surface area). In (3.1), F is expressed in Newton when A is given in m^2 and v is given in m s^{-1}. For a sphere of radius R, $A = \pi R^2$. For a flat disk, $C_D = 1$. For a spherical ball, $C_D = 0.5$ at low ball speeds. A sphere experiences less air resistance than a flat disk because a sphere allows air to flow smoothly around its surface to the rear side of the sphere. Even so, doubling the speed of a ball increases the drag force four times. The momentum transferred to each air molecule is doubled when the ball speed is doubled, and the ball collides with twice as many air molecules each second, so the force on the ball increases four times.

It is interesting to compare the drag force on a baseball with the weight of the ball. For a baseball of diameter 74 mm, (3.1) gives $F = 0.0013v^2$ (N) when $C_D = 0.5$. The weight of a 145 g baseball is $mg = 1.42$ N (0.32 lb). The drag force is equal to the weight when $v = 33$ m s^{-1} (73.9 mph). However, C_D decreases below 0.5 at ball speeds above about 30 mph. Measurements show that the drag force is equal to the weight at a ball speed of about 90 mph (40.2 m s^{-1}), not 73.9 mph, in which case C_D is about 0.34 at speeds around 90 mph.

At high ball speeds the flow of air around the ball becomes turbulent and then the drag coefficient itself decreases as the ball speed increases. The total force on the ball increases with ball speed, because of the v^2 term in (3.1), but if C_D drops as v increases then the force on the ball increases more slowly than it otherwise would. The change in C_D with ball speed has been carefully measured for golf balls, but has still not been measured accurately for baseballs or softballs. The effect in golf is sufficiently important that it makes a big difference to the flight of a golf ball if the ball is perfectly smooth or has dimples. The effect of the dimples is to reduce C_D to about 0.2 at high ball speeds. The reduction in C_D at high ball speeds is referred to as the "drag crisis," the reduction itself depending on the smoothness or roughness of the ball surface [7, 18]. The reduction is large for smooth golf balls, does not occur at all for rough tennis balls, and is relatively small for baseballs and softballs.

If someone were to drop a baseball out of high flying helicopter, the ball would accelerate as it fell, and the speed would increase at a rate of 9.8 m s^{-1} every second. Eventually, the drag force of the air would be so large that it would be equal to the weight of the ball. At that point, the drag force acting up on the ball would be equal to the force of gravity acting down, and there would be zero total force on the ball. The ball would then fall at a constant speed, called the terminal velocity. The terminal velocity of a baseball is about 90 mph as we have just seen.

When a pitcher throws a ball in an approximately horizontal direction, the drag force acts backward along the ball path while gravity acts vertically down (Fig. 3.2). The two forces don't cancel since they act in different directions. Gravity acts to pull the ball downward so the ball follows a curved path. As it does so, the ball slows down along that path due to the drag force. If the pitcher were to throw the ball twice as fast, then the drag force would increase four times but it would act on the ball for only half the time. As a result, the fractional or percentage decrease in speed of the ball, over the 60 ft path to the batter, remains the same (about 10%). The change in ball speed over the 60 ft path is twice as big when the ball is thrown twice as fast, but the percentage change remains the same. Another way of explaining this result

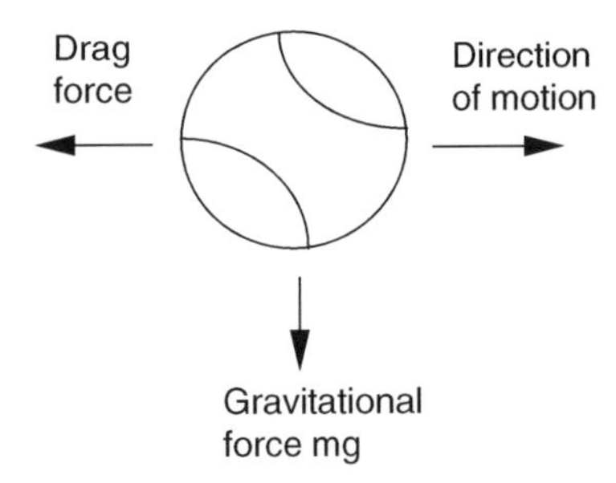

Fig. 3.2 The drag force on a ball acts in a backward direction, slowing it down

is to consider the energy of the ball. If the ball speed doubles then its kinetic energy increases four times, so it needs four times the force to slow it down by the same fractional amount (given that work = force times distance).

One of the most famous experiments of all time was reputed to have been conducted by Galileo around 1600. Whether he actually did the experiment is uncertain since he never actually said he did, despite the fact that he did write up most of his other work. The experiment was to drop two balls simultaneously from the leaning tower of Pisa. The experiment may well have been done by someone else previously, but it demonstrated that a light ball falls at the same speed and with the same acceleration as a heavy ball. If one takes into account the drag force, then the light ball would actually take a fraction longer. Galileo or anyone else at the time would not have been able to measure the difference accurately, nor could they be certain that the balls were released at exactly the same time. The point of the experiment was not to measure these small differences, but to show that if one ball is ten times heavier than another, it does not fall ten times faster.

3.5 Pressure Difference on a Ball

The drag force on a baseball can be explained in several different ways. One way is to consider the effect of air pressure on the ball. Even when the ball is at rest, the air exerts a pressure $p = 14.7$ lb per square inch everywhere on the surface of the ball. The total force on each side of the ball is $F = \pi R^2 p$ where $R = 1.45$ in. is the radius, so $F = 97$ lb. The total force in any one direction is zero since there is a force of 97 lb on one side and an oppositely directed force of 97 lb on the other side. However, when the ball is moving through the air, the air pressure on the upstream side is greater than the air pressure on the downstream side. At a ball speed of 90 mph, the difference in force on each half of the ball is the same as the weight of the ball, or 0.32 lb. The drag force at 90 mph can be explained by a 0.32% increase in air pressure at the front of the ball, or a 0.32% decrease in pressure at the rear of the ball, or a 0.16% increase at the front and a 0.16% decrease at the back.

These numbers make sense in terms of what we know about air. One cubic centimeter of air, about the volume of the top end of your thumb, contains 3×10^{19} molecules of oxygen and nitrogen. That's 3 followed by 19 zeros. Each molecule is moving around at an average speed of 1,085 mph, bumping into other molecules at a rate of 5 billion collisions/s, and colliding with the baseball. It is the continuous bombardment of the surface of the ball that results in a force of 14.7 lb on every square inch. If the ball is traveling at 90 mph, then the average speed of an air molecule striking the front surface effectively increases. The force exerted on the ball surface increases with the molecular speed squared. If the speed of each molecule is doubled then each molecule transfers twice the momentum to the ball in each collision. The squared factor arises because the number of molecules hitting the surface each second is also doubled.

Normal atmospheric pressure is given by the relation $p_A = \rho v_m^2/3$ where ρ is the air density and v_m is the average speed of an air molecule. The pressure exerted against a surface by air moving at speed v is given by $p = p_A + \rho v^2/2$. The fractional increase in air pressure at 90 mph is therefore about 90 squared divided by 1,085 squared, an increase of 0.7%. The shape of the ball reduces the effect by about half since the surface of the ball is not everywhere at right angles to the path of the ball. A 0.32% drop in air pressure across the ball is therefore about the expected amount at a ball speed of 90 mph, and it results in a net force of about 0.32 lb acting to decelerate the ball as it travels through the air.

3.6 Effects of Spin on the Trajectory

A ball that is struck or pitched with spin experiences three different forces on its way to its destination. There is the usual force of gravity pulling it toward the earth, there is the usual backward force due to air resistance, and there is an additional force that deflects the ball in a direction at right angles to its path. The additional force is known as the Magnus force, named after one of the early scientists who discovered its existence. Some simple demonstrations of the Magnus force are described in Project 2.

The Magnus force does not alter the speed of the ball but it acts at right angles to the path of the ball and at right angles to the spin axis. For example, consider the situation shown in Fig. 3.3a where a ball is pitched in a horizontal direction, is traveling right to left, and is spinning about a vertical axis in a clockwise direction when viewed from above. In that case, the Magnus force acts in a horizontal direction, causing the ball to curve to the right when viewed by the pitcher (or to the left when viewed by the batter). If the ball was spinning counter-clockwise, it would curve in the opposite direction.

The origin of the Magnus force can be explained by imagining that the ball spins clockwise about a fixed axis, and that the air is flowing past the ball, as shown in Fig. 3.3b. The air is deflected downward by the ball since it is dragged in a clockwise direction by friction between the ball and the air, with the result that there is an equal and opposite force exerted by the air on the ball.

The effect of the Magnus force is to change the curvature of the trajectory. The trajectory of a ball curves downward due to the downward gravitational pull of the

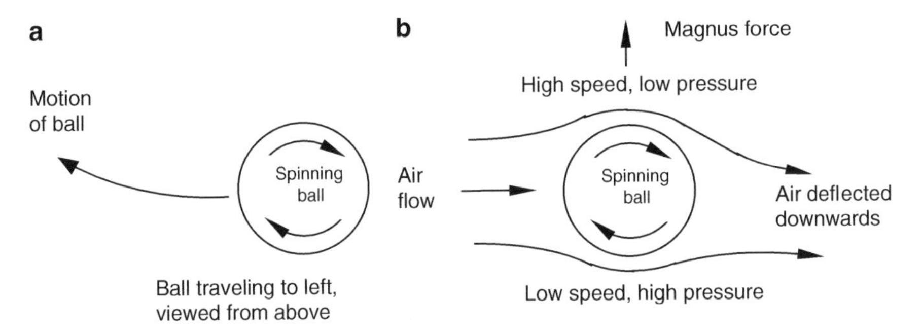

Fig. 3.3 (**a**) A ball traveling to the *left* and spinning clockwise when viewed from above, *curves* to the *right*. (**b**) Air flowing past a spinning ball at rest is deflected downward in this diagram, with the result that the air exerts an upward force on the ball known as the Magnus force

earth, but the Magnus force can increase or decrease that curvature depending on the direction of the Magnus force. If the ball is spinning about a horizontal axis at right angles to the path of the ball, then it has topspin or backspin. With topspin, the Magnus force acts vertically downward, and the curvature increases. With backspin, the Magnus force acts vertically upward, opposing the force of gravity, so the curvature decreases and may even reverse sign. That is, if the ball is spinning fast enough, it can even curve upwards toward the sky instead of downwards toward the earth.

If a ball is pitched with topspin or backspin and is headed over the middle of the home plate, then it will pass straight over the middle of the plate. However, if a ball is pitched with sidespin (that is, spinning about a vertical axis), and if it starts out in a direction that would carry it over the middle of the home plate, then it will curve in a horizontal direction and pass over a different point as it crosses the home plate.

The effect of the Magnus force on a softball struck at 100 mph is shown in Figs. 3.4 and 3.5. The ball can come off the bat with a small or a large amount of topspin or backspin, depending on whether the ball is struck near the middle or toward one edge (of the ball or the bat). Figure 3.4 shows the trajectories of a ball launched at 40° with different amounts of spin. Figure 3.5 shows the trajectories of a ball launched with a large amount of backspin, 40 rev s^{-1}, but at different angles above the horizontal. The question of greatest practical interest concerns the distance traveled by the ball.

Figure 3.4 shows that a softball launched 40° above the horizontal travels the greatest distance when it is struck with backspin, at about 10 rev s^{-1}. When the ball is launched with topspin, the Magnus force acts downward on the ball and the ball lands short. The Magnus force actually acts in a direction perpendicular to the path of the ball, so it has a forward component on a topspin ball as the ball is rising and a backward component as the ball is falling. If the ball is launched with a large amount of backspin, it rises to a greater height due to the extra lift provided by the Magnus force, but the ball then gains additional speed as it falls and strikes the ground at a relatively steep angle. Too much backspin can cause the ball to

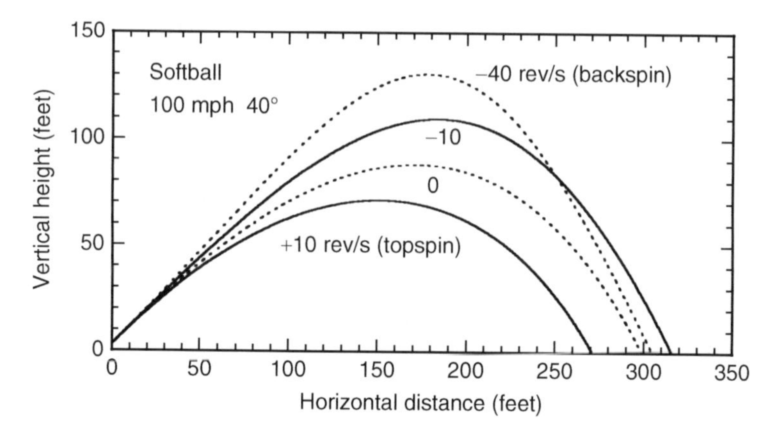

Fig. 3.4 Trajectories of a softball hit from a height of 3 ft at 100 mph, at 40° to the horizontal, with different amounts of spin. The ball travels farthest if it has a small amount of backspin

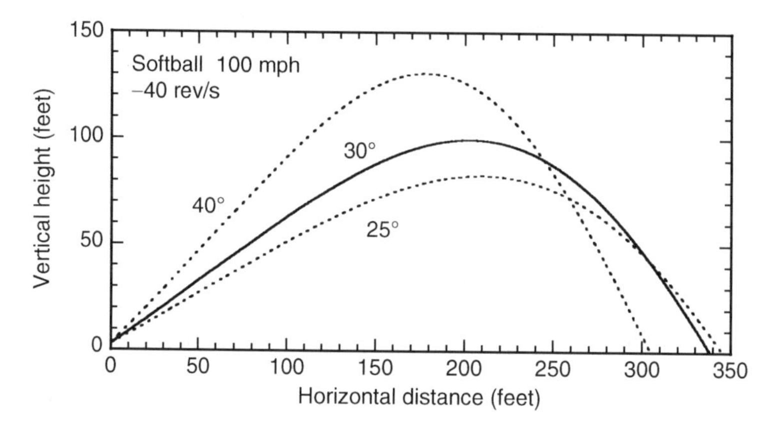

Fig. 3.5 Trajectories of a softball hit from a height of 3 ft at 100 mph, with backspin at $-40\,\text{rev s}^{-1}$, at various angles to the horizontal

travel a smaller horizontal distance than a ball launched with just a small amount of backspin. The optimum amount of backspin depends on the launch angle, as explained in the following paragraphs.

Figure 3.5 shows that a softball launched at 100 mph spinning backwards at $40\,\text{rev s}^{-1}$ (2,400 rpm) will travel the maximum horizontal distance when launched at about 25°. At $40\,\text{rev s}^{-1}$, the Magnus force acting upward on the ball is slightly greater than the force of gravity acting down on the ball, so the ball curves slightly upwards at first. The ball is slowed by the drag force, as well as by the gravitational force while the ball is rising, resulting in a decrease in the Magnus force during the flight of the ball. When the Magnus force drops below the gravitational force, the ball starts to curve downwards.

The horizontal distance traveled by the ball before it falls back to earth is given by the horizontal speed multiplied by the time that the ball remains in the air. The horizontal speed is largest when the ball is launched at a low angle. The time in the air is largest when the ball is launched at a high angle. If the air had no effect on the ball then the horizontal distance would be greatest at a launch angle of 45°. The effect of the drag force in slowing the ball is largest when the ball spends a long time in the air. Consequently, there is an advantage in reducing the time in the air and maximizing the horizontal speed by launching the ball at an angle less than 45°. For a softball or a baseball struck without spin, the flight distance is a maximum when the launch angle is about 39°.

If the ball is launched with backspin, then the ball travels farthest when the launch angle is less than 39°, by an amount that increases as the ball spin increases. One effect of backspin is to increases the amount of time the ball spends in the air. There is an additional effect of backspin while the ball is rising, and it is due to the fact that the Magnus force is not a vertical force. Rather, the Magnus force acts in a direction perpendicular to the path of the ball, so it has a large backward component when the ball is rising steeply. As a result, the horizontal speed of the ball is reduced significantly by the Magnus force, particularly when the ball is launched at a high angle and is spinning rapidly. After the ball reaches its maximum height and starts falling, the Magnus force has a forward component and it acts to accelerate the ball in the horizontal direction. However, by that time, the ball speed has decreased and the Magnus force is weaker. The net result is that the Magnus force reduces the horizontal speed of the ball by an amount that is proportional to the launch angle. That is why, in golf, the longest drives are those where the ball is launched at a relatively low angle. For a baseball or a softball spinning backwards at 40 rev s^{-1}, the ball travels farthest when it is launched at an angle of about 25°.

3.7 Pop-Ups

It is common in baseball and softball for a batter to strike the bottom of the ball rather than the middle of the ball, in which case the ball flies almost straight up in the air as a "pop-up." The ball spins furiously, at around 100 rev s^{-1}, and is difficult to catch since it is difficult to predict where the ball will land. Examples of this type of trajectory are shown in Fig. 3.6, for a ball struck at 75 mph with 100 rev s^{-1} of backspin. The trajectory of the ball is quite sensitive to the launch angle. The Magnus force acts at right angles to the path of the ball so it acts substantially in the horizontal direction if the ball is rising almost vertically. The Magnus force acts to the left while the ball is rising, and to the right as it is falling, so the ball can trace out a loop at the top of its trajectory, confusing anyone trying to catch the ball. The reason that such a ball is so difficult to catch is described in more detail in [27]. Typically, the catcher runs to where he thinks the ball will land, only to find that the ball has a mind of its own and it veers off in a direction away from the catcher.

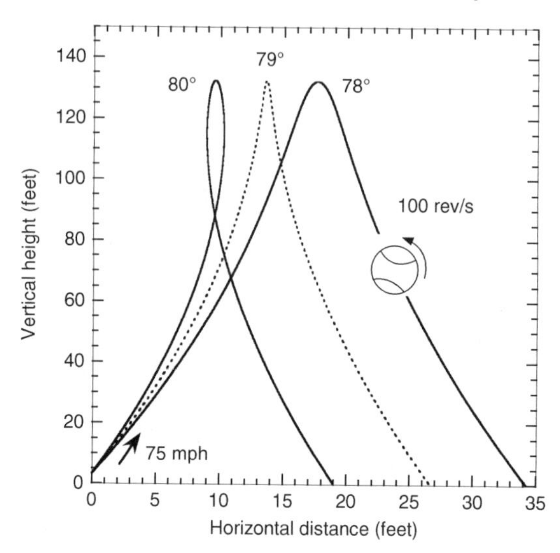

Fig. 3.6 Trajectories of a baseball hit from a height of 3 ft at 75 mph, with backspin at $-100\,\mathrm{rev\,s^{-1}}$, at various angles to the horizontal

3.8 Effects of Weather and Altitude

The trajectory of a ball through the air is affected by the lift and drag forces acting on the ball, both forces being directly proportional to the density of the air. It was assumed when calculating typical trajectories in this chapter that the density of the air was $1.21\,\mathrm{kg\,m^{-3}}$ since this is a typical value. In fact, the density of the air varies slightly with temperature, pressure, altitude, and humidity. Small changes can therefore be expected in the calculated trajectories under conditions where the temperature, pressure, humidity or altitude are not typical. An especially interesting case concerns Denver, Colorado, which is at an altitude of 1 mile or 5,280 ft and where the air density is about 82% of that at sea level. The drag and lift forces on a ball in Denver are therefore about 18% smaller than at other locations near sea level.

At sea level, a baseball pitched over a distance of 60 ft slows down by about 10% through the air, meaning that the average speed over the whole 60 ft is about 95% of the pitched speed. If it slowed down from 100 mph to 90 mph then the average speed would be 95 mph. A baseball pitched in Denver slows down by about 8% through the air, so its average speed over the whole 60 ft is about 96% of the pitched speed. The ball therefore arrives about 1 mph faster, which is only a small effect, especially considering the fact that the pitcher can vary his pitch speed by around 10% from one pitch to the next. Nevertheless, observant batters might detect that fast balls arrive about 1 mph faster at Denver.

A more significant effect results from the Magnus force which is also 18% smaller at Denver. A curveball is pitched with topspin and curves downward at a greater rate than a ball pitched without spin. In Denver, a curveball will not drop as fast and drops about 4 in. less than at sea level. On the other hand, a fastball (with backspin) follows a straighter path at sea level and will drop about 4 in. more at Denver [29].

The variation of air pressure with altitude can be expressed in the form

$$p = p_o(1 - 0.0226A)^{5.258}, \tag{3.2}$$

where A is the altitude in km, p_o is the air pressure at sea level and p is the pressure at altitude A. At Denver, $A = 1.609$ km (1 mile) so $p = 0.823p_o$. On weather maps, air pressure is corrected to reflect the pressure at sea level, otherwise it would seem that Denver was always in the middle of a low pressure zone. If the temperature in Denver is the same as that at sea level, then the air density in Denver is 82% of that at sea level, since air pressure is proportional to air density at any given temperature.

Despite the fact that water is denser than air, humid air is less dense than dry air since water molecules are lighter than air molecules. Air contains 78% nitrogen, 21% oxygen, and only about 2% water vapor (depending on the temperature and humidity). The amount of water vapor in the air is quite small, even at 100% humidity. At 100% humidity, where the air is said to be saturated with water vapor, the pressure of the water vapor is only 611 Pa at 0°C (32°F), 2,338 Pa at 20°C (68°F), and 4,242 Pa at 30°C (86°F). Total air pressure is 101,325 Pa at "standard pressure" so water vapor makes up only about 4% of the total pressure even on a hot, humid day.

Standard pressure is just the average air pressure at sea level, corresponding to an average reading of 29.9 in. of Hg on a weather chart, equivalent to an actual pressure of 14.7 lb in.$^{-2}$ or 1013.2 hPa = 101.32 kPa. Weather maps and barometer gauges show that the air pressure can be higher or lower than the average, ranging from about 28 in. Hg to about 30.6 in. Hg. Since the air pressure can vary by about 9% from one region to another or from 1 week to the next, the air density can also vary by about 9% simply due to differences in air pressure at sea level. Even greater variations in density can occur when extreme pressure variations are combined with extreme variations in temperature and humidity, as indicated in Fig. 3.7. The highest air density in Fig. 3.7 is 1.34, and the lowest is 1.08, a difference of 24%, although it is unlikely that any team would encounter such wide weather extremes in any given season.

The density of humid air, in kg m^{-3}, can be found from the ideal gas law for a mixture of gases and is given by the formula

$$\rho = \frac{p_d}{R_d T} + \frac{p_v}{R_v T}, \tag{3.3}$$

where p_d is the pressure due to the dry air (in Pa), p_v is the pressure due to the water vapor (in Pa), T is the temperature in degrees Kelvin, $R_d = 287.0$ and $R_v = 461.5$. $p_v = Hp_{\text{sat}}$, where H is the relative humidity (e.g., $H = 0.6$ means 60% humidity) and p_{sat} is the water vapor pressure at temperature T when the air is saturated (i.e., when $H = 1$). The total air pressure is $p_d + p_v$, being the sum of the partial pressures. p_{sat} in Pascals is given by the formula

$$p_{\text{sat}} = 610.8 \times 10^S, \quad \text{where} \quad S = \frac{7.5T_C}{T_C + 237.3}, \tag{3.4}$$

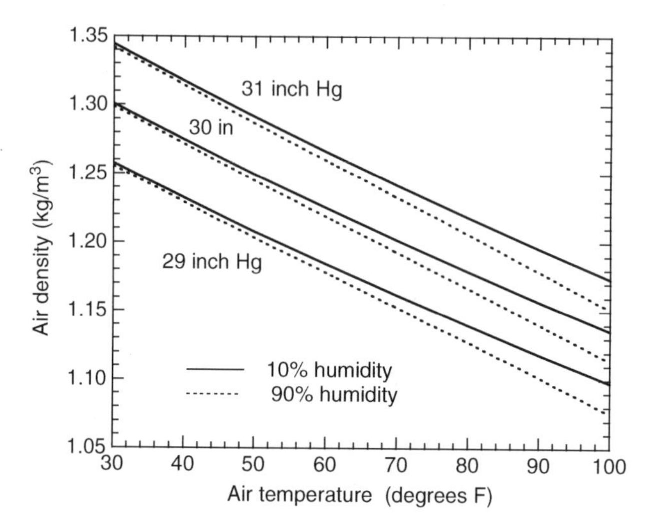

Fig. 3.7 Density of air as a function of air temperature for three typical values of air pressure at sea level or slightly above, each at 10% or 90% humidity. One mile up, the air pressure is about 24.5 in. Hg. The density of air decreases as the temperature increases, so hot air rises and cold air sinks in any region where there is a difference of temperature

and where T_C is the air temperature in Celsius. On a very hot day, when $T_C = 40°C$ (104°F), $p_{sat} = 7{,}375$ Pa. If the humidity is say 60% then $p_v = 5{,}162$ Pa.

Since p_v is relatively small even on a humid day, the main effect on air density is that due to the temperature or pressure of the air. Suppose we ignore moisture in the air, and assume standard air pressure, then $\rho = 352.9/T$ according to (3.3). Since 0°C corresponds to $T = 273$ K, the density of dry air is 1.293 kg m^{-3} at 0°C (32°F) and standard pressure. At 20°C (68°F) or 293 K, $\rho = 1.204$ kg m^{-3} at standard pressure. The air density, therefore, decreases by 7% when the temperature rises by 20°C.

3.9 Effect of Wind

The drag force on a ball depends on the relative speed, v, of the ball and the surrounding air. If the air is at rest then v in (3.1) is the speed of the ball. If the ball is at rest in a wind tunnel then v is the speed of the air. If a ball is traveling at 80 mph into a 10 mph headwind then v is 90 mph. If the ball is traveling at 80 mph assisted by a 10 mph tailwind, then v is 70 mph. If the ball is traveling at 80 mph from north to south and the wind is traveling at 10 mph from east to west then the ball will continue to travel southward but it will curve slightly to the west. In the latter case, the calculation of the separate forces in the north and west directions is slightly tricky. The relative speed in that case is 80.6 mph. Relative to the air, the ball is traveling at 80 mph south and 10 mph west, the vector addition of these components giving a ball speed of 80.6 mph. The drag force can then be calculated for a ball speed of 80.6 mph in still air, and it acts in a direction opposite the direction of motion of the ball. Since the drag force is proportional to v^2, the force on the ball in the westward direction is a lot bigger than one would expect for a light 10 mph breeze acting on a stationary ball. The drag force in the north direction is only slightly larger than that on an 80 mph ball.

Consider the case of a baseball struck at 100 mph without spin and launched 40° above the horizontal. If $C_D = 0.5$ then the ball travels 310 ft, as shown in Fig. 3.1. If $C_D = 0.4$ the ball would travel 343 ft. A 20% decrease in the drag force therefore increases the range of the ball by 33 ft or by about 10% in this case. If $C_D = 0.4$ and the ball is traveling into a horizontal 10 mph headwind then the drag force is about the same as that on a ball struck at 108 mph in still air, at least at the start of the trajectory. The drag force increases by about 17% due to the headwind and it changes direction slightly to be more in line with the horizontal. Trajectory calculations show that the ball then travels only 302 ft. If the ball was assisted by a horizontal 10 mph tailwind then it would travel 383 ft. Wind can therefore make a big difference to the range of the ball, especially if it reverses direction during the course of a game. The range is offset slightly by the fact that the pitch speed is also affected. If the batter has the wind behind him, then the ball will arrive from the pitcher at a slightly lower speed and come off the bat at a slightly lower speed, so the net effect of the wind is not quite as large as we have just calculated.

Appendix 3.1 Trajectory Equations Without Air Resistance

The trajectory of a ball launched at $36\,\mathrm{m\,s^{-1}}$ is shown in Fig. 3.8. The ball reaches its maximum height at $t = 2.04$ s and falls back to the ground, where $y = 0$, at $t = 4.08$ s. During those 4.08 s, the ball travels a horizontal distance $x = 30 \times 4.08 = 122.4$ m. If an outfielder throws the ball not from a height $y = 0$ m

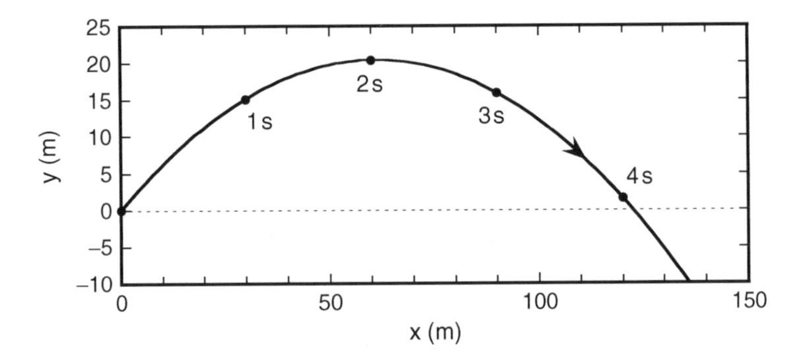

Fig. 3.8 The trajectory of a ball launched at a speed of $36\,\mathrm{m\,s^{-1}}$ at 33.7° to the horizontal, ignoring air resistance. Each *dot* shows the ball position at 1 s intervals. This particular trajectory was also considered in Sect. 1.2

but from a height of 2 m then the ball would travel slightly further before hitting the ground. To take that effect into account we could still start the ball at $y = 0$ but let it land at $y = -2$ m. In practice, air resistance will slow the ball and it will land well before it reaches the 122 m mark.

To work out the trajectory of the ball launched as shown in Fig. 3.8, we can use the equation

$$s = ut + \frac{1}{2}at^2,$$

where s = distance traveled, u = initial speed at launch, t = time after launch and a = acceleration. The speed, v, at any time after launch is given by $v = u + at$.

There is no force and no acceleration in the horizontal direction. The distance, x, in the horizontal direction is therefore given by $x = ut = 30t$ since $u = 30\,\mathrm{m\,s^{-1}}$ in the horizontal direction and $a = 0$. At $t = 0$, $x = 0$. At $t = 1$ s, $x = 30$ m. At $t = 2$ s, $x = 60$ m.

The acceleration in the vertical direction is given by $a = -g = -9.8\,\mathrm{m\,s^{-2}}$ where g is the acceleration due to gravity. The ball is rising upwards and slowing down in the vertical direction so a is negative. In the vertical direction, $u = 20\,\mathrm{m\,s^{-1}}$ and the distance, y, traveled after time t is $y = ut - 0.5gt^2 = 20t - 4.9t^2$. At $t = 0$, $y = 0$. At $t = 1$ s, $y = 15.1$ m. At $t = 2$ s, $y = 20.4$ m.

The maximum range of a ball launched at a fixed speed can be calculated as follows

Suppose that a ball is launched at speed v at an angle θ above the horizontal, starting at $x = y = 0$ when $t = 0$. The horizontal launch speed is then $v_{xo} = v\cos\theta$ and the vertical launch speed is $v_{yo} = v\sin\theta$. The x and y coordinates are given by

$$x = v_{xo}t \quad \text{and} \quad y = v_{yo}t - \frac{1}{2}9.8t^2$$

The horizontal and vertical speeds at any time t are given by

$$v_x = v_{xo} \quad \text{and} \quad v_y = v_{yo} - 9.8t$$

The ball reaches its maximum height at a time t_m when $v_y = 0$, so $t_m = v_{yo}/9.8$. At this time, the values of x and y are

$$x_m = \frac{v_{xo}v_{yo}}{9.8} \quad \text{and} \quad y_m = \frac{v_{yo}^2}{19.8}$$

The ball takes the same time to fall from its maximum height as it takes to rise to its maximum height. The ball, therefore, lands at time $t = 2t_m$ after traveling a horizontal distance R given by

$$R = 2v_{xo}t_m = \frac{2x_o y_o}{9.8} = \frac{v^2\sin(2\theta)}{9.8},$$

where $\sin(2\theta) = 2\sin\theta\cos\theta$. R is commonly known as the range of the ball. If it is launched at $y = 0$ and lands at $y = 0$ then R is a maximum when $\sin(2\theta) = 1$ or when $\theta = 45°$. In that case, $R = v^2/9.8$.

Appendix 3.2 Measurement of Drag Force

It is difficult to measure C_D accurately for baseballs and softballs using flight time information since air drag has a relatively small effect on the flight of these balls, at least over small distances. Figure 3.3 shows a much bigger effect over large distances since the drag force then acts for a relatively long time and slows the ball considerably. By way of example, consider a baseball pitched at a horizontal speed of 80 mph (117.4 ft s^{-1}) over the 60 ft horizontal distance from the pitcher to the batter. In the absence of air resistance, the ball would travel at 117.4 ft s^{-1} the whole way, taking 60/117.4 = 0.51 s to arrive. The effect of air resistance can be estimated from (3.1), giving

$$m\mathrm{d}v/\mathrm{d}t = -F = -\frac{1}{2}C_D\rho v^2 A = -0.0026 C_D v^2 \quad (3.5)$$

for a 37 mm diameter baseball, where m is the mass of the ball (in kg), v is its horizontal speed (in m s^{-1}) and the air density is taken as 1.21 kg m^{-3}. The effects of spin and vertical motion have been ignored so that we can estimate the flight time in the horizontal direction. For a baseball, $m = 0.145$ kg, and then (3.5) can be written as

$$\mathrm{d}v/\mathrm{d}t = -Kv^2 \quad \text{where} \quad K = 0.0026C_D/0.145 = 0.0179C_D \quad (3.6)$$

The solution of (3.6) is

$$v = \frac{v_o}{(1 + Kv_o t)}, \quad (3.7)$$

where v_o is the initial speed out of the pitcher's hand and $v_o t$ is the horizontal distance the ball would travel after a time t if it continued at speed v_o. Equation (3.7) shows how the velocity of the ball decreases with time, and indicates that the ball will slow down by about 10% over the 60 ft distance from the pitcher to the batter if C_D is about 0.4.

An alternative and slightly more convenient solution of (3.6) can be found in terms of the distance, s, traveled by the ball. Since $v = \mathrm{d}s/\mathrm{d}t$, we can write (3.6) as $\mathrm{d}v/\mathrm{d}t = v\mathrm{d}v/\mathrm{d}s = -Kv^2$, in which case

$$\mathrm{d}v/\mathrm{d}s = -Kv, \quad (3.8)$$

which has the solution

$$v = v_o \mathrm{e}^{-Ks}, \quad (3.9)$$

and which allows us to calculate the speed of the ball after it travels a distance s. If we eliminate v using (3.7) we find that

$$t = \frac{(e^{Ks} - 1)}{K v_0}, \tag{3.10}$$

which gives the time taken by the ball to travel the distance s.

If we take the distance from the pitcher's release point to the home plate as 16.76 m (55 ft) then $Ks = 0.300 C_D$ and hence $v/v_o = 0.91$ if $C_D = 0.3$ while $v/v_o = 0.89$ if $C_D = 0.4$. The ball, therefore, arrives at a speed of about 90% of the launch speed, regardless of the actual pitch speed and regardless of the exact value of C_D.

Suppose that the ball is pitched at 80 mph (117.4 ft s^{-1}). It crosses the home plate at 72.8 mph (106.8 ft s^{-1}) if $C_D = 0.3$ or at 71.2 mph (104.5 ft s^{-1}) if $C_D = 0.4$. A very accurate measurement of both the pitch speed and the arrival speed is, therefore, required to distinguish the two cases. The best estimates of C_D to date indicate that C_D is about 0.4 for a low speed pitched baseball (around 70 mph) and C_D is about 0.35 for a high speed pitched baseball (around 90–100 mph).

Appendix 3.3 Measurement of Lift Force

The Magnus force on a spinning ball is commonly described as a lift force since it sometimes acts in a vertical direction. In golf, the Magnus force acts vertically upward when the ball is traveling horizontally with backspin, so it indeed acts as a lift force. Lift forces are described by the relation

$$F = \frac{1}{2} C_L \rho v^2 A, \tag{3.11}$$

where C_L is called the lift coefficient. If the ball is not spinning, or if the spin axis is parallel to the direction of motion of the ball (a gyroball) then $C_L = 0$. If the ball is spinning at angular velocity ω, and if the spin axis is perpendicular to the direction of motion of the ball, then a good fit to the experimental data for baseballs [26] is

$$C_L = \frac{1}{[2.32 + 0.4(v/v_{spin})]}, \tag{3.12}$$

where $v_{spin} = R\omega$ is the peripheral speed of the ball and R is its radius. A typical value of C_L is about 0.2, depending on the spin of the ball. For example, suppose that a ball is pitched at 90 mph (40.2 m s^{-1}) and is spinning at 2,000 rpm (209 rad s^{-1}). For a baseball, where $R = 37$ mm, $v_{spin} = 7.73$ m s^{-1} so $v/v_{spin} = 5.2$, in which case $C_L = 0.23$. From (3.11), the Magnus force is then 0.97 N when $\rho = 1.21$ kg m^{-3}. The force of gravity on a baseball is 1.42 N.

The measured lift coefficient for a spinning tennis ball is similar to that for a spinning baseball but slightly lower. For example, Stepanek [13] found that $C_L = 0.25$ for a tennis ball when $R\omega/v = 0.5$, whereas Nathan [26] found that $C_L = 0.33$ for a baseball when $R\omega/v = 0.5$. In Nathan's experiment, a pitching machine was used to project balls at high speed, and the flight of the ball was recorded with ten high speed video cameras operating at 700 frames s^{-1}.

When the spin is relatively small, then C_L is proportional to v_{spin}/v and the lift force is proportional to the spin and also proportional to v (rather than v^2). The vertical or horizontal break of a pitched ball will therefore increase as the spin increases, but it doesn't necessarily increase as the pitch speed increases. If the ball is thrown twice as fast then it takes only half the time to reach the batter. Even though the Magnus force might double, if it acts for only half the time then the break of the ball will be only half as big, the break being proportional to the force and to the time squared. The only advantage of throwing a faster ball in that case is that it gives the batter less time to think about it. However, a pitcher is likely to spin the ball faster when he throws it faster, in which case the break in a high speed pitch will be slightly smaller than the break in a low speed pitch. Further discussion concerning the break of pitched balls is given in Sect. 4.3.

Appendix 3.4 Trajectory Equations with Lift and Drag

Consider a ball rising at angle θ to the horizontal at speed v. The x and y components of the velocity are $v_x = v\cos\theta$ and $v_y = v\sin\theta$. The forces acting on the ball, if it has backspin, are shown in Fig. 3.9. The drag force $F_D = 0.5C_D\rho Av^2$ acts backward and the lift force $F_L = 0.5C_L\rho Av^2$ acts at right angles to the path of the ball. The equations describing the trajectory of a ball through the air are then

$$m\mathrm{d}v_x/\mathrm{d}t = -F_D\cos\theta - F_L\sin\theta \tag{3.13}$$

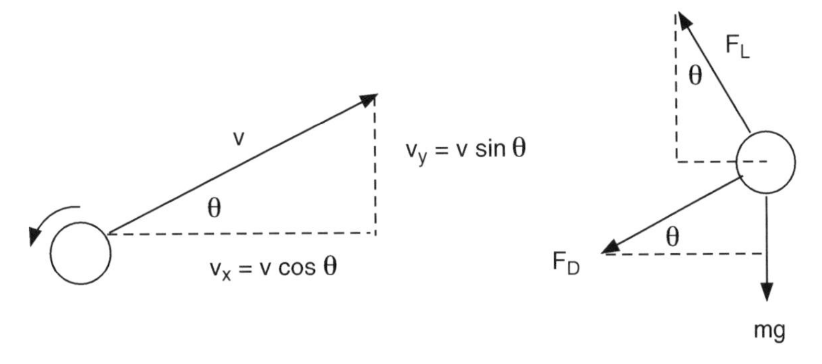

Fig. 3.9 The velocity components and forces acting on a ball rising upward at angle θ to the horizontal with backspin

and

$$m\mathrm{d}v_y/\mathrm{d}t = F_{\mathrm{L}}\cos\theta - F_{\mathrm{D}}\sin\theta - mg, \tag{3.14}$$

which can be expressed in the form

$$\mathrm{d}v_x/\mathrm{d}t = -kv(C_{\mathrm{D}}v_x + C_{\mathrm{L}}v_y) \tag{3.15}$$

and

$$\mathrm{d}v_y/\mathrm{d}t = kv(C_{\mathrm{L}}v_x - C_{\mathrm{D}}v_y) - g, \tag{3.16}$$

where $k = 0.5\rho\pi R^2/m$, R being the ball radius. If the ball has topspin then the signs in front of C_{L} need to be changed. Equations (3.15) and (3.16) need to be solved numerically in general. A good check on the numerical accuracy is to consider the vertical drop of a ball without spin, in which case $v_x = 0$ and then (3.16) can be solved analytically.

References

1. F.L. Verwiebe, Does a Baseball Curve? Am. J. Phys. **10**, 119–120 (1942)
2. R.M. Sutton, Baseballs Do Curve and Drop! Am. J. Phys. **10**, 201–202 (1942)
3. L.J. Briggs, Effect of Spin and Speed on the Lateral Deflection Curve of a Baseball; and the Magnus Effect for Smooth Spheres. Am. J. Phys. **27**, 589–596 (1959)
4. N.F. Smith, Bernouilli and Newton in Fluid mechanics. Phys. Teach. **10**, 451–455 (1972)
5. R.G. Watts, E. Sawyer, Aerodynamics of a knuckleball. Am. J. Phys. **43**, 960–963 (1975)
6. B.L. Coulter, C.G. Adler, Can a body pass a body falling through the air? Am. J. Phys. **47**, 841–846 (1979)
7. C. Frohlich, Aerodynamic drag crisis and its possible effect on the flight of baseballs. Am. J. Phys. **52**, 325–334 (1984)
8. R. Mehta, Aerodynamics of sports balls. Ann. Rev. Fluid Mech. **17**, 151–189 (1985)
9. H. Erlichson, Is a baseball a sand-roughened sphere? Am. J. Phys. **53**, 582–583 (1985)
10. C. Frohlich, Comments on "Is a baseball a sand-roughened sphere?" Am. J. Phys. **53**, 583 (1985)
11. P.J. Brancazio, Looking into Chapman's homer: The physics of judging a fly ball. Am. J. Phys. **53**, 849–855 (1985)
12. A.F. Rex, The effect of spin on the flight of batted baseballs. Am. J. Phys. **53**, 1073–1075 (1985)
13. A. Stepanek, The aerodynamics of tennis balls – The topspin lob. Am. J. Phys. **56**, 138–142 (1986)
14. R.G. Watts, R. Ferrer, The lateral force on a spinning sphere: Aerodynamics of a curveball. Am. J. Phys. **55**, 40–44 (1987)
15. R.G. Watts, S. Baroni, Baseball-bat collisions and the resulting trajectories of spinning balls. Am. J. Phys. **57**, 40–45 (1989)
16. D.T. Kagan, The effects of coeficient of restitution variations on long fly balls. Am. J. Phys. **58**, 151–154 (1990)
17. L.W. Alaways, M. Hubbard, Experimental determination of baseball spin and lift. J. Sports Sci. **19**, 349–358 (2001)
18. G.S. Sawicki, How to hit home runs: Optimum baseball bat swing parameters for maximum range trajectories. Am. J. Phys. **71**, 1152–1162 (2003)

19. G.S. Sawicki, M. Hubbard, W.J. Stronge, Reply to Comment on How to hit home runs: Optimum baseball swing parameters for maximum range trajectories. [Am. J. Phys. **11**, 1152–1162 (2003)], Am. J. Phys. **73**, 185–189 (2005)
20. G.S. Sawicki, M. Hubbard, W.J. Stronge, Erratum: Reply to Comment on How to hit home runs: Optimum baseball swing parameters for maximum range trajectories. [Am. J. Phys. **11**, 1152–1162 (2003)], Am. J. Phys. **73**, 365 (2005)
21. R.K. Adair, Comment on How to hit home runs: Optimum baseball bat swing parameters for maximum range trajectories. [Am. J. Phys. **11**, 1152–1162 (2003)], Am. J. Phys. **73**, 184–185 (2005)
22. J.M. Pallis, R.D. Mehta, in *Balls and Ballistics*, Ch5, ed. by M. Jenkins. Materials in Sports Equipment (Woodhead, Cambridge, 2003), pp. 100–126
23. J.E. Goff, Heuristic model of air drag on a sphere. Phys. Educ. **39**, 496–499 (2004)
24. J. Rossman, A. Rau, An experimental study of Wiffle ball aerodynamics. Am. J. Phys. **75**, 1099–1105 (2007)
25. R. Cross, Aerodynamics of a party balloon. Phys. Teach. **45**, 334–336 (2007)
26. A. Nathan, The effect of spin on the flight of a baseball. Am. J. Phys. **76**, 119–124 (2008)
27. M.K. McBeath, A.M. Nathan, A.T. Bahill, D.G. Baldwin, Paradoxical pop-ups: Why are they difficult to catch? Am. J. Phys. **76**, 723–729 (2008)
28. E.R. Meyer, J.L. Bohn, Influence of a humidor on the aerodynamics of baseballs. Am. J. Phys. **76**, 1015–1021 (2008)
29. A.T. Bahill, D.G. Baldwin, J.S. Ramberg, Effects of altitude and atmospheric conditions on the flight of a baseball. Int. J. Sports Sci. Eng. **3**, 109–128 (2009)

Chapter 4
Pitching Trajectories

4.1 The Basics

The basic task of the pitcher is to throw the ball so that it passes over the home plate at a height between the batter's knees and shoulders. In baseball, the strike zone extends from the knees to a point midway between the shoulders and the top of the pants. In fast pitch softball the strike zone is essentially the same, although the upper part of the zone is defined as the bottom of the sternum. In slow pitch softball, the zone extends to the shoulders.

The object of the exercise is to pitch the ball in a manner that makes it as difficult as possible for the batter to strike the ball well. The pitcher can do this by varying the speed of the ball, or the height of the ball as it passes over the plate, or the magnitude and direction of the spin imparted to the ball. The pitcher can even deliberately throw the ball outside the strike zone, hoping that the batter might think the ball will curve into the strike zone. All pitched balls curve vertically downward toward the ground due to the effect of gravity. The effect of ball spin is to change the trajectory of the ball so that it curves at a different rate. If a ball has topspin it will fall toward the ground at a faster rate. With backspin, the trajectory is straighter and the ball falls to the ground at a slower rate. With sidespin, the ball curves in a horizontal direction (as well as in the vertical direction), and it can curve inward toward the batter or outward away from the batter depending on the direction of spin.

The distance between the pitcher and the batter is about 60 ft in baseball, about 43 ft in fast pitch softball and about 50 ft in slow pitch softball. The field dimensions are shown in Appendix 4.1. Pitchers throw overhand in baseball and underhand in softball. In fast pitch softball, the ball is pitched at speeds up to about 70 mph using a "windmill" style of pitching, starting with the ball at one hip, rotating the arm quickly in a full circle and then releasing the ball as it passes the hip.

In slow pitch softball, the ball is released underhand at a slow speed and must reach a maximum height between 6 and 12 ft above the ground before falling back down over the home plate. Most pitchers prefer to pitch at a low speed, around 30 mph, so that the ball drops from a height of about 12 ft. If the ball is pitched at around 45 mph, then it follows a straighter path and may not reach the required height of at least 6 ft. A faster ball is usually but not always easier to hit since it follows a straighter path.

R. Cross, *Physics of Baseball & Softball*, DOI 10.1007/978-1-4419-8113-4_4,

4.2 Some Pitched Ball Trajectories

If a pitcher could pitch a ball at 1,000 mph (incredibly fast!) then the ball would follow a path from the pitcher to the batter that is essentially a straight line. At normal pitching speeds, the ball travels in a curved path. The faster the ball, the straighter the path, and the slower the ball the more curved is the path. A ball pitched at about 30–40 mph in slow pitch softball needs to be launched at an angle about 20–30° above the horizontal so that it crosses the plate at the correct height, the ball reaching maximum height about half way between the pitcher and the batter.

In fast pitch softball, the ball travels at about 50–70 mph and it needs to be launched at angle of about 5 to 10° above the horizontal to cross the plate at the correct height. If the ball is launched without spin then it reaches maximum height after traveling about 30 ft and then falls downward as it heads toward the plate. If the ball is launched with backspin then it reaches maximum height closer to the plate. If the ball is spinning fast enough then it might reach its maximum height after it crosses the plate, in which case the batter will see the ball rising all the way. Such a pitch is called a rise ball.

In baseball, the ball is launched at high speed from about shoulder height and needs to be launched in an almost horizontal direction to cross the plate at the correct height. Curve balls are thrown slightly above the horizontal (since topspin makes the ball curve downward faster than a ball without spin) while fast balls are thrown slightly below the horizontal. The ball reaches its maximum height near the pitcher's hand and falls through a distance of about 3 ft all the way to the batter. Thrown with backspin, and launched at the same speed and angle, the ball will fall through a smaller distance, but it is not humanly possible to spin the ball fast enough so that it rises upward as it approaches the batter.

Slow Pitch Softball

Figure 4.1 shows some relatively slow and some relatively fast trajectories for slow pitch softball, where the ball is launched either at 29 mph or 45 mph so that it passes over the plate between the batter's knees and shoulders. For each trajectory, the ball was launched from a height of 3 ft, without spin. The position of the ball is shown at intervals of 0.1 s. If the ball is launched at 29 mph, then the launch angle must be between 32.4° and 35.8°. If the ball is launched at angle greater than 35.8° then it will rise to a height greater than 12 ft. If it is launched at an angle less than 32.4° then the ball will cross the plate below the batter's knees.

Two other trajectories are shown in Fig. 4.1, both for a launch speed of 45 mph. In this case, the ball must be launched at an angle between 12.6° and about 14°. If the ball is launched at an angle less than 12.6° then the ball will not reach a maximum height of at least 6 ft. If the ball is launched at an angle greater than about 14° then it will pass over the batter's shoulders.

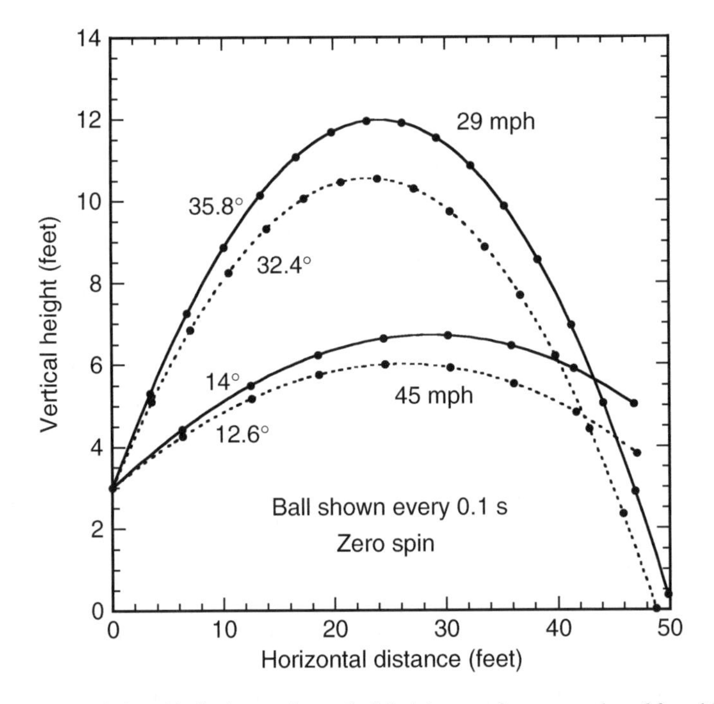

Fig. 4.1 In slow pitch softball, the maximum ball height must be greater than 6 ft and less than 12 ft above the ground. The ball must therefore be pitched at a speeds between about 29 mph and 45 mph. The trajectories here were calculated for a ball pitched without spin

It is clear from these results why pitchers generally prefer to launch the ball at low speed. First, there is a greater range of launch angles to choose from, and second, the ball drops at a steeper angle, making it harder to hit [1].

Fast Pitch Softball

Fast pitch softball requires more skill than slow pitch softball and is a more competitive sport, especially at college level. Slow pitch is a more popular recreational sport. Figure 4.2 shows two 50 mph trajectories and two 70 mph trajectories. In each case, one of the trajectories is for a ball pitched without spin, and the other trajectory is for a ball pitched with backspin, at 10 rev s^{-1} (or 600 rpm). With backspin, the ball reaches its maximum height closer to the batter since the aerodynamic force on a ball spinning backward acts upward, opposing the downward pull of gravity. Pitched at around 70 mph and with enough backspin, the ball reaches its maximum height as it crosses the home plate. Consequently, the ball rises the whole time, and is then called a rise ball.

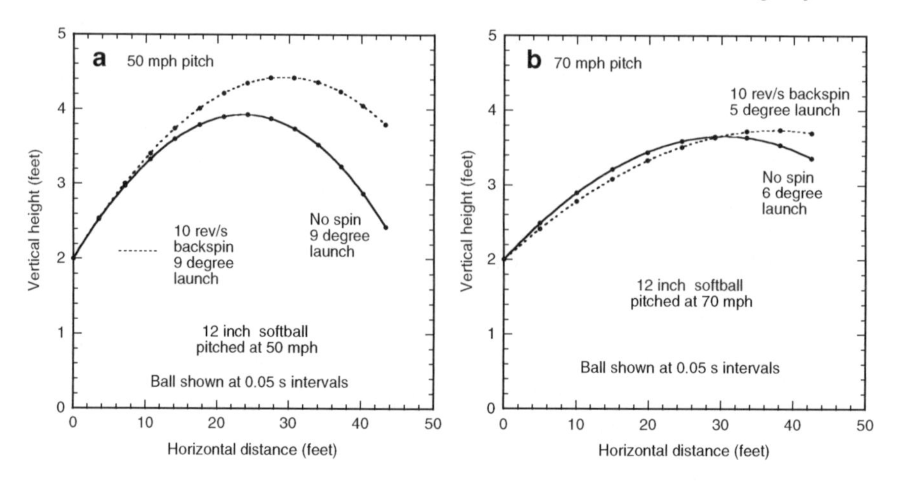

Fig. 4.2 Typical ball trajectories in fast pitch softball, assuming that the ball is launched at (**a**) 50 mph or (**b**) 70 mph, either without spin (*solid lines*) or with backspin, at 10 rev s^{-1} (*dashed lines*). Pitched at 70 mph with sufficient backspin, the ball rises all the way to the batter

Pitched at 50 mph, the ball takes about 0.65 s to reach the batter. At 70 mph, the ball takes about 0.45 s to reach the batter. Both pitches are difficult to hit. The slower ball drops more steeply, while the batter has less time to respond to the faster ball. If the slower ball is pitched with topspin, the ball falls even more sharply.

Baseball

The strike zone in baseball is defined by the rules to be an area above the home plate about 17 in. wide, about 26 in. high and centered about 32 in. above the plate. The pitcher releases the ball about 53 ft from the batter, at a launch height about 6 ft above the ground, at a speed of about 80–90 mph. The pitcher releases the ball about 5 ft in front of the rubber and the batter strikes the ball near the front edge of home plate. For an experienced pitcher, this is not a difficult task. The hard part is to maintain accuracy at high speed, especially when applying spin to the ball.

The path of a baseball pitched from a height of 6 ft at 90 mph in a horizontal direction is shown in Fig. 4.3 for three different values of ball spin, including zero spin. The middle path labelled "no air" is one where the effect of air resistance was ignored, as if the ball was thrown in a vacuum. The vacuum path is easy to calculate since in that case the ball travels at a horizontal speed of 90 mph all the way to the batter, without slowing down. The ball takes 0.402 s to travel the 53 ft distance. During that time, the ball falls through a vertical distance of 2.6 ft, and therefore crosses the plate at a height of 3.4 ft. The effect of the air is to slow the ball from 90 mph to about 81 mph as it crosses the plate, so the ball takes longer to reach the plate, about 0.425 s. In air, the ball falls for a longer time so it falls through a slightly greater height when thrown without spin.

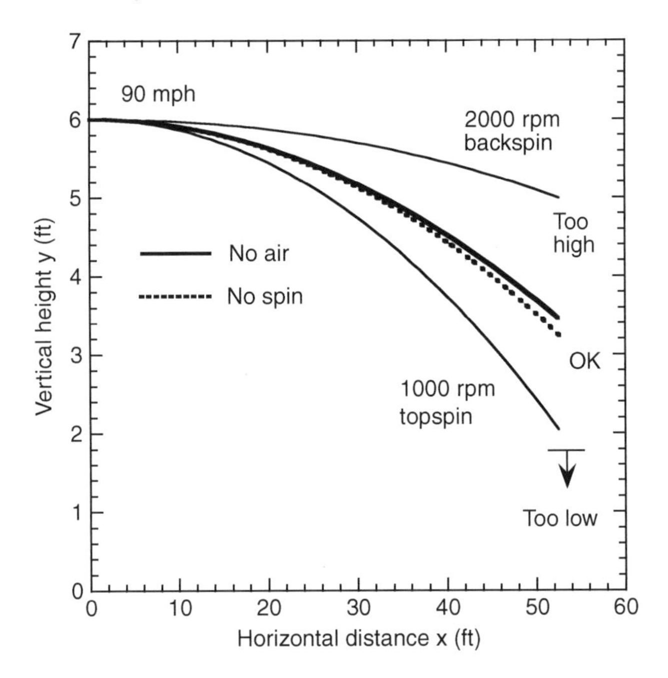

Fig. 4.3 Paths of a baseball pitched at 90 mph in a horizontal direction with topspin or backspin, or without spin. In a vacuum (no air) the ball would cross the plate, 53 ft away, at a height of 3.4 ft. In air, the ball crosses the plate at a height of 3.2 ft if it is pitched with no spin. A spin of 1,000 rpm corresponds to 16.7 rev s^{-1}

The most natural, and the fastest throwing action in baseball is one where the pitcher's arm is inclined slightly to the vertical and the fingers pull down along the back of the ball, giving it backspin. If the pitcher's fingers pull down along the front of the ball, or rise up the back of the ball, the ball is given topspin and the pitch speed is usually slower. The easiest ball to pitch in softball is one where the ball simply rolls out of the hand while the fingers rise up the back of the ball. The combined effect of the air and the ball spin is to introduce an additional force on the ball, known as the Magnus force. If the ball has topspin and is traveling horizontally then the Magnus force acts vertically down on the ball, assisting gravity, so the ball falls through a greater distance on its way to the batter. If the ball has backspin, then the Magnus force acts vertically upwards, opposing gravity, so the ball falls through a smaller distance on its way to the batter.

Golf balls are struck with backspin so they can spend a longer time in the air before falling to the ground, and therefore travel further. A baseball pitcher uses backspin for a different reason. It is simply to confuse the batter. If the pitcher mixes up topspin and backspin in successive pitches, and if the batter can't pick the difference in the spin direction or the pitch speed, then the ball will sometimes drop quickly and sometimes appear to rise quickly. The ball always falls as it approaches the batter, but it can rise above or fall below the expected path, depending on the spin direction and on the amount of spin.

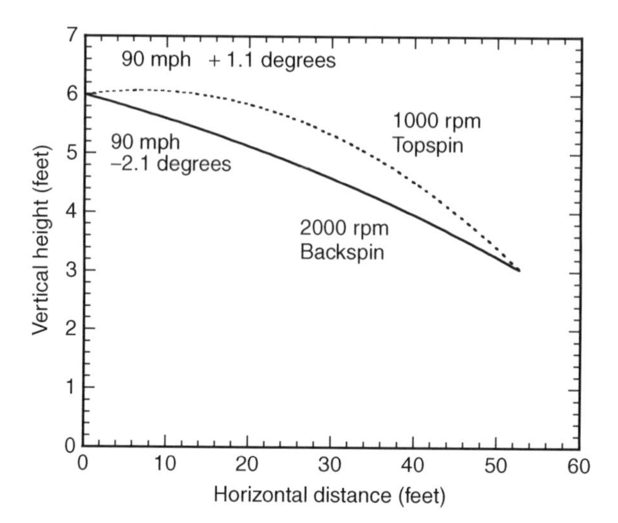

Fig. 4.4 Paths of two balls pitched at 90 mph to cross the home plate at a height of 3 ft. The ball with topspin is launched at an angle of 1.1° above the horizontal. The ball with backspin is launched 2.1° below the horizontal

It would appear from Fig. 4.3 that a ball pitched with topspin drops much faster than a ball pitched with backspin, and would therefore be more difficult to strike. A better comparison is made in Fig. 4.4 where we have plotted the paths of two 90 mph balls pitched so that they cross the plate at the same height. The ball launched with topspin does indeed drop faster but by a smaller amount than shown in Fig. 4.3.

The difficulty of pitching the ball through the allowed strike zone can be calculated in terms of the ball trajectory. If the ball is aimed too high or too low, then it won't pass through the zone. At a launch speed of 90 mph, a fastball thrown from a height of 6 ft with backspin at 2,000 rpm needs to be launched at an angle between 1.3° and 3.7° below the horizontal, as shown in Fig. 4.5. If a curveball is thrown at 80 mph with 1,000 rpm of topspin, then the launch angle must be between 0.5° and 2.8° above the horizontal. In both cases, the launch angle can vary over a range of only about 2.3°. Regardless of the pitch speed and the spin of the ball, the allowed range of launch angles is only about 2.3°.

4.3 The PITCHf/x System

A baseball game viewed on TV shows clearly that the batter either connects with the ball or misses, but the event usually happens too fast for the viewer to get an accurate view of the bat and ball trajectories. Slow motion cameras help to slow down the action, but the view is still two-dimensional and it is still difficult to determine the precise trajectories followed by the ball and the bat. Modern technology has

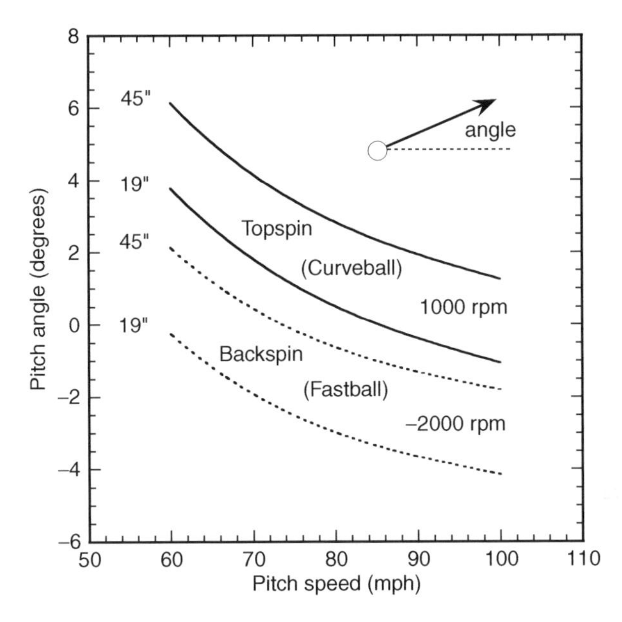

Fig. 4.5 Maximum and minimum pitch angles above the horizontal required for the ball to cross the plate within the allowed strike zone, between 19 in. and 45 in. above the ground, assuming that the ball is pitched from a height of 6 ft and the plate is 53 ft from the pitcher's release point. Curve balls are usually thrown a few degrees above the horizontal and fast balls are usually thrown a few degrees below the horizontal

come to the rescue by capturing multiple images of the batter and the ball from two or three cameras positioned around Major League ballparks. The viewer is not presented with all that information directly. Instead, it is processed by a computer, then displayed on the TV screen in a manner that the viewer can more easily interpret.

PITCHf/x is a system devised to track the path of a pitched baseball. It was developed by a company called Sportsvision and the data can now be downloaded for every MLB game at www.mlb.com/mlb/gameday. The information from the cameras is displayed numerically in terms of the pitched ball speed, the break in inches toward or away from the batter in the horizontal direction, and the break in the vertical direction. The information is also displayed visually in a rectangular box depicting the strike zone, using a white dot to show where the ball crossed in relation to the strike zone.

The break of a pitched ball is a measure of the spin imparted to the ball. If the ball was pitched without spin, then it would cross a point at the front of the home plate at a certain vertical height z_1 and at a certain horizontal distance x_1 from the center of the plate. If the ball actually crosses at height z_2 and horizontal distance x_2, then the break due to the spin of the ball would be $z_2 - z_1$ in the vertical direction and $x_2 - x_1$ in the horizontal direction, both expressed in inches. MLB/Gameday defines the break as the "movement" of the ball, but the word "break" is used in this book since it is more commonly used by physicists. MLB/Gameday defines break to mean something slightly different. The physics definition of break is explained graphically in Fig. 4.6.

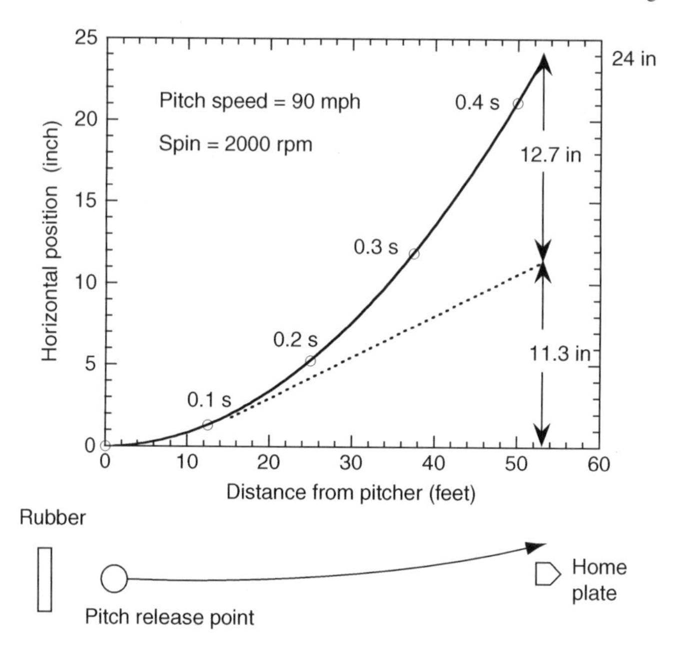

Fig. 4.6 A 90 mph pitched ball with 2,000 rpm of sidespin breaks by 24 in. over 53 ft. However, PITCHf/x records the break as 12.7 in.

The break in PITCHf/x is not measured over the whole distance from the pitcher's release point to the home plate. Rather, it is measured from a start point 40 ft from the point of home plate, the reason being that the break is smaller over the reduced distance and is more like the break that batters are accustomed to seeing. From measurements of the speed and angle of the ball out of the pitcher's hand and beyond, an estimate is made of the speed and trajectory angle at the 40 ft point, to predict where the ball would cross the plate if it was thrown with zero spin and was subject only to the force of gravity. In that manner, the break due to the spin is combined with the small break due to the drag force to quote the break of the ball in the vertical and horizontal directions.

There are several different ways that the break in the vertical direction could be defined. In the absence of any vertical force on the ball, the ball would follow a straight line path in the vertical direction from the pitcher to the home plate. When gravity is the only vertical force on the ball, and the ball is pitched in a horizontal direction at 90 mph, the ball crosses the home plate at a point about 2.6 ft lower since the ball curves downward, as shown in Fig. 4.3. The difference in the two heights in this case would be the break due to gravity. When there is no ball spin, the vertical forces acting on the ball are those due to gravity and the drag force. For the example shown in Fig. 4.3, the ball crosses the home plate at a height of 3.4 ft when there is no air and when gravity is the only force on the ball. If the ball is subject to a drag force through air, then the ball crosses the home plate at a height of 3.2 ft. The break due to the drag force is therefore about 0.2 ft. If the ball is pitched with 1,000 rpm of topspin, then the ball crosses the plate at a height of 1.9 ft, so the break due to the extra spin is 1.3 ft. If the ball is pitched with 2,000 rpm of backspin, then the ball crosses the plate at a height of 5.0 ft, so the break due to the backspin is −1.8 ft. The break values here refer to the break over a 53 ft path. The break over the last 40 ft is smaller but that is how it is measured in the PITCHf/x system. The break of the ball due to its spin, in both the vertical and horizontal directions, therefore provides the viewer with an indication of the type of ball that was pitched and the amount of spin imparted to the ball.

Estimating the Break of a Spinning Ball

Pitchers spin the ball to confuse batters. The spin axis can point in any direction the pitcher choses, so the ball can rise or fall vertically above or below the natural fall due to gravity, and it can simultaneously curve horizontally into the batter or away from the batter. The amount of curvature due to spin, or the break, is typically about 20 in. in baseball. It can be 20 in. in any direction, either up, down, left, right, or at say 45° to the horizontal, in which case the ball breaks about 14 in. vertically and 14 in. horizontally.

Suppose that the ball is spinning about a vertical axis. In that case, the ball curves downward in the usual way due to gravity, but it also curves in a horizontal direction due to the spin. The break is just the horizontal distance traveled by the ball, due to its spin, as it travels from the pitcher to the front edge of the home plate. The break from the pitcher to the 40 ft position is quite small, only about 2 in., so at first sight it might seem that the PITCHf/x break would be only about 2 in. less than the actual break. However, the PITCHf/x break is considerably smaller than the actual break, since a straight line drawn from the 40 ft position of the ball, when extended tangentially to the home plate, can result in an extra 11 or 12 in. in the measured break.

The Magnus force on a baseball pitched at 90 mph with 2,000 rpm of sidespin is about 0.97 N or about 68% of the weight of the ball. The horizontal acceleration of the ball is then 6.7 m s^{-2}, assuming its mass is 0.145 kg. Without spin, the ball would travel in a straight line path in the horizontal (x) direction. At 40.2 m s^{-1} (90 mph or 132 ft s^{-1}), the ball takes 0.401 s to travel 16.15 m (53 ft) in a vacuum, or 0.426 s in air. The Magnus force causes the ball to curve horizontally in say the y direction, at right angles to the x direction. The horizontal distance, y, traveled by the ball in time t is given by $y = 0.5at^2$ where a is the horizontal acceleration, which is 6.7 m s^{-2} in this case. The ball therefore breaks by a horizontal distance $y = 0.61$ m after 0.426 s, or by 24 in. In the PITCHf/x system, the break is measured from a start point 40 ft (12.19 m) from the home plate, in which case the ball takes only 0.310 s to travel from that point to the front of home plate. Starting from the direction of the ball at the 40 ft mark, the ball breaks by $0.5 \times 6.7 \times 0.31^2 = 0.322$ m $= 12.7$ in., as shown in Fig. 4.6

The ball takes only 0.116 s to travel the first 15 ft, and breaks by 1.8 in. during that time. If the ball stopped spinning at that point and then continued on its way in a perfectly straight line, it would cross the home plate with a break of 11.3 in. But it doesn't stop spinning and breaks another 12.7 in. by the time it arrives at home plate, for a total break of 24 in.

What would happen to the break if the pitcher threw the ball at 80 mph instead of 90 mph, keeping the spin fixed at 2,000 rpm? Will the break be more, or less or the same? At 80 mph, the ball takes 0.480 s to travel 53 ft, and the Magnus force on the ball drops to 0.80 N. The horizontal acceleration is then 5.5 m s^{-2} so the break is $0.5 \times 5.5 \times 0.48^2 = 0.63$ m $= 24.9$ in. The break is slightly larger than for a 90 mph pitch. The Magnus force is proportional to the ball speed squared, the time taken to travel 53 ft is inversely proportional to the ball speed, and the break is proportional to the travel time squared. All these factors cancel out. The small increase in the break at 80 mph is due to the fact that the Magnus force is also proportional to the lift coefficient (defined in Appendix 3.3) which increases slightly as the ball speed decreases if the spin remains the same.

There are three interesting conclusions we can draw from these calculations. The first is due to the fact that the ball accelerates in the horizontal direction. The batter sees the ball break by only 1.8 in. in the first 15 ft as if the ball is breaking slowly, as indeed it is. But then it breaks by another 22 in. on the way to the batter because it is accelerating. The batter therefore gets the impression that the ball travels in a relatively straight line at the start, prepares to strike the ball on that basis, and then discovers when it is too late that the ball suddenly swerves away from his predicted path.

The second point of interest is that the break due to the spin does not depend strongly on the speed of the ball. At any given spin, the break will be approximately the same, at least for pitch speeds commonly found in baseball.

The third point of interest is that PITCHf/x data on ball break can be used to estimate the spin of the ball and the direction of the spin axis. PITCHf/x determines the position of a pitched ball as it travels from the pitcher to the batter, but there is not sufficient image resolution to determine the spin. The ball is just a small white dot on the screen. Nevertheless, if the ball breaks by say 12.7 in. in the PITCHf/x system, then the spin will be about 2,000 rpm or more, depending on the ball speed and the direction of the spin axis. The spin might actually be more than that. If the spin axis is tilted forward or backward then the tilt has no effect on the break. A gyro ball has its axis parallel to the direction of motion of the ball, in which case the Magnus force is zero, the break is zero, and it might appear that the ball is not spinning at all. The estimated spin, when using PITCHf/x data, is just the minimum spin that the ball needs to produce the measured break. PITCHf/x cannot be used to determine the forward or backward tilt of the ball, but the vertical and horizontal values of the break determine the inclination of the spin axis in the vertical plane. The vertical break can be used to calculate the amount of topspin or backspin, the horizontal break can be used to calculate the amount of sidespin, but the spin about an axis parallel to the path of the ball remains unknown.

Some Technical Considerations

The success and popularity of the Pitchf/x system is based not only on theoretical considerations but also on solutions of a number of technical problems. A few of the technical issues involved in measuring ball trajectories are as follows.

Suppose that a ball is pitched at 90 mph and is filmed using a video camera viewing at right angles to the path of the ball. That is, the pitcher is on the left of screen and the batter is on right of screen. The camera sees the ball travel across the screen at 132 ft s^{-1}. If the camera operates at 30 fps then it takes one image or one "photo" of the ball every 1/30 s. The ball travels 4.40 ft between each photo, and 12 or 13 such photos can be taken as the ball travels from the mound to the home plate.

Each photo takes a certain time to take, depending on the exposure time. The longer the exposure time the brighter the image but the more blurred the image will be. Suppose that the exposure time is 1/1,000 s. During one exposure, the ball travels 0.132 ft = 1.58 in., about one ball diameter. Sportvision claims an accuracy of 1/2 in. in measuring the position of the ball. In that case, the exposure time needs to be shorter than 1/3,000 s and there needs to be a sufficient number of pixels in the camera. If there were only 60 pixels across the screen, then each pixel would record an image that represented one foot of the ball's path. To resolve 1/2 in., the camera would ideally need to have $60 \times 24 = 1{,}440$ pixels across the screen. Standard resolution in most digital video cameras is 640 pixels across the screen. That is sufficient in this case since it is not just one but many ball images that are recorded, allowing the trajectory to be determined with greater resolution than one whole pixel width.

Suppose that the ball travels 4.4 ft between each photo and there is an error or uncertainty of 1/2 in. in the position of each ball. Using any two consecutive photos, the speed could be $30 \times 4.48 = 134.4$ ft s^{-1} or $30 \times 4.32 = 129.6$ ft s^{-1}, so speed = 132 ± 2.4 ft s^{-1} = 90 ± 1.6 mph. Since the ball slows down during the flight and since there are 13 or 14 images over the flight distance, the recorded ball positions can be fitted by a smooth curve to give the ball speed to within about 1 mph at each position during its flight. Slightly greater accuracy can be achieved by de-interlacing the camera images to record the position of the ball at intervals of 1/60 s. Video cameras actually record 60 frames s^{-1}, but every second frame is normally combined or interlaced with the first to provide a smoother image.

One other technical consideration with the PITCHf/x system concerns the prediction of the path when the ball is pitched with zero spin. The observed path, in the vertical and the two horizontal directions, can be described to sufficient accuracy over the short pitched distance by curve fitting three separate parabolas to estimate the acceleration in each of the three directions. The break in the vertical direction is calculated by comparing the fitted parabolic path with one where the acceleration due to gravity is subtracted out, but no correction is made for the acceleration due to the drag force, either in the vertical or the horizontal directions. As a result, the break currently measured with PITCHf/x is that due to the combined effects of spin and drag rather than that due to the spin alone. This feature does not introduce a significant error since the break due to the drag force is typically only a few inches.

4.4 Curveballs, Fastballs and Other Oddballs

Baseball and cricket share two things in common. In both sports, the ball is struck with a bat. And both sports have a richly documented variety of pitched ball types. In cricket, there are inswingers, outswingers, reverse swingers, legspinners, legbreaks, offbreaks, flippers, doosras, bouncers, zooters, and googlies. Some of those terms refer to the flight through the air and some refer to the bounce off the pitch before the ball reaches the batter. In baseball there are curveballs, fastballs (two-seam and four-seam), cutters, sliders, sinkers, changeups, knuckleballs, screwballs, and gyroballs plus a few others [2–4] (Fig. 4.7). It is difficult to find clear definitions of each pitch type in terms of physical properties, such as the inclination of the spin axis, partly because the forward or backward tilt of the spin axis is difficult to measure.

Cricket balls and baseballs have almost the same weight and diameter and both have a leather cover. The main difference concerns the stitching used to attach the cover. In cricket, there are several parallel rows of stitches running around the equator. The stitching used in a baseball follows the same curved path as the seam in a

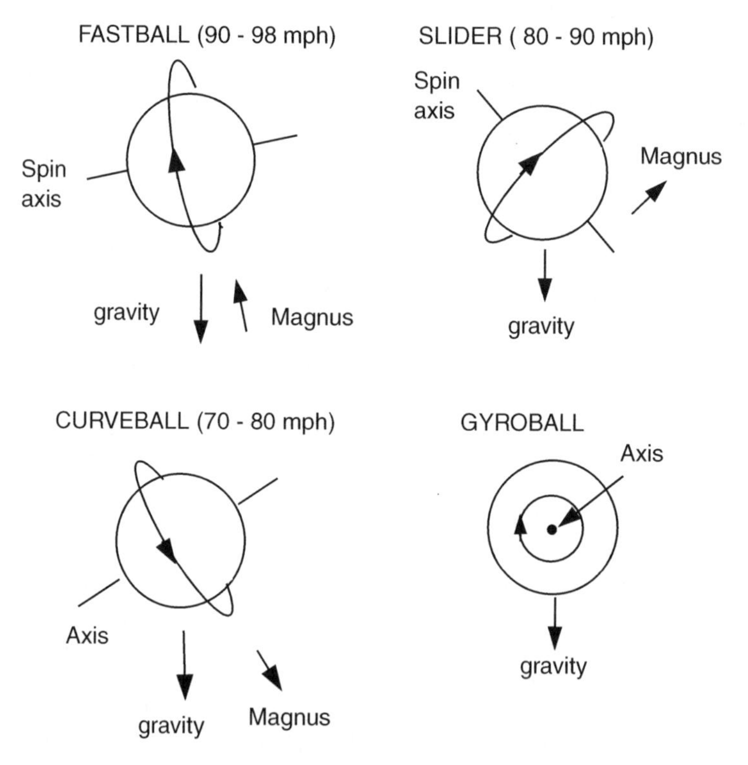

Fig. 4.7 Directions of ball spin and the Magnus force on various pitched balls, as viewed by the batter and pitched by a right hand pitcher. Gravity pulls the ball downward in all cases. The Magnus force acts in a direction perpendicular to the spin axis, in the directions shown, causing the ball to break left or right as well as up or down. There is no Magnus force on a gyroball

tennis ball. The stitching makes a big difference to the flight of the ball through the air. A cricket ball can be spun in such a way that the stitching always faces the batter or is inclined at some fixed angle to the path of the ball. The air then flows smoothly over one side of the ball but is tripped into turbulence by the stitching on the other side of the ball. A spinning baseball cannot maintain the stitching in any fixed position, although the ball can be spun in such a way that the batter sees either two or four lines of stitches each revolution. Sometimes, a batter might detect a red dot on the approaching ball. If that happens, then it means that the ball is rotating about an axis that passes through the stitching. A point on the axis remains on the axis as the rest of the ball rotates, so the batter then sees a particular stitch as a red dot.

Pitchers classify different ball throws by the way they hold and release the ball, emphasizing the importance of holding the seams with the fingers to get a better grip on the ball, and the importance of wrist and finger action to generate the required spin direction. If the fingers pull down on the back of the ball, then the ball will be given backspin, spinning about a horizontal axis. If the fingers pull down the side of the ball, the then ball will spin about an axis pointing in the same direction as the pitched ball. The result is a gyroball and it won't curve left or right at all. If the fingers push forward on the right side of the ball, then the ball will spin about a vertical axis and the ball will curve to the left as a slider, breaking away from a right-handed batter. If the fingers push forward on the left side of the ball, then the ball will curve to the right as a screwball, breaking away from a left-handed batter. If the fingers push both forward and down on the right side of the ball then the ball will spin in the direction that the fingers move, and the spin axis will be tilted forward. The ball will curve to the left due to rotation about the vertical axis. To see how this works, hold a ball in your hand and rotate it about a vertical axis. Then tilt the axis forward and rotate the ball again. Any given point on the right side of the ball will rotate both forward and downward.

Physicists classify different throws in terms of the resulting spin imparted to the ball. The spin axis of the ball can be selected by the pitcher to point in almost any direction. To explain the flight path of the ball, it helps to consider just the three primary axes, one being vertical (the z axis) and two being horizontal. The horizontal x axis points from the pitcher to the batter, and the other horizontal axis (y) is perpendicular to the other two. A ball spinning around the x axis is called a gyroball. A ball spinning around the y axis can rotate either with topspin or with backspin. A ball spinning around the z axis has sidespin and is deflected horizontally by the Magnus force. To describe the trajectory of a ball spinning about any other axis, we can consider separately the forces on the ball in the x, y and z directions.

A spinning ball in flight is subject to three different forces:

(a) The force of gravity acts vertically down on the ball, in the $-z$ direction, and causes it to curve downward into the ground. This force is given by $F = mg$ so it remains constant throughout the flight and is independent of the speed or the spin of the ball.
(b) The drag force is due to air resistance and acts backward along the path of the ball, essentially in the $-x$ direction, causing the ball to slow down as it makes its way to the batter. The drag force is proportional to the ball speed squared.

(c) The Magnus force acts in a direction that is at right angles to the path of the ball and at right angles to the spin axis. If the spin axis happens to be parallel to the path of the ball, as in a gyroball, then there is no Magnus force.

One way to remember the direction of the Magnus force is to remember that a golf ball is struck with backspin, so it spins around the y axis, and the Magnus force acts primarily in an upward direction on the ball, in the $+z$ direction, opposing the force of gravity. At least, that is case while the ball is traveling horizontally. While the ball climbs upward the Magnus force is directed slightly backward, at right angles to the path of the ball (as shown in Fig. 3.9). The Magnus force helps to hold the ball in the air for a longer time and it therefore travels farther before it lands. If a pitched ball spins about a vertical axis, then it curves in the horizontal direction (in the $+y$ or $-y$ direction). It is the same effect as that on a spinning golf ball, but it helps to turn your head sideways to convince yourself that it is the same effect.

Suppose that a ball is pitched in the $+x$ direction and spins about an axis that is perpendicular to the x direction but tilted at 10° to the horizontal. Then the spin is basically topspin or backspin, depending on the spin direction, but it also has a small amount of sidespin. The ball will therefore behave essentially like a ball with topspin or backspin but it will curve slightly in the $+y$ or $-y$ direction. If the axis was tilted at 80° to the horizontal (or 10° to the vertical) then the spin is basically sidespin, and the ball would curve in the $+y$ or $-y$ direction. But it would also have a small amount of topspin or backspin, causing it to curve slightly in the $+z$ or $-z$ direction. The spin axis is rarely vertical or horizontal since it is difficult for a pitcher to throw the ball that way. If the fingers pull down on the back of the ball at say 20° to the vertical, because of the way the arm is inclined to the vertical, then the spin axis will be at right angles to that, or 20° to the horizontal. That is the usual way a fastball is pitched.

A ball pitched without any spin or with very little spin is called a knuckleball. It can be thrown off the knuckles or off the fingertips, typically at relatively low speed. The ball tends to travel in an erratic path since the flow of air around the ball depends on the alignment of the stitches. If the air flows smoothly around one side of the ball and is turbulent on the other side then, there will be a sideways force that deflects the ball away from the expected path. As the ball rotates slowly, the direction of that force changes. The same effect can be seen even more dramatically by dropping a sheet of paper to the ground. As the sheet twists and turns, the flow of air around the paper causes it to follow an erratic path, in the same way that a leaf flutters as it falls to the ground. The path of a knuckleball is not as erratic as a leaf or a sheet of paper but it is still enough to trouble most batters.

Appendix 4.1 Playing Field Dimensions

The main dimensions of the playing field in fast pitch softball and baseball are shown in Fig. 4.8. The distance between the pitcher and the batter is about 60 ft in baseball, about 43 ft in fast pitch softball, and about 50 ft in slow pitch softball.

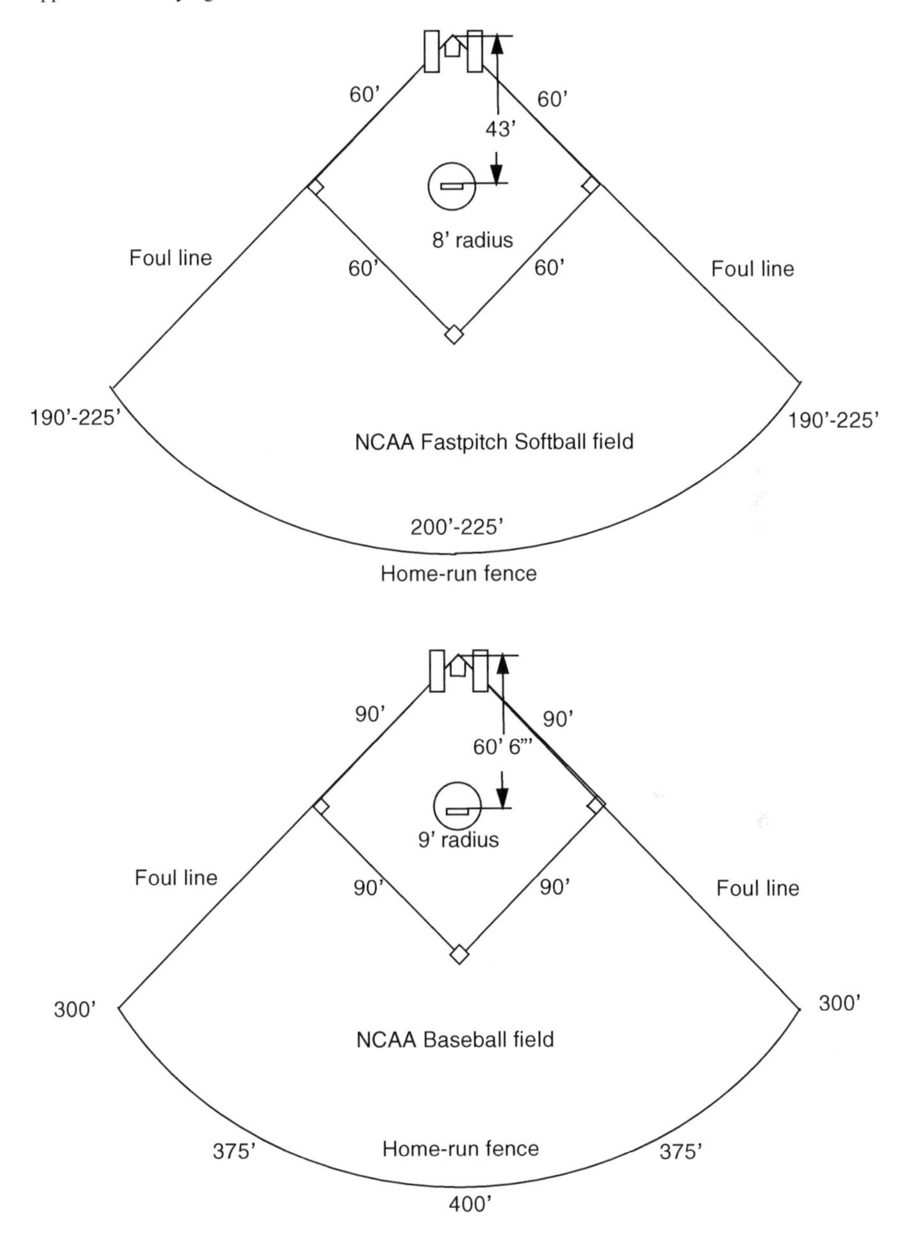

Fig. 4.8 Softball and baseball are played on different size fields as specified by the rules of each game

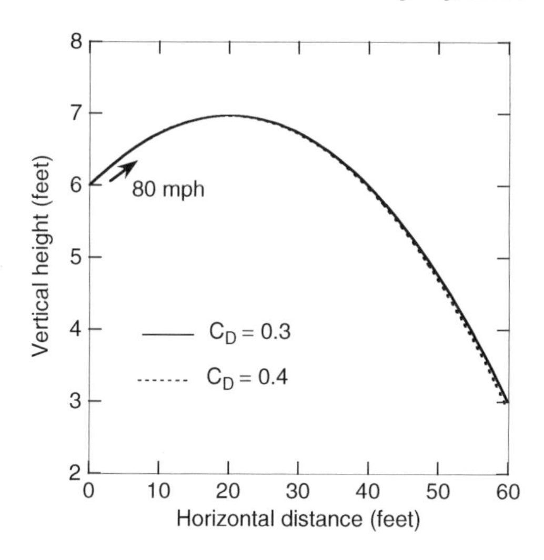

Fig. 4.9 Paths of a baseball pitched without spin at 80 mph, assuming that $C_D = 0.3$ or $C_D = 0.4$. The two paths are almost identical, although the one with greater air resistance (with $C_D = 0.4$) crosses the plate 1 in. lower, 2 mph slower and 0.008 s later

Appendix 4.2 Drag and Lift Coefficients

For the calculations in Figs. 4.1–4.5 we assumed that the drag coefficient for a baseball or a softball is $C_D = 0.4$ and that the lift coefficient is $C_L = 1/[2.324 + 0.40/(R\omega/v)]$, where v is the ball speed (in m s^{-1}) and $R\omega$ is the circumferential speed of the ball (in m s^{-1}). While these coefficients may vary slightly in practice, the actual coefficients will not alter the resulting ball trajectories by any significant amount. As shown in Fig. 4.3, the trajectory of a pitched ball is determined mainly by its launch speed and angle and by the force of gravity acting on the ball. The spin of the ball also plays an important role, but if we change the drag coefficient from say 0.4 to 0.3 then the trajectories shown in Fig. 4.3 will be almost the same. Figure 4.9 shows the effect on a ball launched at 80 mph without spin, when $C_D = 0.3$ or 0.4. The two paths and the time taken to reach the plate are very similar, meaning that measurements of C_D to better than say ± 0.05 are not only very difficult in practice, but do not make much difference anyway.

References

1. T. Wu, P. Gervais, An examination of slo-pitch pitching trajectories. Sports Biomech. **7**, 88–99 (2008)
2. T. Jinji, S. Sakurai, Direction of spin axis and spin rate of the pitched baseball. Sports Biomech. **5**, 197–214 (2006)
3. A.T. Bahill, D.G. Baldwin, Describing baseball pitch movement with right-hand rules. Comp. Biol. Med. **37**, 1001–1008 (2007)
4. D.G. Baldwin, A.T. Bahill, A. Nathan, Nickel and dime pitches. Baseball Res. J. **35**, 25–29 (2007)

Chapter 5
Pitching Mechanics

I became a good pitcher when I stopped trying to make them miss the ball and started trying to make them hit it.

– Sandy Koufax

Pitching mechanics can mean different things to different people. To coaches and players, pitching mechanics refers to biomechanics, to the actions of all the different body segments involved, the correct sequence of those actions, the alignment of the fingers along the ball, injury prevention, and so on. In this chapter, we examine the mechanics of pitching from a physics point of view. That is, we treat pitching as a mechanics problem. There are three main questions that we ask. The first question is: how does a pitcher manage to release the ball at just the right moment to project it at the angle he or she wants to pitch it? The second question is: why can a baseball be thrown faster than a brick? And the third question is: How can a pitcher transfer as much energy as possible from his or her arm to the ball? The second question might seem a little strange at first sight, but the answer is not as simple as you might expect.

5.1 Timing Accuracy Problem

The pitcher must release the ball at a very precise time during his throwing action. If he releases the ball a fraction too early, the ball will head off toward or above the batter's shoulders. If he releases the ball a fraction too late, the ball will head off toward or below the batter's knees. It is interesting to work out just how accurate the timing must be since it also relates to the problem of swinging a bat so that it connects with the ball at exactly the right time. Suppose that the pitcher's hand is rotating in a circular arc of radius 2 ft at a speed of 80 mph $= 117\,\mathrm{ft\,s^{-1}}$ just before releasing the ball. The circumference of that circle is 12.6 ft, so the hand could rotate one full circumference or 360° in $12.6/117 = 0.108$ s. It, therefore, rotates by 1° in $0.108/360 = 0.0003$ s. If the ball is released 0.0003 s late, the hand rotates another 1° and the ball is launched 1° lower than desired. Given that the ball must be launched within a 2° angle, the pitcher has to release the ball within a time period of 0.0006 s. If the hand rotated in a 4 ft radius circle, the pitcher would have 0.0012 s to release the ball, but that is still an extremely short time interval.

R. Cross, *Physics of Baseball & Softball*, DOI 10.1007/978-1-4419-8113-4_5,

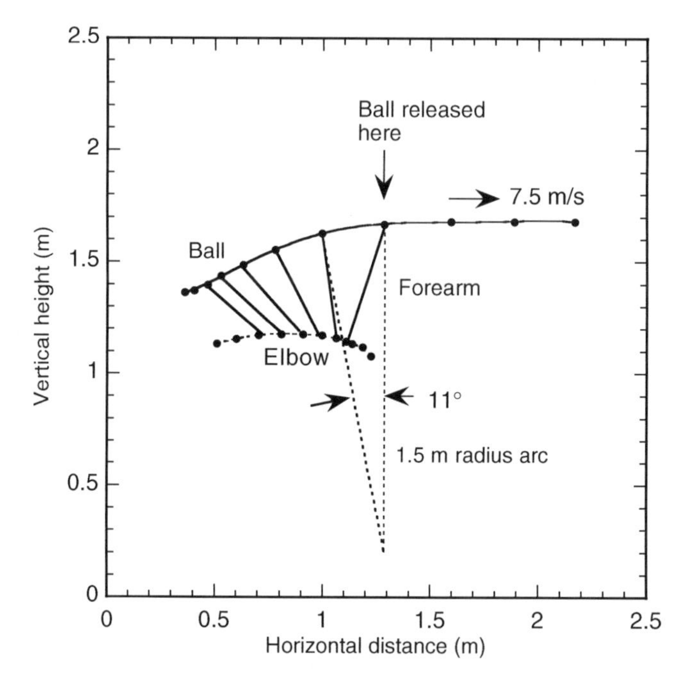

Fig. 5.1 Trajectories of a thrown ball and the elbow while the ball is being thrown at low speed, showing the positions of the ball and the elbow at intervals of 0.04 s, as indicated by the *black dots*

The calculation here seems wrong. Timing within ± 0.0003 s or even ± 0.0006 s seems impossible. What could be wrong with the calculation? To answer this question, the author filmed a low speed throw with a video camera to capture the action. In a high speed throw, the arm rotates so rapidly that the details cannot be captured without using a special high speed camera operating at 200 frames s^{-1} or more. The result of the low speed throw is shown in Fig. 5.1, captured at 25 frames s^{-1}, corresponding to a time interval of 0.04 s between frames. The ball was thrown at only 7.5 m s^{-1} (16.8 mph) but the throw action at high speed is not drastically different, just faster. Just before the ball was released, the ball was rotating in a circular arc of radius 1.5 m (4.9 ft). This is much longer than the length of the forearm, so the forearm could not have been rotating about a fixed axis through the elbow. The elbow itself was moving forward, which acted to straighten out the path of the ball in the hand. In addition, the ball rolled out of the hand, up and along the fingers, which also helped to straighten out the path and gave the ball some backspin.

The 1.5-m arc radius was measured by drawing the dashed lines shown in Fig. 5.1, at right angles to the path of the ball, to see where they intersected. These two lines form an arc angle of 11°. The axis in this case is quite different from the rotation axis of the forearm, which can be estimated by extending lines along the forearm. The thrower was able to launch the ball in a horizontal direction quite accurately, within $\pm 1°$, meaning that the ball was released at a time within 1/11th of that taken to rotate the last 11°, that is, within 0.04/11 = 0.0036 s. Even this accuracy is quite amazing, but that is what is required to pitch a successful ball even at very low speed. At high throw speeds, the timing accuracy must be better than 0.001 s.

An alternative, and much more likely explanation of the required accuracy is that the pitcher doesn't time the ball release at all. Rather, he works out the correct *position* to release the ball, not the correct time. Humans are very good at judging the position of their body parts, even with their eyes closed. For example, you can raise your arm almost exactly vertically or stretch it out horizontally, at least to within a few degrees. With practice, using a protractor to tell you if your practice position is right, you can probably do it within 1° without the protractor. Throwing requires similar positional accuracy. With practice, you will know where to release the ball, rather than when. For example, if someone starts a clock, and tells you to release the ball in 10 s time, you would be lucky to guess the release time to within 1 s. But, with practice, you could guess the correct release position of your hands and arm to within 1 °. The brain has a feedback system that monitors the speed and position of the arm and warns you when you are getting close, but this is physiology, not physics. However, there is a simple test to prove that humans are capable of judging the position of the hand quite accurately. Try swinging your arm at a table or a wall so your hand will smash into it. Instinct will tell you to stop about 1 in. before you get there. Instinct doesn't tell you to stop after 0.476 s. It tells you to stop when you are about 1 in. away, or maybe 2 in. away, depending on how brave you are.

5.2 Physics of Pitching: Without Equations

Throwing is a skill that evolved many thousands of years ago as a means of killing animals for food using rocks, spears and boomerangs. These days we use a different technique to obtain food, primarily by driving to the supermarket. Nevertheless, throwing has survived as a popular sporting activity, especially honed for the purposes of pitching baseballs and softballs and other objects such as a basketball, a football, a javelin, the discus and the shotput. Each of these objects requires a different throwing action but all throwing actions are based on the same basic principles that were used in ancient times to hunt for food.

A ball can be thrown faster when it is thrown overhand since energy is transferred more efficiently from the upper arm to the lower arm. When throwing overhand, the upper arm is swung using muscles attached to the shoulder, starting with the forearm locked nearly at right angles to the upper arm. Muscles surrounding the elbow joint then take over, straightening the whole arm in such a way that the forearm speeds up while the upper arm slows down. That way, energy is transferred from the upper arm to the forearm. Finally, muscles attached to the wrist and finger joints take over to project the object in the hand at whatever speed, spin and direction the thrower desires. The biomechanics of this process is described in [1] and [2].

A pitcher doesn't actually start the throw by swinging the upper arm. There is a wind-up stage before that, followed by an ordered unwinding sequence of events that starts at the legs, works its way up through the hips and torso, and ends in the arm and the hand. Video film of the action of a pitcher shows this sequence clearly. At each step in the chain, energy is transferred from one body segment to the next until the very last step where maximum energy is transferred to the ball.

Not all of the energy used by a pitcher goes into the ball, otherwise the pitcher would come to a dead stop as soon as the ball was thrown. A large fraction of the energy used by a pitcher is used to swing the arm, which is a lot heavier than the ball. Since the arm follows through after releasing the ball, most of the energy used by a pitcher is wasted in the follow through motion of the arm, and only a small part of the throwing energy is transferred to the ball. The problem here is that the ball is relatively light. By contrast, athletes who throw the shotput transfer almost all of their throwing energy to the shotput since their throwing arm comes almost to a stop as soon as the shotput is released. Obviously, there is a tradeoff between the energy that can be given to a ball and the speed at which the ball can be thrown. The physics of this is interesting, as we will see shortly.

In terms of the basic arm and other body segment actions, batting is similar to pitching. The main difference is that a batter uses both arms to swing the bat and holds onto the bat after striking the ball, rather than throwing the bat. The same result would be achieved if the batter actually let go of the bat as the ball was struck, or even just before the ball was struck, but throwing the bat at an opponent after striking the ball is not considered good etiquette and is against the rules. You might like to try it one day to see what happens, on your own when no-one is looking.

Throw Speed Question

Everyone knows that a baseball (or a softball) can be thrown faster than a brick. An interesting question is why this is so. Most people would probably guess that it's because the baseball is lighter. That is indeed part of the answer but it is not the whole answer. The speed of an object thrown by hand depends on three factors. One is the weight of the object. Another is the force exerted on the object. The third factor is the distance over which the force acts on the object.

If a force is applied to an object then the object accelerates. If the force is applied for a short time then the object will accelerate from rest to a relatively small speed. If the same force is applied for a longer time then the object will accelerate to a greater speed and the force will be applied over a longer distance. For example, a vehicle can't accelerate from rest to 100 mph over a distance of only 1 foot. The vehicle needs to travel a much larger distance to reach 100 mph, accelerating at a roughly constant rate over a time period of 10 s or more. Maximum speed of a thrown ball therefore results when the ball is accelerated over as large a distance as possible. In other words, the pitcher needs to stretch back as far as possible at the start of the throw and then release the ball when stretching forward as far as possible.

Throwing a baseball doesn't require much effort at all, at least compared with throwing a brick. Throwing a piece of paper is even easier. The reason is slightly subtle. If you throw each of these objects as fast as you can, then you are exerting as much force as you can on each object. Does that not mean you are exerting the same effort? Not really, since the force you can exert on an object depends on how heavy it is. Try pushing a piece of paper through the air, then try pushing a brick wall. Obviously, the force you can exert on a brick wall is a lot bigger than the force on the piece of paper, even though you might be trying just as hard.

When you push or throw an object with your hand, then muscles in your arm move your hand forward, and your hand moves the object forward. The force exerted by the muscles might be the same, but the force on the object depends on its mass. The force, F, on the object is given by $F = ma$, where m is the mass of the object and a is the acceleration of the object. The acceleration of the object is the same as the acceleration of your hand. The acceleration of your hand depends on both the mass of the object and the mass of your arm. If the object is much lighter than your arm, then the acceleration of the object will be essentially the same, regardless of its mass, but the force $F = ma$ will increase with the mass of the object. You can throw a 2-g peanut just as fast as 1-g peanut, the acceleration of both peanuts being the same, but the force on the 2-g peanut is twice as large. That is why the force you can exert on an object increases as its mass increases, even though the muscle force might be exactly the same.

The following section explains the effect in more detail. If the force on a thrown object was exactly proportional to its mass then all objects could be thrown at the same speed. When the object being thrown is about 100 g or more then doubling the mass increases the force by a factor of only about 1.8. That is the subtle reason why heavy objects can't be thrown as fast as light objects.

5.3 Physics of Pitching: With Equations

When a force F is applied to an object over a distance D then the work done by the force is given by FD. That is, work = force × distance. The result is an increase in the kinetic energy of the object. The kinetic energy of an object is given by $mv^2/2$, where v is the speed of the object. The speed at which an object can be thrown is therefore given by

$$FD = \frac{1}{2}mv^2$$

When throwing a baseball, D is about 2 m since the pitcher starts from a point well behind the shoulder and releases the ball at a point well in front of the shoulder. The average force F applied to the ball is therefore about equal to $mv^2/4$. If we take $m = 0.145$ kg and $v = 40\,\text{ms}^{-1}$ (89 mph) then $F = 58$ N (13.0 lb), which is 41 times larger than the weight of the ball. The weight of the ball is $mg = 0.145 \times 9.8 = 1.42$ N. The acceleration of the ball, in common terminology, is then 41 g's. Even though this is a large acceleration, it is much smaller than the acceleration of the ball when it is struck by a bat.

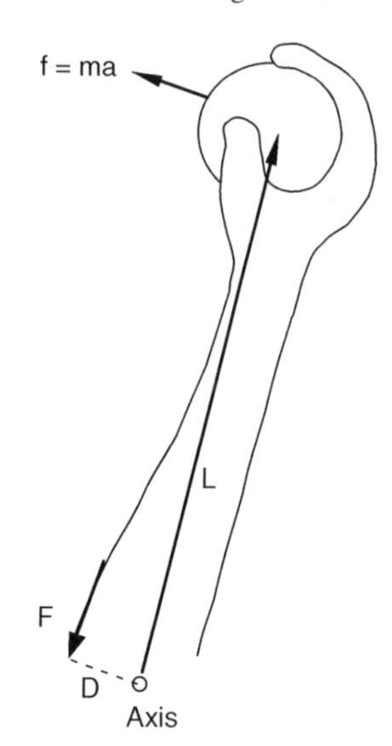

Fig. 5.2 The pitcher exerts a force f on the ball by contracting the arm muscles so they exert a force F on tendons attached to bone near the elbow. The elbow itself is also propelled forward by muscles in the upper arm, but the essential physics can be understood by assuming that the elbow rotates about a fixed axis

The relation between force and ball speed shows that a small increase in throw speed requires a large increase in the required force. Throwing the ball twice as hard results in an increase in ball speed of only 41%. To throw the ball twice as fast, the pitcher would need to exert four times the force on the ball.

The essential physics involved in throwing can be understood in terms of the diagram shown in Fig. 5.2. To simplify matters, suppose that a ball of mass m is located at the end of an arm of mass M and length L. The other end of the arm rotates about a fixed axis. This model could describe throwing with the forearm only, with the elbow resting on a table. The forearm rotates when a person contracts the arm muscles in such a way that a force F is applied along the arm at a distance D from the axis. Muscle is attached to bone via stretchy tendons which look and act like thick strings. You can see or feel them under the skin when you tense your muscles, especially those in your wrist.

Multiplying F by D gives the torque $\tau = FD$ applied to the arm about the axis. This torque causes the arm to rotate at a speed that is proportional to the torque. The rotation speed also depends on the mass of the arm, the mass of the ball and the length of the arm. If M, m, and L are all relatively small, then the arm will rotate rapidly. If M, m, and L are all relatively large then the arm will rotate more slowly. Mathematically, we describe the resistance to rotation in terms of the moment of inertia (MOI) of the arm plus the MOI of the ball. The arm is a complicated shape so the actual MOI is hard to calculate, but it will be about the same as that of a uniform rod of mass M and length L. That is MOI(arm) $= ML^2/3$ when the axis is at one end of the rod. The MOI of the ball about the same axis is mL^2. The equation describing arm rotation is then

$$\tau = FD = (m + M/3)L^2\alpha, \tag{5.1}$$

where α is called the angular acceleration of the arm. It is like ordinary linear acceleration along a straight line, but it is measured in terms of the change in the rotation angle of the arm rather than the change in distance.

The force of the muscles acting on the arm is not the same as the force on the ball. The ball itself is rotating at the end of the arm and has a linear acceleration $a = L\alpha$ along a circular arc of radius L. The force f on the ball is given by $f = ma$. Using (4.1) we find that

$$f = \frac{mFD}{(m + M/3)L} \tag{5.2}$$

The force on the ball is proportional to the muscle force, as one would expect, but is it bigger or smaller than F? Consider a baseball of mass $m = 0.145$ kg, an arm of mass $M = 2$ kg and length 0.35 m, and let $D = 0.03$ m. Then $f = 0.0153F$. In this case, f is 65 times smaller than F. If the ball was a golf ball of mass 0.04 kg, then f would be 206 times smaller than F. If the ball had a mass of say 1 kg, similar to that of a bat, then f would be 19 times smaller than F. The force exerted on the ball increases with the mass of the ball but it remains much less than the force needed by the muscles to accelerate the arm.

An interesting result is that heavy balls can be thrown almost as fast as light balls, provided that the mass of the ball is much less than the mass of the arm. An experiment [3] was conducted in 2003 to examine this effect, using balls which varied in mass from 57 g (a tennis ball) to 3.4 kg (a lead brick). Despite the factor of 60 difference in mass, the tennis ball could be thrown only 2.4 times faster than the lead brick. The result is not especially surprising, since most of the effort of the thrower is required just to rotate the arm. Adding a small mass to the hand, which has a mass typically about 0.5 kg, makes only a small difference to the throw speed.

5.4 Double Pendulum

A better model of pitching and batting, taking into account the upper arm as well as the forearm, is one that describes the motion of a double pendulum. A single pendulum is pivoted about a fixed axis at one end and swings back and forth at the other end. A double pendulum is just two single pendulums joined end to end. The upper arm and forearm are indeed joined end to end at the elbow and each rotate about a different axis. The upper arm rotates about an axis through the shoulder and the forearm rotates about an axis through the elbow. The upper and lower parts of the leg are also joined end to end like this, and can also be described as a double pendulum. Consequently, actions such as walking, running, and kicking a ball can also be understood in terms of the motion of a double pendulum.

You can make a single pendulum by tying a length of string to a fixed support at the top end, and then tying a weight at the bottom end. If you pull the weight aside and then let go, the weight will swing back and forth at a rate that depends on the length of the string, but is typically about one cycle each second. To make a double pendulum, you tie a second length of string to the weight and then tie another weight to the bottom end of the second string. The two weights don't need to be the same and the two lengths of string can be different. If you now pull the bottom weight aside and let go, the motion of the two weights will be more complicated than you might expect. The motion is not just that of two separate pendulums because the bottom pendulum pulls back and forth on the top pendulum and vice-versa. The result can be totally unexpected. For example, the top pendulum can come to a stop for a while while the bottom pendulum swings wildly, then the bottom pendulum stops and the top pendulum swings wildly.

You can also make a single pendulum using a wood or metal rod with a hole drilled in the top end, without any weight attached to the bottom end. By passing a thin rod through the hole as an axis, the pendulum will swing back and forth just like a length of string with a weight on the bottom end. This type of pendulum, without a weight on the bottom end, is called a physical pendulum. The weight of the pendulum is distributed along its whole length rather than being concentrated at the bottom end. If you then swing another rod at the bottom end of the top rod you will have a model of the upper arm at the top and a forearm at the bottom, with an elbow joint joining the two rods and a shoulder joint at the top. The easiest way to make the elbow joint is to drill holes through the two rods and join them with a loose-fitting bolt or a short loop of string.

A neat experiment is to make a double pendulum and then pull it aside so that the upper arm is horizontal and the lower arm is almost vertical and above the horizontal, as shown in Fig. 5.3. If you then release the pendulum, the lower arm starts to fall vertically and pushes down on the upper arm, causing the upper arm to rotate faster than it would if it was a single pendulum. Both arms swing around until both reach the bottom of their circular path, at which time the upper arm has slowed down considerably and the lower arm is rotating rapidly. This happens naturally without any force acting on the system other than the force of gravity. The resulting motion

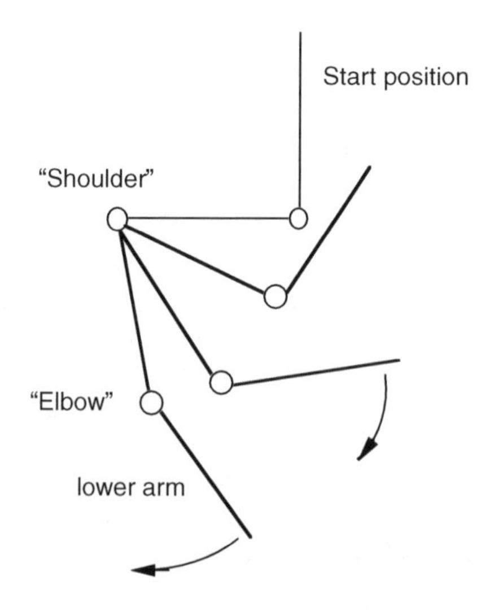

Fig. 5.3 A double pendulum consists of an upper arm and a lower arm joined end to end. The combination provides a good model of the upper and lower arm or the upper and lower parts of the leg. It also provides a good model to analyze the swing of a bat. The two arms start out at right angles when pitching a ball or swinging a bat or a golf club, as shown here, then straighten out just before throwing or impacting the ball

is very similar to that of the arm actions when throwing a ball or swinging a bat or when swinging a golf club or a tennis racquet. In all these actions, the lower arm is locked at about a right angle to the upper arm at the start, the upper arm speeds up first and then slows down as the forearm speeds up, and both sections swing around to be approximately in line when the forearm is swinging at maximum speed.

Swinging a bat or club or racquet is slightly different because the implement in the hand acts as a third segment attached to the forearm. In other words, the arrangement is really a triple pendulum and only the first two or the last two segments behave like a double pendulum. In fact, the forearm slows down just before impact with the ball so that maximum speed is transferred to the implement in the hand and hence to the ball. If we want to get really serious about the pendulum model, then there is yet another segment involved. The hand adds another segment, joined to the forearm about an axis through the wrist.

Energy Efficiency in the Double Pendulum

Another difference with respect to the mechanical pendulum model, when throwing a ball or swinging a bat, is that gravity plays only a minor role. The muscles in the arm provide a much bigger force than gravity and swing the arm much faster than gravity can. Nevertheless, the resulting motion of the upper arm and the forearm is very similar to the mechanical model since it is the natural way that a double pendulum behaves. It is natural because it is efficient. That is, maximum speed results with minimum effort when swinging in the manner of a double pendulum. A striking implement by itself can't generate any extra energy, but it can be used as an extra link in the chain to increase the transfer of energy from the player's arms to the ball.

A double pendulum acts to improve energy transfer from one segment to the next. To illustrate this point, we can consider the effect of the wrist on throwing a ball. Suppose that we strap a pitcher's forearm to a vertical pole so that he can only throw the ball using his wrist. Suppose that he can throw a 140-g baseball at a speed of $3\,\mathrm{m\,s^{-1}}$ this way. The kinetic energy of the ball is then 0.63 J so the work done on the ball by the wrist is 0.63 J. Now suppose that we strap the pitcher's upper arm to a horizontal surface so he can throw the ball using both his forearm and his wrist. If he can throw the ball at $20\,\mathrm{m\,s^{-1}}$ with his forearm alone then he can throw it at $23\,\mathrm{m\,s^{-1}}$ using a final flick of the wrist. At $20\,\mathrm{m\,s^{-1}}$, the kinetic energy of the ball is 28.0 J. At $23\,\mathrm{m\,s^{-1}}$, the kinetic energy is 37.0 J, an increase of 9.0 J. In other words, the pitcher's wrist works $9/0.63 = 14$ times more effectively when it works together with the forearm to propel the ball, at least if he waits until the ball is traveling at $20\,\mathrm{m\,s^{-1}}$ before he flicks his wrist. If he flicks his wrist first and then swings his forearm, there is no gain in energy given to the ball.

The extra energy given to the ball is taken from the forearm. When the pitcher flicks his wrist, an extra force is applied to the ball and an equal and opposite force is applied to the forearm. As a result, the ball speeds up and the forearm slows down. That way, energy is taken from the forearm and transferred to the ball. Another way to explain this result is that the force on the ball due to the wrist action acts over a longer distance when the forearm is used to throw the ball. Consequently, the work done on the ball (work = force × distance) is greater.

It is clear from this example that maximum energy is transferred to a ball or a bat when a pitcher or batter activates muscles in the correct sequence, with appropriate time delays, so that energy is transferred efficiently from one segment to the next. A similar sequence of actions occurs when swinging a golf club or a tennis racquet, with the result that a golfer or a tennis player can hit a stationary ball much faster than he or she can throw it. In baseball and softball, batters can hit a stationary ball no faster than pitchers can throw it, the problem here being that baseballs and softballs have a low COR, meaning that a large amount of energy is dissipated in the ball when it is struck by a bat. A much smaller amount of energy is dissipated in the ball when the ball is thrown.

References

1. N. Zheng, G.S. Fleisig, S. Barrentine, J.R. Andrews, in *Biomechanics of Pitching*, ed. by G. Hung, J. Pallis. Biomedical Engineering Principles in Sports (Kluwer, Dordrecht, 2004), pp. 209–256
2. G.S. Fleisig, R. Phillips, A. Shatley, J. Loftice, S. Dun, S. Drake, J.W. Farris, J.R. Andrews, Kinematics and kinetics of youth baseball pitching with standard and lightweight balls. Sports Eng. **9**, 155–163 (2006)
3. R. Cross, The physics of overarm throwing. Am J. Phys. **72**, 305–312 (2004)

Chapter 6
Swinging a Bat

By the time you know what to do, you're too old to do it.

– Ted Williams

I've always swung the same way. The difference is when I swing and miss, people say, He's swinging for the fences. But when I swing and make contact people say, That's a nice swing. But there's no difference, it's the same swing.

– Sammy Sosa

6.1 The Basics

Most people can walk or run successfully, without tripping or stumbling, even without any special coaching. Similarly, most people can swing a bat without needing anyone to show them how. The problem is, everyone has trouble connecting with the ball. In fact, some people claim that hitting a baseball is the single most difficult task of all sporting tasks [1]. More often than not the batter misses the ball completely even though the swing itself might be sensational. Hitting a hole in one in golf or sinking a 60-foot putt are actually more difficult but striking a moving ball with a bat is definitely not easy. A good batter can swing a bat perfectly well with his eyes closed but will only rarely connect with the ball. The task of hitting a ball is one that involves watching the ball carefully, it involves good hand-eye coordination, and it requires lots of practice. Some people never get the hang of it, and some are a lot better than others.

It is not immediately obvious why it is so difficult to hit the ball. Golfers and tennis players don't have this problem. It would be very embarrassing if a golfer or a tennis player missed the ball. So, why do batters have so much trouble? Golf is easier since the ball just sits on the ground waiting to be struck. Tennis is easier since the head of a racquet is much bigger than the barrel of a bat. The main problem with hitting a baseball or a softball is that the barrel of the bat is considerably smaller than the head of a tennis racquet. We can't place all the blame on the pitcher for throwing the ball so fast or with a nasty curve since tennis players are faced with exactly the same problem.

The best cricket batter in the world, by a long way, was Don Bradman, who played for Australia in the 1930s and 1940s. He had a slightly unorthodox style and was never coached. He taught himself how to hit a ball at a young age by hitting a golf ball onto a corrugated water tank in his back yard, using a shortened broom stick. The ball bounced off the tank at odd angles, so he needed very good footwork, as well as very good eyesight and very good hand-eye coordination. He kept

R. Cross, *Physics of Baseball & Softball*, DOI 10.1007/978-1-4419-8113-4_6,

practising until he hardly ever missed the ball. For him, hitting a 90 mph cricket ball with a cricket bat was almost as easy as walking, and he could hit the ball wherever he wanted to hit it.

A cricket ball is about the same size and weight as a baseball, but a cricket bat is about twice as wide as the barrel of a baseball bat. Even so, a common problem in cricket is that the batter sometimes contacts the very edge of the ball with the edge of the bat. There are usually several fielders lying in wait behind the batter to catch the ball. If the bowler in cricket (the pitcher) is especially fast, the opposing team might spread out four or five fielders behind the batter, all waiting to catch an edged ball.

Good batters can swing at a baseball sitting at rest on a tee and never miss. The problem with hitting a moving ball is, therefore, trying to predict where it will be when the bat arrives to meet it. A batter might have a perfect swing, but if he predicts that the ball will be 2 in. higher or lower than it actually is, he might miss the ball completely. Furthermore, the batter also needs to predict when, as well as where, the ball is going to arrive. If his prediction is out by a split second, he can swing too early or too late.

We will examine in more detail the problem of making contact with the ball in the next chapter. In this chapter, we will concentrate on the somewhat simpler task of swinging the bat, and we will examine the task through the eyes of a physicist. A coach, or a biomechanist, would view the task differently. A coach will tell you how to place your feet, how and where to hold the bat, and how you need to swing each body segment, including the legs, the hips, the arms and the elbows and wrists. All of those actions are obviously important, but we want to focus on the final result. That is, we will examine the effect of all those actions on the bat, without being too concerned about how the effect comes about. The basic questions we ask are these: what is it, precisely, that the batter does when he swings a bat, and how does he do it?

For example, is the batter trying to rotate the whole bat as fast as possible or is he trying to get the whole bat to move forward, toward the ball, as fast as possible? We are also interested in what each arm is doing. Does the left hand push while the right hand pulls? In what direction does each hand push or pull? Is it along the handle, at right angles to the handle, or is at some other angle to the handle? These are not the sorts of questions that a batter or a coach might ask, but they are fundamental from a physics point of view. A coach might tell a batter to use the wrists more aggressively, or to "throw your hands at the ball," but the physics of the problem might indicate that an entirely different action is needed. Some day in the future, physicists, coaches, and biomechanists might all get their heads together to work on these issues. So far, it hasn't happened.

The actions involved in swinging a bat are not intuitively obvious and are not the actions you might expect. For example, consider the start of the swing when the bat is starting its forward motion. The batter starts rotating his hips and shoulders, keeping his arms and the bat locked in position with respect to his upper body. The result of this action is that the batter pushes on the handle in the opposite direction to the direction of motion of the handle. That's really weird, but it happens every time. It's almost like pushing on a door knob to pull the door toward you, but there is a difference. When a batter pushes on the handle, the barrel moves away from the batter but the handle moves toward the batter. Even though the batter pushes on the handle, he also exerts a twisting or turning force to rotate the bat. When opening a door, a person first twists the handle and then pulls or pushes the handle to open the door. When swinging a bat, the batter pushes the handle and twists the handle (using both hands) at the same time. The end result is that barrel moves in the direction of the push force but the handle moves in the opposite direction.

What happens near the end of the swing, just before the batter strikes the ball? Does the batter push or pull on the handle or does one hand push while the other hand pulls? The answer here is also surprising. If you watch a batter in action, and if the batter is swinging as fast as he or she can, then you will see that the batter is leaning backward and pulling the handle toward his or her body as hard as possible, with both hands. That is, the force on the handle is almost at right angles to the direction of motion of the handle.

6.2 Film of a Swing

The nature of bat swinging is illustrated in Fig. 6.1. The positions of the bat here were measured by filming a batter using a video camera mounted about 10 ft above his head. A batter usually starts off the swing with the bat near or above one shoulder

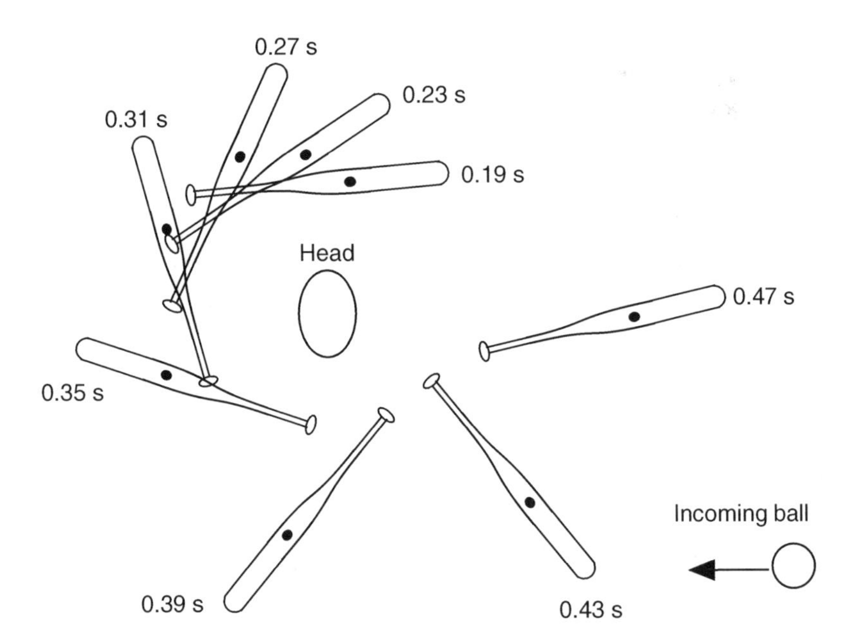

Fig. 6.1 Overhead view of a typical swing of a baseball bat showing the position of the bat every 0.04 s, starting 0.19 s into the swing. The knob end of the bat rotates at relatively low speed in an approximately semi-circular path, while the tip of the bat rotates at higher speed along a path that spirals outward

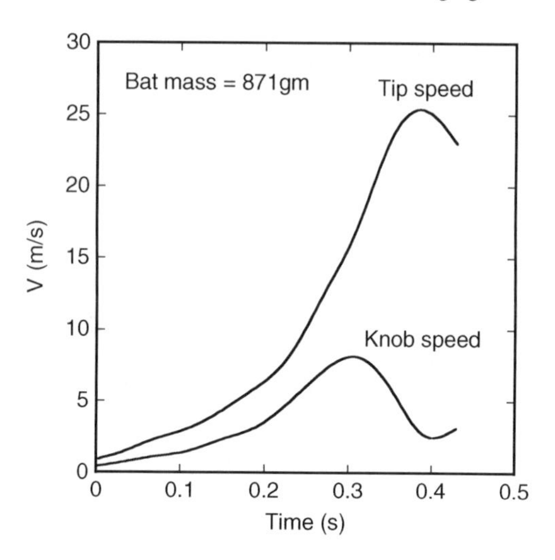

Fig. 6.2 Speed of the knob and the tip of the bat for the swing shown in Fig. 6.1. Just before impact with the ball, the knob end slows down (and so do the batter's arms) as the barrel end speeds up

and then swings it around through 180° or more so that it impacts at right angles with the incoming path of the ball. In Fig. 6.1, the bat started off at almost zero speed. After 4/10 of a second, the tip was traveling at $26\,\mathrm{m\,s^{-1}} = 58\,\mathrm{mph}$, as shown in Fig. 6.2. Not a particularly fast swing, but one that is relatively common in the game of baseball or softball. Despite the fact that it is easy to swing a bat like this, it is not easy to explain what the batter needs to do to achieve this result. There is probably not a batter alive who could explain what he does when he swings a bat, at least in terms of the forces on the bat.

A surprising result, shown in Fig. 6.2, is that the batter did not swing his arms as fast as possible during the whole swing. The knob speed started to decrease before the tip of the barrel reached its maximum speed, meaning that the batter decreased the swing speed of his arms before the bat reached its maximum speed. A similar effect is observed when golfers swing a golf club. The arms of a golfer slow down before the club head reaches its maximum speed. There is a simple explanation for this effect. The object of the exercise is not to swing the arms as fast as possible but to swing the bat or the club as fast as possible. That is best done by first accelerating the arms and then by transferring the energy in the arms to the bat or club. The action is the same as that of a double pendulum, as described in the previous chapter. To transfer energy from the arms to the bat, the arms must slow down so that the bat can reach maximum speed [2].

We will analyze the swing in Fig. 6.1 in some detail in this chapter, not because it was a particularly good swing but rather to highlight some of the physics issues. One issue is that the swing took 0.4 s. The last 180° of the swing took only 0.15 s. Some batters complete the last 180° of the swing in only 0.10 s. There is, therefore, an issue as to when the swing actually starts. In Fig. 6.1, we started counting when the bat was as far back as it went and when the bat then started to rotate forward. There was some movement of the bat even before this time.

Whenever film of a baseball game is taken showing both the pitcher and the batter in view at the same time, the film shows clearly that the batter starts his first move before the pitcher releases the ball. As the pitcher raises his front foot, the batter crouches down slightly by bending at the knees, like a tiger getting ready to pounce. Just before the pitcher releases the ball, the batter starts to lift his own front foot. The batter steps forward into the shot as soon as the pitcher releases the ball. By the time the batter plants his front foot on the ground, the ball is already half way to its destination and the batter has commenced his final rapid swing. The ball takes only 0.4–0.5 s to arrive in the hitting zone after the pitcher releases the ball. The batter is in motion the whole time, sizing up the situation and getting ready to slog the ball. When he is finally ready, he swings the bat with an acceleration that is 40 times greater than that of a Ferrari at a Grand Prix. A Ferrari can accelerate to 60 mph in 4 s. A batter can accelerate a bat to 60 mph in 0.1 s.

The batter in Fig. 6.1 was filmed in the laboratory for convenience and didn't actually strike a ball. That might have influenced his swing technique and timing to a small extent, but it didn't alter the physics of what did happen. In fact, the swing was almost a carbon copy of the one analyzed in considerable detail by Adair in his book [3]. The question we now ask is, what does a batter actually do when he swings a bat? A batter grabs hold of the bat handle with both hands. What force does he apply to the handle with each hand, and in what direction do those forces act?

6.3 Effect of a Force Acting on an Object

To work out the forces on a bat, it will help to first consider the effect of forces in a more general way. Figure 6.3 shows the effect of a vertical force on a baseball and the effect of horizontal forces on a block of wood. Figure 6.4 shows the effect of various forces on a bat.

If a ball is held at rest in the hand and then dropped vertically, the force of gravity acts in a vertical direction through the middle of the ball, and the ball falls vertically in a straight line, as shown in Fig. 6.3a. In Fig. 6.3b, the ball is thrown horizontally and follows a curved path, despite the fact that the same force of gravity acts on the ball as in Fig. 6.3a. In Fig. 6.3b, the ball travels at constant speed in the horizontal direction since there is no horizontal force on the ball (apart from the small horizontal force due to air resistance). Simultaneously, the ball accelerates in the vertical direction due to the vertical force of gravity on the ball. The result is that the ball follows a curved, parabolic trajectory.

Now suppose we apply a horizontal force, F, to the middle of a rectangular block of wood, as shown in Fig. 6.3c, and the block is free to slide on a horizontal table. The block will then move in a straight line path, accelerating along the way. If F is applied near one end of the block, as in Fig. 6.3d, the block will still slide along the table as before, but it also rotates. The rotation rate depends on the applied torque and on the length and weight of the block. The torque, τ, is given by $\tau = Fd$ where d is the distance between middle of the block and the point where F is applied. In Fig. 6.3c, $d = 0$, so the torque is zero and there is no rotation.

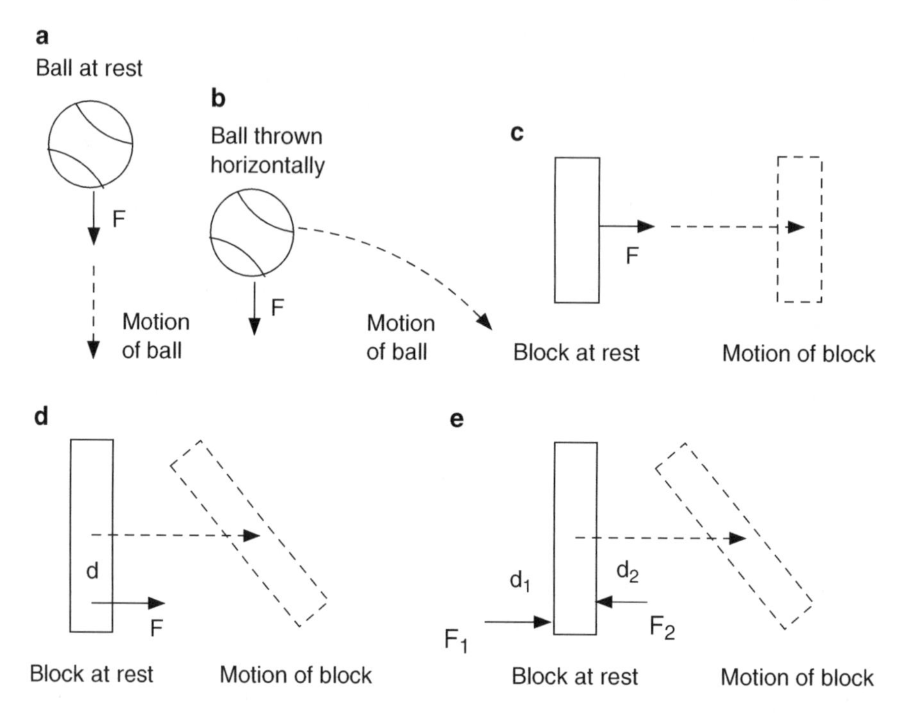

Fig. 6.3 Motion of a ball and a block of wood when subject to various forces. The result depends on whether the object started out from rest or whether it was moving when the force was applied, and it also depends on where the force is applied

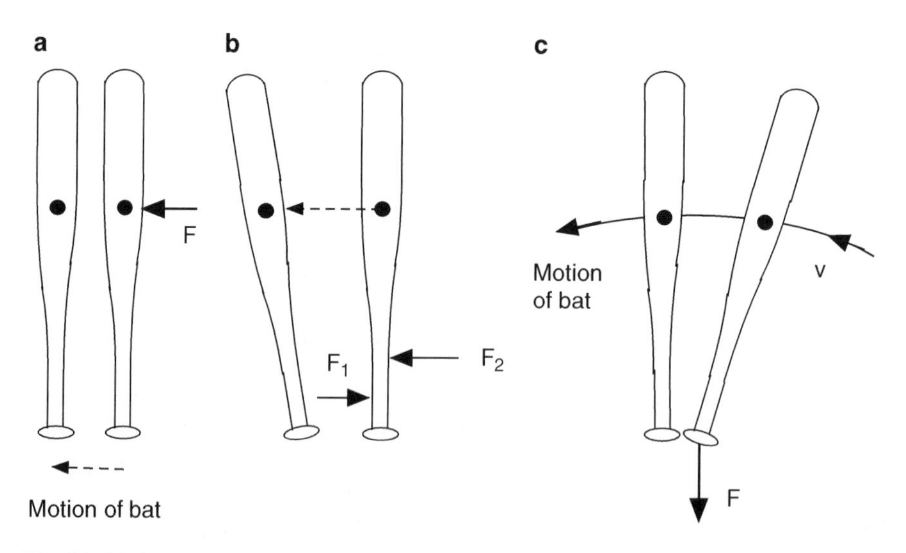

Fig. 6.4 Motion of a bat subject to various forces

Figure 6.3e shows the result when two forces are applied near one end of the block, a situation that is more like that when a batter swings a bat with two hands. If F_1 is bigger than F_2 then the block will move to the right, but if F_1 is smaller than F_2 then the block will move to the left. The block also rotates. There is a torque $F_1 d_1$ which, on its own, would cause the block to rotate counter-clockwise. The torque $F_2 d_2$, on its own, would cause the block to rotate clockwise. The final result depends on the total torque $F_1 d_1 - F_2 d_2$ which could be positive or negative or zero depending on the values of the F's and the d's. The block will rotate in a clockwise direction if $F_2 d_2$ is larger than $F_1 d_1$.

6.4 Forces Acting on a Bat

Figure 6.4 shows the results of applying various forces to a bat. The centre of mass of the bat is shown by a black dot, and it is located about 10 in. from the end of the barrel. Figure 6.4a shows the effect of applying a force F at right angles to a bat, in line with its center of mass (CM). If the mass of the bat is m then the acceleration of the bat is given by $F = ma$ or $a = F/m$. If the bat is initially at rest then the whole bat accelerates to the left without rotating. You can try this yourself by pushing a ruler (or a pencil) along a table. In that case, the CM of the ruler is probably in the middle, so you will need to push in the middle to avoid any rotation. Actually, you might need to cheat a little and push with two fingers, since friction with the table might cause one end to get stuck and then the ruler will rotate, which will ruin the whole experiment. Physics experiments are often like that. Scientists set out to observe or measure something and might discover that something else is just as important if not more so. That is how things often get discovered but it can be frustrating for a beginner.

Figure 6.4b shows the effect of applying a force at right angles to the handle, using both hands. One hand exerts a force F_1 to the right and the other exerts a force F_2 to the left. Provided that F_2 is larger than F_1 then the whole bat will accelerate to the left. But how do we make sure that the bat also rotates counter-clockwise, as shown in Fig. 6.4b? Given that F_2 is larger than F_1, won't the bat rotate clockwise, in the wrong direction? To ensure that the bat rotates in the correct direction, the torque due to F_1 must be bigger than the torque due to F_2. Such a result will be achieved if F_2 is only slightly bigger than F_1 and if F_2 acts along a line that is closer to the bat CM than F_1.

You can try this with a pencil or a ruler on a table. Your brain will tell you what to do without even thinking about the physics of it. You don't need two hands to rotate a pencil or a ruler in the manner shown in Fig. 6.4b. You can rotate a pencil or a ruler using just one hand, swinging it through the air like a small table-tennis bat. In that case, the two separate forces are supplied by different parts of the same hand. The part of the hand near the index finger supplies the force F_2 and the part closest to the little finger supplies the force F_1. Alternatively, you can push and rotate a ruler on the table using just the thumb and the index finger of the same hand.

Figure 6.4c shows a situation where the bat CM is moving in a curved path at speed v, and the whole bat is simultaneously rotating counter-clockwise about an axis through the CM. To move in a curved path like this, there must be a force on the bat acting at right angles to the curved path, similar to the situation shown in Fig. 6.3b for the curved ball path. In Fig. 6.4c, the force F on the handle acts in a direction that is almost along the handle but it is at a slight angle to the handle. That is exactly what the batter needs to do to hit the ball as fast and as far as possible. The force is directed towards the batter's body and causes the bat CM to follow a curved path. In addition, the force generates a torque on the bat causing the whole bat to rotate in such a way that the barrel speeds up as it approaches the ball. Simultaneously, the handle slows down.

Near the end of the swing, the bat rotates so fast that the handle pushes backward against the hands, causing the forearms to slow down. As we saw in the previous chapter, if one segment slows down then the next segment speeds up. At least, this is the case if the wrists are relaxed, as they are near the end of the swing. At the start of the swing, the batter needs to lock his wrists when he accelerates his forearms, otherwise the bat would rotate backwards.

6.5 How Big is the Force on a Bat?

A bat needs to rotate in an approximately circular path, accelerating along that path. If an object undergoes circular motion at constant speed then it can do so only if there is a force acting at right angles to the path, pointing in a direction toward the center of the circle. Without that force, the object would fly off in a straight line tangential to the circular path. The force in this case is supplied by the batter who needs to pull the bat towards the center of the bat path, somewhere near his chest. That's why a batter leans backward slightly when swinging a bat. A force directed towards the center of a circle is called a centripetal force. In the case of a bat swung at around 58 mph (26 m s^{-1}) the centripetal force is about 200 lb, greater than the weight of the batter and about 100 times greater than the weight of the bat.

The centripetal force required to swing a bat is given by $F = mv^2/R$ where m is the mass of the bat, v is the speed of its center of mass and R is the distance from the CM to the center of the circular path. The speed of a bat, at the impact point on the bat has been measured for many players and varies typically from about 50 mph (22 m s^{-1}) to about 90 mph (40 m s^{-1}). The impact point on a bat is about 27 in (0.70 m) from the knob end of the handle, while the CM is about 23 in (0.58 m) from the knob. The CM is slightly closer to the center of the circular path and therefore travels at a slightly lower speed than the impact point.

For example, if $m = 1$ kg, $v = 25$ m s^{-1} and $R = 0.65$ m then $F = 625/0.65 = 962$ N. This is quite a large force, equal to the weight of a 98 kg (216 lb) mass. Even though the bat mass is only 1 kg, the force needed to swing it around in a circle at 25 m s^{-1} is 98 times larger than its weight.

A bat is not normally swung at constant speed in a circular path since it accelerates along the way. That means there needs to be a force acting at right angles to the bat, in a direction along the path, so the CM can accelerate along that path. Suppose the bat CM accelerates from rest to a maximum speed of 56 mph (25 m s^{-1}) in 0.15 s. The average acceleration is then $25/0.15 = 167$ m s^{-2} and the average force on the bat acting along the circular path is $F = ma = 1 \times 167 = 167$ N (38 lb).

It seems that all the batter needs to do is to push the bat sideways with a force of about 38 lb and to pull it toward his chest with a force that increases with time up to about 200 lb. But there is something else he needs to do. To hit the ball with the barrel, the batter also needs to rotate the bat through an angle of at least 180° and perhaps 270°. The batter starts the swing with the knob pointing roughly toward first base or even further around toward the catcher. Contact with the ball is made when the bat is aligned at right angles to path of the ball. To rotate the bat in this manner, the batter also needs to exert a torque on the handle.

There are two ways that a batter can exert a torque on the bat, and both of them are used to swing the bat at different stages of the swing. One way is to push with one hand and pull with the other, as shown in Fig. 6.4b. The other way is to pull with both hands at a slight angle to the bat, as shown in Fig. 6.4c. The situation in Fig. 6.4b is used at the start of the swing and the situation in Fig. 6.4c is used near the end of the swing.

6.6 Close Inspection of the Swing in Fig. 6.1

There is a great deal of information contained in Figs. 6.5 and 6.6, and it allows us to determine the forces and torques acting on the bat, at least in the horizontal plane. Batters normally swing a bat in a plane that is inclined to the horizontal, given that the tip of the bat starts at a point above the shoulders and drops to about waist level when the bat collides with the ball. To obtain the results in Figs. 6.1, 6.5, and 6.6, the batter was asked to swing in a horizontal plane, although he started off with the bat near shoulder height, a bit closer to the camera. That is why the bat in Fig. 6.6 appears to be a bit longer at the start of the swing than later on. Ideally, several cameras should be used to get a full three-dimensional view of the swing, but that would have complicated the experiment considerably. Alternatively, a single camera could be used, viewing in a direction perpendicular to the swing plane, as is sometimes done to view the swing action of a golfer.

By plotting the x and y coordinates of the bat CM as a function of time, the velocity V_x in the x direction and the velocity V_y in the y direction were determined to calculate the speed $V = (V_x^2 + V_y^2)^{1/2}$. The bat CM moved in a curved, spiral path of radius R, where R increased with time since the batter started the swing with his elbows bent and finished the swing with both arms relatively straight. The length of the dashed lines in Fig. 6.5 shows the value of R at each instant during the swing. Early on, R was about 0.6 m, and near the end of the swing R was about 0.8 m.

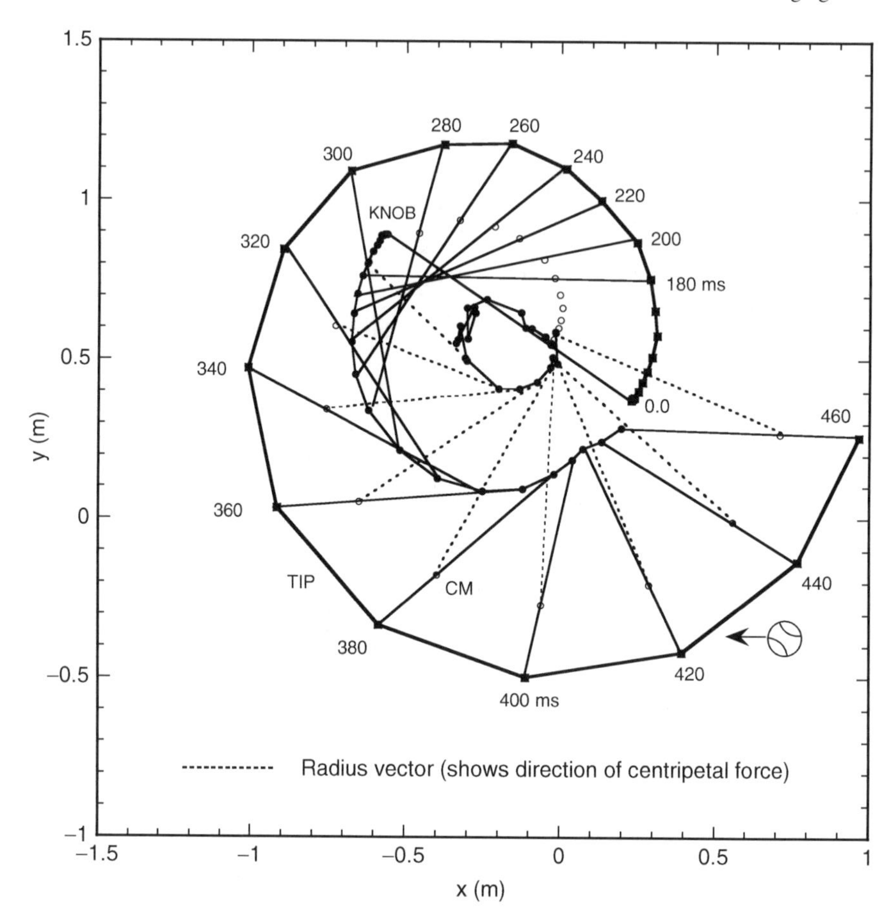

Fig. 6.5 The swing in Fig. 6.1 shown at intervals of 20 ms = 0.02 s. The *outer spiral* shows the path of the tip of the barrel. The *middle semi-circle* shows the path of the knob end of the bat. *Small, open circles* show the position of the bat center of mass (CM). The *inner circle* of *black dots* shows the instantaneous center of curvature of the path followed by the CM. The centripetal force on the bat is directed from the CM to the center of curvature, as indicated by the *dashed lines*

By plotting a graph of V vs. time, the acceleration $\mathrm{d}V/\mathrm{d}t$ along the spiral path of the bat CM could be determined, and so could the centripetal acceleration V^2/R perpendicular to the spiral path. Multiplying by the bat mass M then gave the force $M\mathrm{d}V/\mathrm{d}t$ acting along the path of the CM and the force MV^2/R perpendicular to the path.

The force acting along the path acts to increase the speed of the bat CM, and the force perpendicular to the path causes the bat to follow a curved path rather than a straight line path. The two forces are shown in Fig. 6.7. Both of the force components were calculated from changes in position of the bat CM, but the forces are actually applied at the handle end of the bat. An alternative plot is shown in Fig. 6.8 where we show the total force, F, acting at the handle end, and the angle, θ,

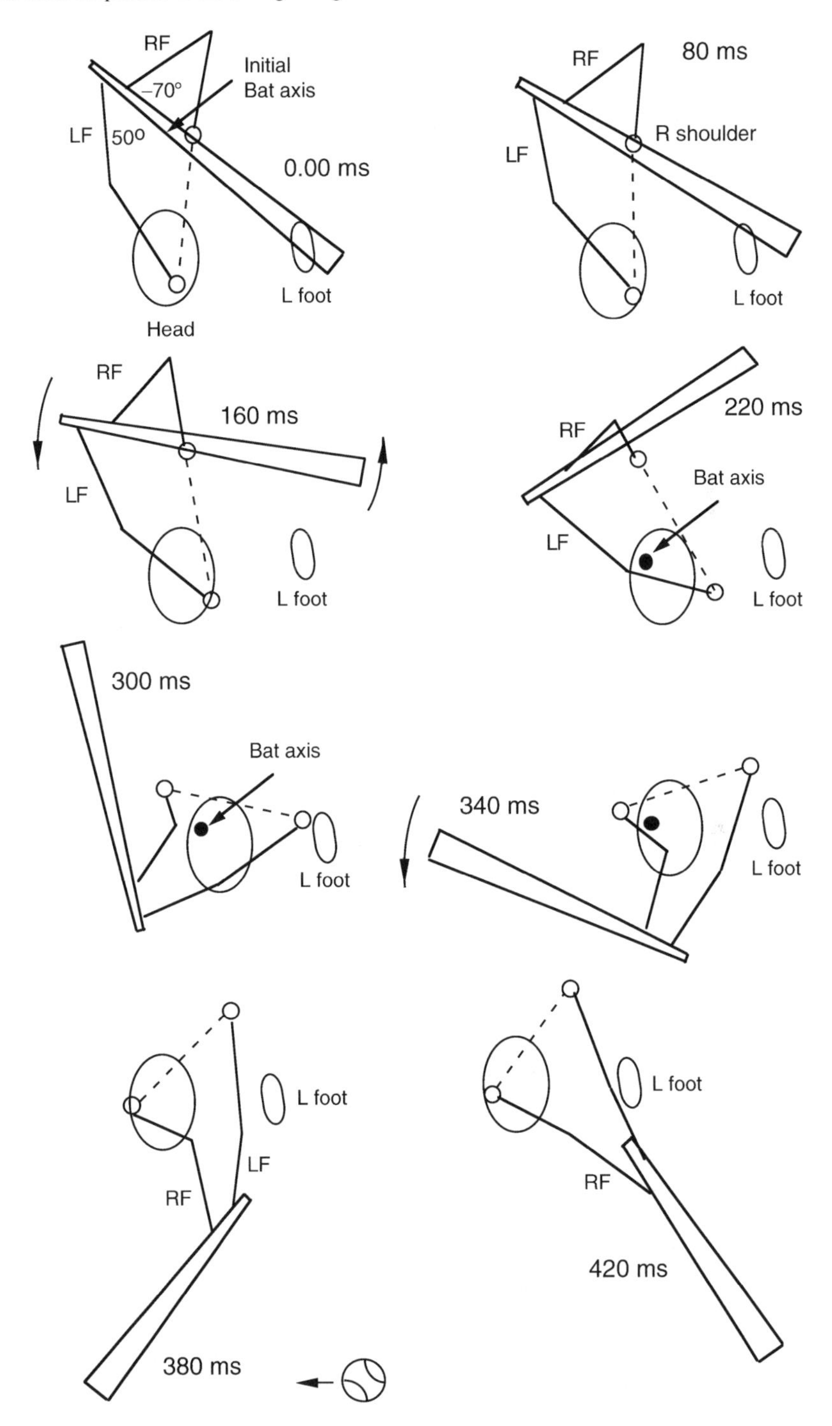

Fig. 6.6 View of the swing shown in Fig. 6.1 at eight selected times, showing the positions of the bat, the four arm segments, the batter's head and shoulders, and his stationary left foot. *LF* Left forearm, *RF* Right forearm

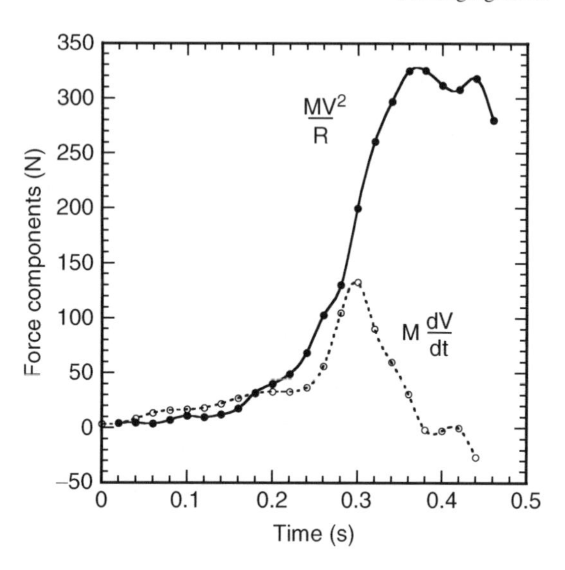

Fig. 6.7 The force components $M\mathrm{d}V/\mathrm{d}t$ and MV^2/R acting on the bat

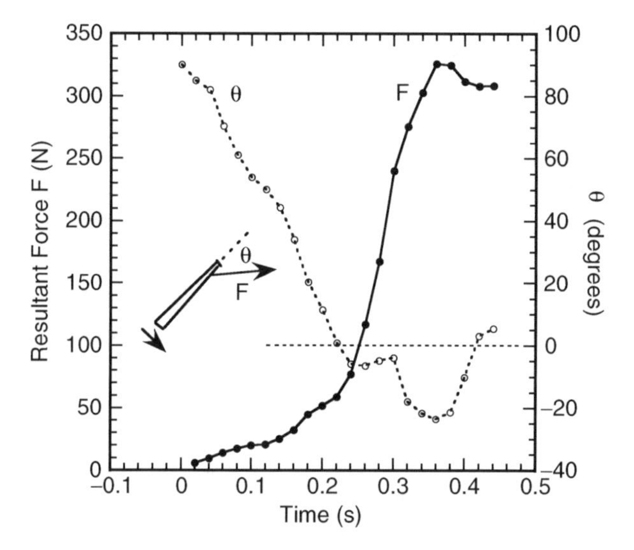

Fig. 6.8 The total force, F, on the bat and the direction of that force, measured in terms of the angle θ shown in the inset

between the line of action of F and the long axis of the bat. At the start of the swing, $\theta = 90°$, meaning that the batter exerts a force at right angles to the handle. As the bat swings around, θ drops to zero and remains close to zero from $t = 0.2$ to $t = 0.3$ s. During this time, the batter pulls in a direction that is essentially along the handle. From $t = 0.3$ to $t = 0.4$ s, θ is negative, meaning that the batter pulls at an angle of around 10–20° to the long axis of the bat and he pulls slightly backward.

Towards the end of the swing, the centripetal force is much larger than the other component, so the batter pulls on the handle in a direction almost parallel to the dashed lines shown in Fig. 6.5.

6.7 Rotation of the Bat

The motion of the bat can be regarded as consisting of two separate motions. One is the motion of the bat CM, which follows a spiral path, starting from a point near the batter's right shoulder, and ending in front of the batter when the batter makes contact with the ball. On top of that motion is a rotation of the whole bat around an axis through the bat CM. In Fig. 6.6, the whole bat rotates through an angle of about 180° from $t = 0$ to $t = 340$ ms, and almost 360° from $t = 0$ to $t = 420$ ms.

The object of the exercise, assuming the batter wants to hit the ball at high speed, is to allow the bat to line up almost at right angles to the path of the incoming ball at a time when the impact point on the barrel is traveling at maximum speed.

The forces shown in Figs. 6.7 and 6.8 are not the only forces acting on the bat. In addition, the batter exerts a torque on the bat to make it rotate. If the batter exerts equal and opposite forces on the handle, using both hands, then those forces will have no effect on the motion of the bat CM but they will cause the bat to rotate, as indicated in Fig. 6.4b. In Fig. 6.4b, we show the two forces as F_1 and F_2. If these two forces are equal and opposite then there is no net force on the bat but there is still a torque or turning force. Two equal and opposite forces acting in this way are known as a "couple."

The rate of rotation of the bat is determined by the total torque acting on the bat. That torque can conveniently be regarded as consisting of three separate parts that all add up to give the total torque. The batter just does what he needs to do, so the three separate parts are not part of three separate actions. The three separate parts just help us to understand what is happening.

The total torque on the bat was determined from its rate of rotation and the known moment of inertia of the bat. The total torque was relatively small, less than 6 Nm throughout the whole swing. The torque arising from the two force components, MV^2/R and $M\mathrm{d}V/\mathrm{d}t$, was also calculated. Subtracting these two components from the total torque gives the torque arising from the couple, C, exerted by the batter. The results are shown in Fig. 6.9.

The three components of the torque are all much larger than the total torque, meaning that they all add up to give a small net torque. At the beginning of the swing, τ_A is very small since the bat is rotating at low speed and hence the centripetal force on the bat is very small. The centripetal force increases rapidly as the bat accelerates, leading to a big increase in τ_A, but then τ_A rapidly drops to zero near $t = 0.42$ s since the centripetal force is then directed along a line from the bat CM to the knob end (as indicated in Fig. 6.5). The peak value of τ_A is 75 Nm.

The value of τ_B remains negative throughout the swing, meaning that the force component $M\mathrm{d}V/\mathrm{d}t$ acts in the correct direction to accelerate the bat, but it acts in

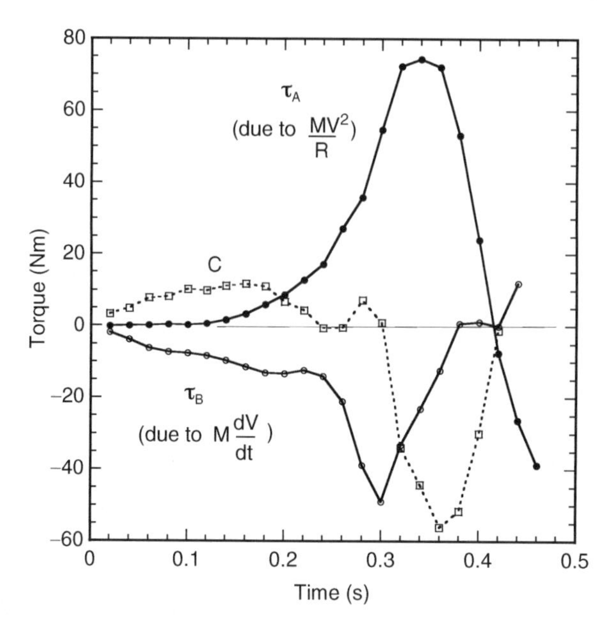

Fig. 6.9 The three torque components τ_A, τ_B and C, where τ_A is the torque due to the centripetal force on the bat, τ_B is the torque due to the force component $M\,\mathrm{d}V/\mathrm{d}t$ and C is the couple exerted by the batter's two arms (one arm pushing and the other arm pulling, or both arms pushing in opposite directions)

the "wrong" direction in terms of bat rotation. It is for that reason that the batter must apply a small positive couple near the start of the swing to make sure the bat rotates in the correct direction. He does that by keeping his wrists locked so that barrel of the bat does not get left behind when he first applies a force to the handle.

The situation near the start of the swing can be explained with reference to Fig. 6.4b. If $F_1 = 0$ then the force F_2 would cause the bat CM to move and accelerate to the left, but the torque due to F_2 would cause the bat to rotate in the wrong (clockwise) direction and the tip of the barrel would move to the right. To overcome this problem, the batter needs to apply an additional force F_1. The combined effect of F_1 and F_2 can be described as a net force $F = F_2 - F_1$ acting to the left plus a couple that is large enough to rotate the bat in the correct (counter-clockwise) direction.

As the bat accelerates, the couple required to rotate the bat in the correct direction decreases, because the torque due to the centripetal force increases. The latter torque, labelled τ_A in Fig. 6.9, acts in the correct direction to rotate the bat in a counter-clockwise direction. When the bat is moving fast enough, the batter can relax his wrists completely and allow his hands to rotate freely about an axis through the wrist. However, the torque due to the centripetal force rises to such a large value that the batter must then apply a negative couple to prevent the bat rotating too fast. That is why, in Fig. 6.9, a negative couple of almost 60 Nm was applied by the batter near the end of the swing. Without that couple, the bat would rotate much too fast and the bat might end up pointing straight at the pitcher rather than at right angles to the path of the incoming ball.

The large negative couple near the end of the swing arises naturally, without the batter doing anything special to the bat, apart from hanging onto the handle firmly. The situation shown in Fig. 6.6 at 300 or 340 ms indicates that if the bat were allowed to rotate at high speed then the handle would push firmly on the batter's left hand and pull out of his right hand. To prevent this happening, the batter pushes on the handle with his left hand and pulls with his right hand, thereby generating the large negative couple required to counteract the large positive torque arising from the centripetal force on the bat. The batter simultaneously pulls the bat towards his chest to provide that centripetal force. If he didn't, the bat would tend to follow a straight line path headed toward first base. By pulling the bat towards his chest, the bat follows a circular or spiral path and is headed toward the pitcher when the bat meets the ball. Further details of the mechanics involved, including a mathematical model of the swing, are given in [4].

6.8 Wrist Torque

Much has been written, especially in relation to the golf swing, about the action of the wrists. When swinging a bat or a club, the wrists are used at the beginning of the swing to hold the bat or club at an angle of about 90° to the forearms. By locking the wrists in this manner, the bat or club, as well as the forearms, can swing around like a solid, rigid object. As the bat speeds up, the wrists relax. By the time the bat collides with the ball, the wrists have allowed the bat to line up with the forearms, as indicated at the end of the swing in Fig. 6.6. It might appear that the wrists are actively causing the bat to rotate, but it is primarily the centripetal force that causes the bat to rotate. The wrists are not strong enough to generate rapid rotation of the bat, nor are they strong enough to prevent the rotation. The bat, therefore, causes the hands to rotate about an axis through the wrist, not the other way around.

The strength of the wrists can be measured by holding a bat in a horizontal position with one hand and by hanging a weight at the far end. A person can easily hold the bat by itself in this manner, but when a weight is added to the barrel end, it becomes more difficult to hold the bat in a horizontal position. The wrist torque needed to support a 0.9 kg bat in a horizontal position, when the bat center of mass is 0.45 m from the wrist, is 4 Nm. If a person can support an additional 2.2 lb (1 kg) weight located 0.7 m from the wrist, then the additional torque exerted by the wrist is 6.9 Nm, giving a total torque of 10.9 Nm. With two wrists, a batter can exert a maximum wrist torque of about 22 Nm on the bat.

In Fig. 6.6, the maximum couple exerted on the bat is about 60 Nm. That couple is too large to be provided by wrist action alone, and must be supplied by the equal and opposite forces exerted on the handle by each arm. Consequently, the wrists play only a small role at the end of the swing, although they are often used by batters when they roll one wrist over the top of the other during the follow-through.

6.9 Rotation Axes Again

The rotation axis of a bat is not as easily identified as one might expect. The rotation axis of the bat is shown in Fig. 6.6 at several different times during the swing. At the start of the swing, the bat rotates about an axis in the handle near the batter's right shoulder. That axis can also be identified in Fig. 6.5 by the intersection point of images of the bat at times $t = 0$, 180 and 200 ms. At later times, the bat axis moves to a point outside the bat, above the batter's head (see Fig. 6.6). In the latter case, the bat axis is not defined as the intersection point of sequential images of the bat. Rather, every point in the bat rotates about the axis in a circular orbit, although the radius of the orbit is different for different points. You can see how this arises if you rotate the letter L about an axis through the bottom, right hand corner of the letter. The axis is not at the intersection of subsequent images of the vertical part of the letter. The axis would correspond to the intersection of subsequent images of the vertical part of the letter only if the actual axis was located somewhere along the vertical part of the letter.

6.10 Summary of Forces Acting on a Bat

When a batter swings a bat, there is only one force of any significance acting on the bat. That is the force exerted by the batter. The force of gravity also acts on the bat but it is much smaller than the force exerted by the batter. To understand how the batter influences the motion of the bat, it is useful to divide the total force acting on the bat handle into four separate components. A similar situation would arise if four people were lifting a heavy load. Each person would exert a separate force, but there is only one total force on the load, which is the sum of the four separate forces. Similarly, when swinging a bat there is only one total force on the bat, arising from four separate components, as shown in Fig. 6.10. The four components are:

(a) A "push" force F_a exerted by the batter at right angles to the bat,
(b) A "push" force F_b exerted by the batter at right angles to his foream,
(c) A "pull" force F_c exerted by the batter in a direction along the bat, and
(d) A "pull" force F_d exerted by the batter along his forearm.

A batter usually uses both arms but it is simpler to imagine that the combined effect of both arms is equivalent to that of a single forearm.

The formula for each force component is relatively simple, but it helps to consider that the motion of the bat consists of two parts. First, we suppose that the bat handle moves at the same speed as the batter's hand, at speed V_h, and the hand moves in a circular arc of radius R_A. If the hand is accelerating the bat along that arc then the force along the arc (in a direction perpendicular to the arm) is $F_b = M\mathrm{d}V_h/\mathrm{d}t$, where M is the mass of the bat. There is also a centripetal force $F_d = MV_h^2/R_A$ acting toward the center of the arc.

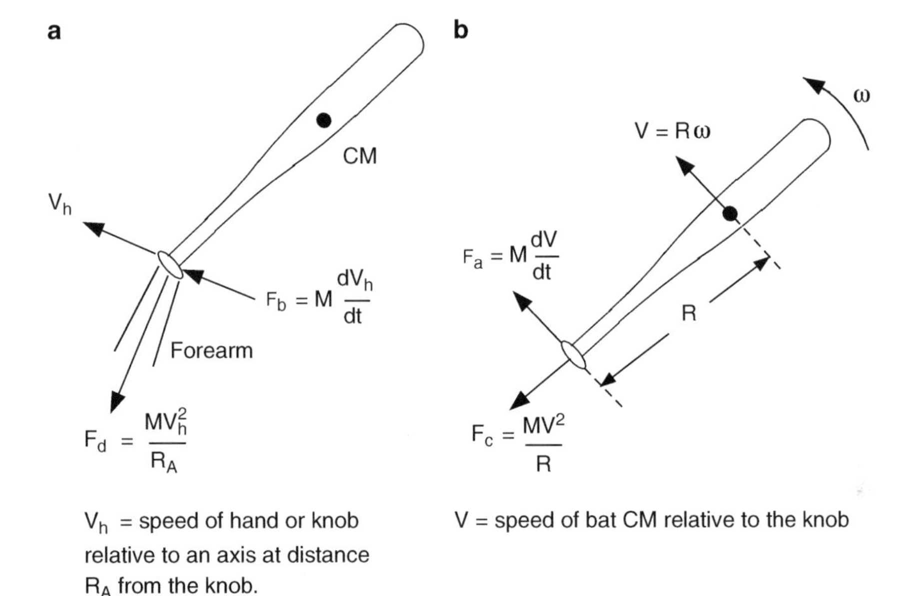

Fig. 6.10 The total force acting on a bat can be regarded as the sum of four different components

In reality, the whole bat moves in such a way that different parts of the bat move at different speeds. Suppose that the bat center of mass is located at a distance R from the knob end. Relative to the knob, the bat CM rotates about an axis through the knob at speed $V = R\omega$ where ω is the angular velocity of the bat. To rotate the bat in this manner, the batter needs to exert an additional force $F_a = M\mathrm{d}V/\mathrm{d}t$ at right angles to the bat and needs to exert an additional pull force $F_c = MV^2/R$ in a direction along the bat.

The biggest force component by far is the pull force F_c. It starts off being relatively small but increases rapidly during the swing to a maximum value of around 200 lb or more at the time of impact with the ball. This component has no effect on the bat speed and does not result in any torque being applied to the bat. In that respect, the biggest force component doesn't appear to do anything useful at all. However, it *does* allow the bat to rotate, so it allows the batter to swing the bat in a circular arc from behind his back to meet the ball as it crosses the plate. That's actually quite useful.

The three other force components all help to rotate the bat by exerting a torque on the bat. F_b causes the speed of the knob end of the bat to increase and F_a allows the barrel end to speed up relative to the knob end. That is, it allows V to increase. A detailed discussion of the effects of the various force components is given in [2].

References

1. P. Kirkpatrick, Batting the ball. Am. J. Phys. **31**, 606–613 (1963)
2. R. Cross, A double pendulum swing experiment: In search of a better bat. Am. J. Phys. **73**, 330–339 (2005)
3. R.K. Adair, *The Physics of Baseball*, 3rd edn. (Harper Collins, NY, 2002)
4. R. Cross, Mechanics of swinging a bat. Am. J. Phys. **77**, 36–43 (2009)

Chapter 7
Contacting the Ball

During my 18 years I came to bat almost 10,000 times. I struck out about 1,700 times and walked maybe 1,800 times. You figure a ballplayer will average about 500 at bats a season. That means I played seven years without ever hitting the ball.

– Mickey Mantle

7.1 Introduction

One of the problems faced by a batter is that the ball approaches the batter at high speed, up to 90 mph. Batters are somewhat sluggish in responding to this situation. It takes a certain time before they react, and when they do their arms and legs are too heavy to respond instantly. It doesn't help that the bat itself is quite sluggish, except of course when actual contact with the ball is made and then a heavy bat is an asset rather than a liability.

Another problem facing batters is that the bat is smaller in diameter than the ball. If the batter just dangled the bat over the plate, the chance of the ball striking the bat would be pretty slim. We can compare this problem with the situation in tennis. Tennis players almost always make contact with the incoming ball because the racquet head is about 11 in. wide. It is very rare for a tennis player to miss the ball completely. Not only that, a tennis player can choose to return the ball down the line or across court, meaning that he is able to start his final swing action precisely at the right time. He can also quickly vary the angle of the racquet head using his wrist since modern tennis racquets weigh only about 300 g. A baseball bat is about three times heavier and can be swung properly only by using both hands.

One of the differences between a very good tennis player and an average player is that very good players almost always hit the ball near the middle of the strings. Average players strike the ball all over the strings, which is part of the reason that racquet heads need to be so big. The same effect translates to baseball and softball, in that very good batters miss the ball by a small amount, and less frequently, while average players miss the ball more often and by a larger amount.

To compensate for the small diameter of a bat, the rules of baseball require the pitcher to throw the ball into a relatively small hitting zone. That way, the batter has a reasonable chance of hitting the ball. In tennis, the server whacks down serves at speeds up to 140 mph into a hitting zone that is 13.5 ft wide. The receiver therefore has to dash sideways in either direction to chase down the ball before he can strike it, using either a forehand or a backhand stroke. Baseball and softball are much easier

R. Cross, *Physics of Baseball & Softball*, DOI 10.1007/978-1-4419-8113-4_7,

in this respect. The batter doesn't need to chase after the ball and he needs only one hitting style, essentially a double-handed forehand. Despite these advantages, baseball and softball are games of chance as much as games of skill. Batting is literally a hit or miss event.

7.2 The Timing Problem

A potential problem in striking a ball is the timing problem. The batter has very little time to judge the flight of the ball and to swing the bat so that they both collide at the right time and at the right spot. The timing problem was explored in Chap. 5 in relation to pitching, and is now reconsidered in relation to batting. After a baseball leaves the pitcher's hand, the ball takes between 0.4 and 0.5 s to arrive in the hitting zone, 60 ft from the pitcher. The hitting zone is about 17 in. (0.43 m) deep. Traveling at a speed of around 80 mph (36 m s^{-1}), the ball spends at most only 0.012 s in the hitting zone before being struck or passing through to the catcher. The task of the batter is to make sure the bat arrives in that zone at the same time. If we compare the 0.012 s time interval with other time intervals, the task appears at first sight to be almost impossible.

The time available is much shorter than human reaction times. The very first movements of a top sprinter out of the blocks take about 0.2 s after the starting gun. If someone holds a ruler between your open fingers and then drops it through your fingers, the ruler will fall about 6 in. or more before you can catch it. After 0.1 s, the ruler falls 2 in. After 0.2 s it falls 7.7 in. After 0.3 s it falls 17.4 in. You can test your own reaction time this way. Human reaction times are typically about 0.2 s. That is, it takes about 0.2 s for the brain to process visual or auditory or tactile information received by the eyes or ears or hands and to then send signals to the relevant muscle groups to respond.

Getting the bat to the ball is not just a reaction time problem since the batter has about 0.4 s to react after the pitcher releases the ball. In fact, batters normally start swinging at the ball even before the pitcher releases the ball. Not the full-blooded swing that comes at the end but a relatively slow swing to get things started. If the batter sees the ball coming at him slowly then he can delay his actions appropriately by maintaining a slow swing until the last 0.15 s. If the ball is approaching rapidly then he needs to speed up the swing so the bat will be in the right spot just before the full-blooded part of the swing. In this manner, the batter has continuous feedback from the speed and position of the ball, extending over a period of about 0.3 s, that allows him to adjust the speed of the bat so they both meet within the hitting zone. Even so, when the batter unleashes his full swing power, about 0.15 s before impact, he still has to make sure that the bat and ball meet precisely within that 0.012 s time interval. How can he do that?

Suppose that the batter swings his bat at an average speed of 50 ft s^{-1} (34 mph) for the last 0.2 s of his swing. The speed figure here must be about right because the bat travels about 10 ft in that time. Suppose also that the bat somehow arrives right in the middle of the hitting zone just when the ball arrives. If the batter starts his swing 0.01 s too late or too early then the bat will arrive 0.01 s too late or too early.

Will he miss the ball? At 50 ft s^{-1}, the bat travels 6 in. in 0.01 s. Hitting the ball 6 in. in front or behind the front of the home plate is fine. The batter, therefore, needs to make a prediction of when the ball will arrive and swing the bat so that the bat crosses the plate within about 0.01 s of that time.

If the batter simply held the bat at the correct height in the middle of the hitting zone, and just bunted the ball, then that is where the ball would meet the bat. The fact that the ball spends only 0.012 s over the plate would then be irrelevant. The timing problem arises only when the bat is swung at the ball. What if the batter starts his last powerful swing phase say 0.05 s too early or too late? At an average bat speed of 50 ft s^{-1}, the bat will overshoot or undershoot the middle of the home plate by 2.5 ft. That will result in a foul hit. However, the batter might be able to sense that he is in trouble and swing a fraction faster or slower during the last 0.2 s to help make up for the late or early start to the swing.

As discussed in Sect. 5.1, batters are probably not aware of the timing problem at all. When a batter sees a ball approaching, he or she cannot tell whether it will take say another 0.125 s or another 0.130 s to arrive. People don't have in-built clocks to measure time that accurately. However, a batter will know from experience, based on the position and speed of the ball, when to start swinging at it. Based on past experience, and a mental estimate of the ball speed, a batter will start to swing vigorously when he or she sees the ball a certain distance away, roughly half way between the pitcher and the home plate. The end result, when judged correctly, is that the bat will meet the ball at the correct time, within say 0.01 s. Provided the batter uses his eyes to interpret distances accurately and his brain to estimate the velocity accurately, the timing will be almost perfect.

A batter presumably processes this information instinctively, based on previous experience. For example, information may be stored in the brain that tells a batter that a ball is approaching at a speed of say 8 on a 1–10 scale, and that it is currently at a distance of say 6 on a 1–10 scale. Other information in the brain tells the batter to start swinging at a speed of 7 on a 1–10 scale when the ball gets to a distance of say 4 on the 1–10 scale. Good batters might use a finer scale than average batters, with the result that their timing is more precise, even though they don't consciously or even subconsciously process timing information at all.

That line of thought suggests an interesting experiment. Suppose that a ball machine is available to pitch balls at various speeds and a batter is asked to judge the speed on say a 1–10 scale. It is doubtful that he or she could pick the difference between a 70 mph ball and a 71 mph ball, but it should be easy to pick the difference between a 50 mph ball and a 70 mph ball. The question is, can good batters pick small differences better than average batters? As far as the author is aware, nobody has done that experiment.

7.3 The Height Problem

Batting in baseball (and softball) is much more difficult than striking a ball in tennis because the barrel of the bat is only 2.75 in. in diameter at most. The diameter of the ball is 2.90 ± 0.04 in., slightly larger than a tennis ball. The batter will miss the

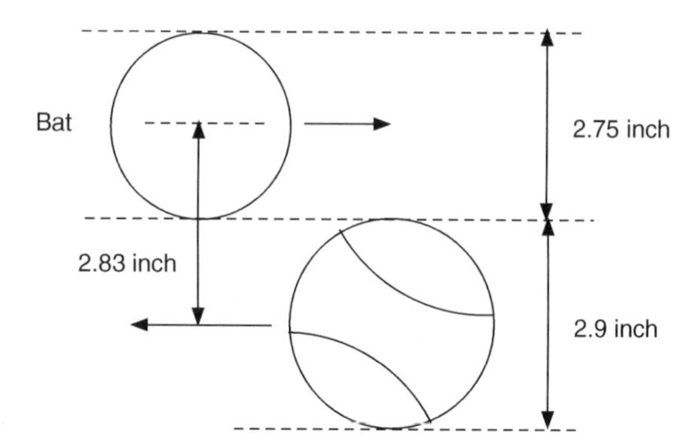

Fig. 7.1 The batter will miss the ball if the bat is swung 2.83 in. too high (or too low)

ball completely if the bat is swung 2.83 in. too high or too low (Fig. 7.1). A height error of 2 in. will result in the ball just nicking the bat and passing behind the batter.

The ball does not remain at the same height during its flight. It falls from a launch height of about 6 ft and passes over home base at a height between about 2 ft and 4 ft. It therefore drops at a rate of about 10 ft s^{-1} since the flight time from the pitcher is about 0.4 s. During the last 0.1 s, the ball drops in height by about 1 ft depending on the type of pitch.

Suppose that the ball was always pitched at about the same speed and always dropped by the same amount in a predictable way. A good batter would then be able to hit the ball about 90% of the time. But what if the drop height changed by 3 in. from one pitch to the next on a random basis? Using the same swing as that for the predictable ball, the batter would miss the ball almost every time. However, the batter gathers information about the pitch for the whole 0.4 s of its flight and will see that it is about to cross the home plate at a different height. During the first 0.25 s of the flight he can drop or raise the barrel by about the right amount to correct for the changed trajectory, and then he swings as fast as he can for the last 0.15 s. At that point he is committed to the predicted flight path. The problem here is that once the bat reaches about one quarter of maximum speed, it has too much forward momentum for the batter to be able alter its height by more than 1 in. even if he wanted to.

Suppose that the batter doesn't actually swing the bat forward but instead leaves it hanging over the middle of the plate. As the ball approaches, he can raise or lower the bat quickly so that it at least meets the ball. The ball will simply bounce off the bat at low speed. The force needed to raise a 1 kg bat by 5 cm (2 in.) in 0.15 s is only 3.3 N (0.74 lb). We can compare that force with the force needed for a full-blooded swing. If a 1 kg bat accelerates from 0 to 40 m s^{-1} over 0.15 s then $F = m\mathrm{d}v/\mathrm{d}t = 1 \times 40/0.15 = 267\,\mathrm{N}$, or about 60 lb.

If the bat has already accelerated to 10 m s^{-1} when the batter decides to raise the bat, then the vertical force required can be estimated in terms of the situation shown in Fig. 7.2. The bat needs to rise by say 5 cm in 0.15 s, following a curved arc.

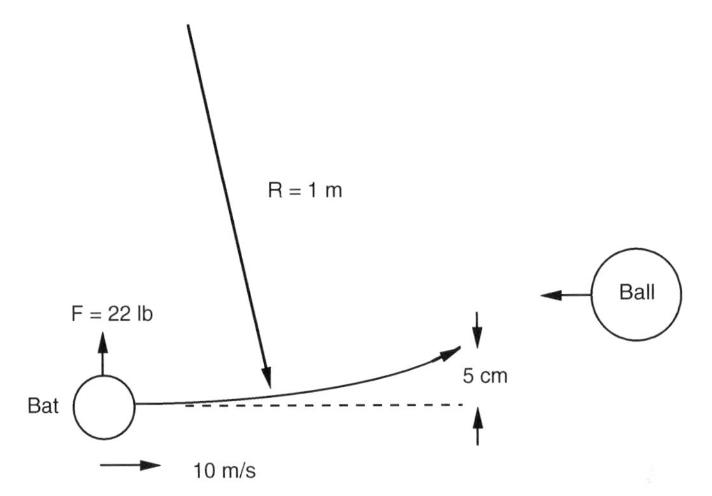

Fig. 7.2 If the batter wants to raise the bat 5 cm while swinging it horizontally at $10\,\mathrm{m\,s^{-1}}$, he needs to exert a vertical force of 22 lb on the bat, rising to 90 lb or more just before contact with the ball

Taking an arc radius $R = 1$ m (39 in.), the required force is $mv^2/R = 100$ N or about 22 lb. If the bat accelerates to say $20\,\mathrm{m\,s^{-1}}$ during that time, then the required force would increase to about 90 lb. In other words, the whole of the batting effort and then some would be needed to raise the bat height by only 5 cm.

If the batter is already committed to a full-blooded forward swing then he can't suddenly change the direction of the force to a vertical force even if he wanted to. It would be like getting a sprinter to jump almost vertically over a high bar half way through a 100 m dash. Even in a hurdling event, the hurdler needs to slow down over a few paces to change the direction of the force of his or her feet on the ground.

The batting problem in baseball and softball is, therefore, quite clear. The batter has only 0.25 s after the ball is pitched to decide where the ball is going to cross the plate. By that time, the batter has already started his swing and is ready to unleash the bat onto the ball. He has estimated the arrival height of the ball to within about one bat width and aims for that spot. He might just as well close his eyes at that time because it is too late to change his mind. The bat is already on its way and accelerating rapidly. There is a very good chance that he will miss the ball completely.

7.4 Predicting the Flight of a Ball

To strike an incoming ball, the batter must be able to predict when and where the ball will pass through the hitting zone so that he can swing the bat toward that spot at the right time. It is interesting to consider how the batter manages to predict the flight path of the ball. The short answer is that it is essentially intuitive, derived from years of practice, but it is more complicated than that. If a batter tries to hit a ball while blindfolded then intuition will be of no help at all. Obviously, the batter somehow uses visual information to track the path of the ball onto the bat. Visual information is part of the story but there is more to it.

A batter also uses spatial perception. A simple example will illustrate the point. You can reach behind your back and touch the tips of your two index fingers without being able to see where they are. Your eyes are not part of this process. Your brain just knows where your finger tips are, just as it knows where each hand is located. If you close your eyes and raise one hand, you just know where your hand is even though you can't see it. You make a conscious decision to raise your hand, then lo and behold it moves upwards. You can't repeat that trick and raise a glass of water to your lips without touching the glass, but the ability to move your own hand just by thinking about it is truly amazing. It can be explained in terms of electrical signals sent to the right muscles which contract accordingly, but that doesn't explain how you manage to generate those particular electrical signals just by thinking about it. If you want to, you can raise one hand, wiggle one finger, lift one leg and sing a song all at the same time. Or you can think about doing it and not actually do any of those things. One part of the brain decides what needs to be done and another part decides if it wants to put those plans into action.

Visual information and spatial perception are both employed when catching a ball. Get someone to throw a ball over your head and then try to catch it without moving your head to look up at the ball. It is not difficult. You can just reach up and pluck it out of the air, provided of course you watch the ball as it approaches so you can predict where and when it will be when it passes over your head. You can even catch a very fast ball this way. In fact, if the ball is moving really fast, you won't have time to raise your head to follow the flight of the ball into your hand. It is essential to keep your eye on the ball for part of the flight but it is not necessary to do so for the whole flight.

Careful observation of the ball provides several clues concerning the flight path. One is that the ball appears to get larger in diameter as it approaches, compared with other stationary objects. If you were to capture the ball on video film, the ball would be a tiny speck when it is 200 ft away, and it would fill the whole screen when it is only 1 in. away. The relative size of the ball and the rate at which the size increases both provide clues as to where the ball is and how fast it is approaching.

Another clue is the change in position of the ball relative to the position of the pitcher. The ball can move to the left or right of the pitcher or it can approach the batter without any sideways motion. At the start of the flight, the ball is located at about the same height as the pitcher's head and moves in the vertical direction relatively slowly. Near the end of the flight the ball drops quickly to a position well below the pitcher's feet. Consequently, the batter sees the ball approach along a certain line and can project that path visually into the hitting zone, as indicated in Fig. 7.3, even if the path is curved. It is easy to predict that the ball will not intercept points A, B, or D, but will it pass through point C or slightly above or below? The closer the path gets to point C the easier it is to predict the subsequent path, meaning that it helps to keep your eye on the ball as long as possible.

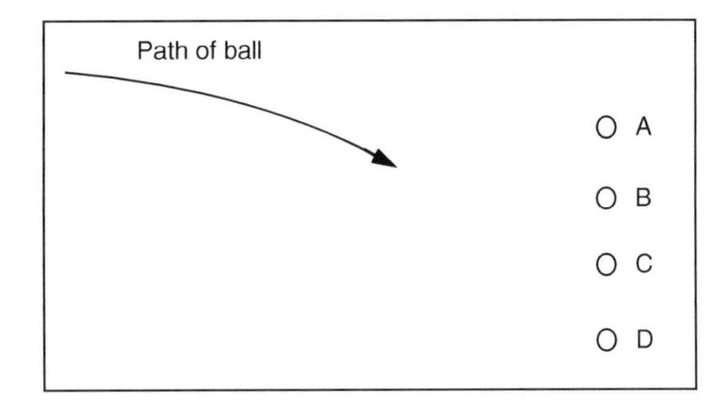

Fig. 7.3 A batter predicts the path of the ball and then swings the bat toward the estimated arrival point. Will this ball pass through C or above C or below C?

7.5 Stereo Vision

A batter uses both eyes to watch the ball [1]. Each eye sees a different ball path relative to the background. If you hold one finger at arm's length and observe your finger with one eye at a time against a distant object then the relative position of your finger will change by about six finger widths. If you move your finger closer to your eyes, then the relative position, with respect to the distant object, will change by a greater amount. This information on its own tells you where your finger is relative to the distant object, and relative to your head, and it will tell you how fast your finger is approaching.

If both eyes are located along a horizontal line and if the batter focusses on the pitcher, then one eye will see an image of the ball to the left of the other image, but both images will be at the same height. Similarly, if the batter focuses on the ball then there will be two images of the pitcher, one slightly to the left of the other. These images provide information on motion of the ball left or right, but the up–down motion is not as well determined. To gather information in both directions, it helps if the batter tilts his head as he swings at the ball.

If the batter moves his head as the ball approaches, then the path of the ball relative to other background objects will change due to motion of the head. For that and other reasons, it helps to keep your head still when you strike a ball.

Figure 7.4 illustrates the type of stereo information received by both eyes when viewing an approaching ball. The images received by the left and right eyes are received as separate images but the brain interprets the combined images as a single image. For example, if you were to focus just on the approaching white ball in Fig. 7.4, then you would see only one white ball but there would be two striped balls in the background. If you were to focus on the striped ball then you would see only one striped ball but you would see two white balls in the foreground.

At time $t = 0$ the white ball is close to the striped ball, slightly to the right and just in front. The striped ball is a few feet in front of the black object, in line with the right eye, and slightly below. You can check this out by holding one finger in front of another and observing separately with the left and right eyes.

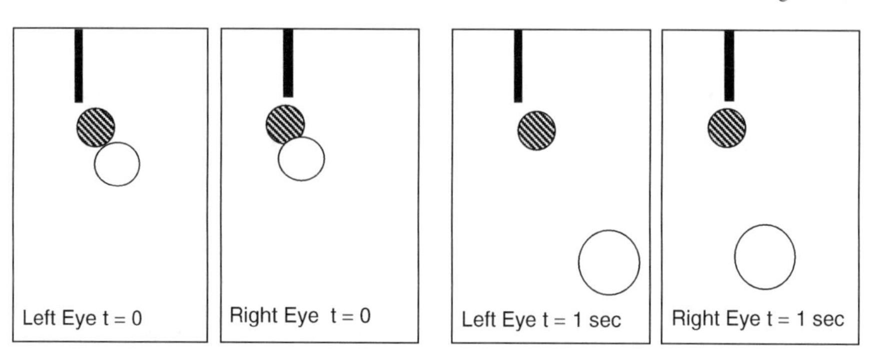

Fig. 7.4 The view of an approaching ball seen through each eye

After 1 s, the white ball has moved closer to the eyes since it is bigger, and it has moved to the right. It hasn't moved a lot closer since it is not a lot bigger. The black and striped objects stayed where they were so they are not moving at all.

The most difficult aspect to interpret is the vertical motion of the ball. At first sight, it appears that the white ball has moved forward and dropped below the level of the striped ball. However, the eyes are a few feet above both balls, looking down on them at an angle. In fact, the drawings in Fig. 7.4 were copied from photographs of a baseball rolling along a horizontal table, with a tennis ball at rest at the far end of the table. Without the image of the table, it is quite difficult to interpret Fig. 7.4 as meaning that the ball approaches along a horizontal surface. Figure 7.4 by itself would be consistent with a ball rolling down an inclined table. A batter receives additional information concerning the height of the ball from a number of different sources, including the height of his eyes above the ground, the location of the pitcher's feet on the ground and the perspective of the whole playing field. Nevertheless, it is difficult to judge ball height as accurately as its horizontal position. If you stand back and look at a table, can you tell whether it is horizontal or sloping forward a fraction? Probably not very well.

It seems that batters have three methods available to judge the height of the ball above the ground. When the pitcher releases the ball, the height of the ball can be compared with the height of the pitcher's body. As the ball gets closer to the batter, the batter loses sight of the pitcher and sees the ball against the background of the playing field. The problem in Fig. 7.4 is that there is no background under the ball, just a patch of white. On a playing field there will be two images of the ball and the playing field, one in each eye. Those images show that the ball is definitely not on the ground, otherwise both images would be the same. The height of the ball off the ground can be judged by the separation of the two background images when the batter focusses on the ball. The size of the ball against the background also provides height information. Obviously, the batter doesn't have time to process all this information by conscious thought, like a physics calculation. He just knows where the ball is, based on previous experience and the sight of the ball. The only problem is that he might see the ball pass over or under the bat because his prediction of the where the ball would be was in error [2–5].

7.6 Psychology of Hitting a Ball

Despite the fact that the physics of hitting a ball is relatively straightforward, there is no player alive who will claim that it easy. Ted Williams said that hitting a ball was the single most difficult thing to do in sport. Yogi Berra said, "Hitting a ball is 90% mental and the other half is physical." Hitting a ball is not just a physics problem. In the real world, it involves biomechanics, it involves lots of practice and it involves the psychology of perception. Perception is a fascinating subject since it involves the way that people perceive and interpret the world through their various senses. Sometimes, that perception is quite different from reality. There are many visual illusions that demonstrate the problem.

A famous example is the "size-weight illusion." When two objects of the same weight but different size are compared by lifting them, the larger object usually feels lighter. Sometimes, it can feel two or three times lighter. The problem is that the brain receives several different signals from different sources and sometimes has trouble knowing which one to trust. Sensors in the fingers and hands respond to pressure and to movement and tell the brain how heavy something is. The eyes tell the brain that one object is bigger than the other. The "experience" or memory sections of the brain knows that big objects are usually heavy. The "conclusions" section of the brain adds all this information together and says "That big object is lighter than I expected" and "That small object is heavier than I expected" so it concludes, incorrectly, that the big object is lighter than the small object. The bigger the size difference, the bigger the apparent weight difference. The illusion persists even when you know that both objects are the same weight. If you don't believe it, then try it with a small steel rod and a large wood rod, both about 300 g.

Similar things must happen when a batter sees a ball approaching. He or she sees the pitcher throw the ball and knows from experience that the pitcher has just thrown the ball at a certain speed. The pitcher could well have disguised the throw and pretended to throw a fast ball but actually threw a slower ball. Or vice versa. The batter knows that slow balls drop further than fast balls and starts to swing accordingly. If the batter is fooled in this way, he or she might then conclude that the ball rose upwards at it approached him when it fact it didn't rise at all. The book "The psychology of baseball" by Mike Stadler [6] has many interesting insights like this.

References

1. T. Bahill, D. Baldwin, J. Venkateswaran, Predicting a baseball's path. Am. Scientist (online at www.americanscientist.org) **93**(3), 218–225 (2005)
2. A.T. Bahill, T. LaRitz, Why can't batters keep their eyes on the ball. Am. Scientist **72**, 249–253 (1984)
3. A.T. Bahill, W.J. Karnavas, The perceptual illusion of baseball's rising fastball and breaking curve ball. J. Exp. Psychol. Hum. Percept. Perform. **19**, 3–14 (1993)
4. A.T. Bahill, D.G. Baldwin, The vertical illusions of batters. Baseball Res. J. **32**, 26–30 (2003)

5. A.T. Bahill, D.G. Baldwin, in *The Rising Fastball and Other Perceptual Illusions of Batters*, ed. by G. Hung, J. Pallis. Biomedical Engineering Principles in Sports (Kluwer, Dordrecht, 2004), pp. 257–287
6. M. Stadler, *The Psychology of Baseball: Inside the Mental Game of the Major League Player* (Gotham, NY, 2008)

Chapter 8
Elastic Properties of Balls

The bodies return one from the other with a relative velocity which is in a given ratio to that relative velocity with which they met. This I tried in balls of wool, made up tightly, and strongly compressed; the balls always receding one from the other with a relative velocity as about 5 to 9. But in balls of glass the proportion was as about 15 to 16.

– Isaac Newton, Axioms or Laws of Motion, in Principia Mathematica, 1687.

8.1 How Does a Ball Bounce?

It is obvious to anyone observing a bouncing basketball that the ball squashes when it bounces, and that it expands back to its original shape when it bounces up off the floor. The same thing happens when any ball bounces, although it usually happens too fast to be seen by eye. A basketball spends about 20 ms (0.02 s) in contact with the floor which is just enough time to catch a glimpse of the ball squashing. For most other balls, the time spent in contact with the floor is so short that the action can only be captured with the aid of a fast camera. For a baseball or a softball, the contact time is only about 1 ms. The very short contact time can be explained with the aid of the simple model shown in Fig. 8.1a, where the ball is regarded as a mass m attached to a spring, of spring constant k. The mass m represents the total mass of the ball, and the spring is added to account for the stiffness of the ball, as defined in Sect. 2.6.

In reality, the mass and stiffness of a ball are distributed throughout the ball. The two are separated in the model to simplify the problem. A better model of the ball would be one where 1,000 masses are connected by 3,000 springs, but that would be too complicated to think about, even though the solution could be obtained using a finite element computer model. One spring and one mass is sufficient to explain the essence of the bounce process, especially considering the fact that the stiffness of a ball is determined mainly by the small region of the ball in contact with the surface. At low impact speeds, a baseball or a softball might compress over a contact region that is only 1 in. wide and about 0.1 in. thick, as shown in Fig. 8.1b, while the rest of the ball remains uncompressed. The stiffness of that small contact region is usually a lot less than the stiffness of the ball as a whole [1]. Even at high impact speeds, high speed film shows that the compression of a baseball or a softball is confined almost entirely to a region about half an inch thick, as shown in Fig. 8.1c, while the rest of the ball remains largely unaffected. Consequently, the bounce of a ball can indeed be modeled quite well by regarding the ball as a mass on a spring, the spring representing the stiffness of the small compressed region at the bottom of the ball.

R. Cross, *Physics of Baseball & Softball*, DOI 10.1007/978-1-4419-8113-4_8,

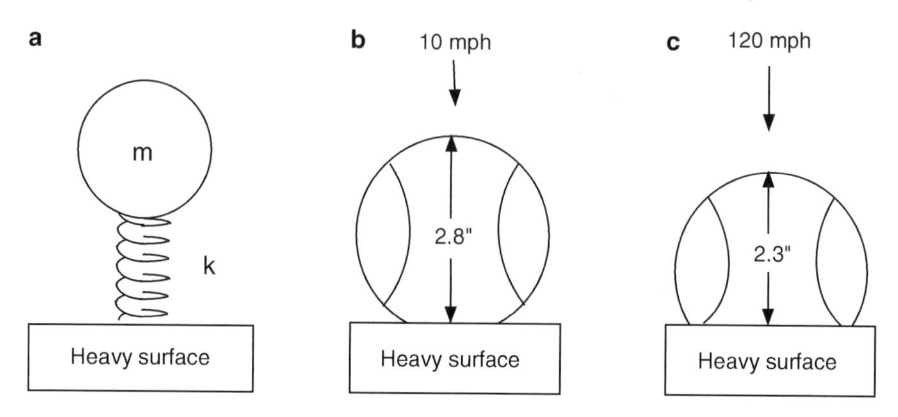

Fig. 8.1 (**a**) A simple model of a bouncing ball. The spring represents the stiffness of the small contact region at the *bottom* of the ball. A 2.9 in. diameter baseball incident at 10 mph on a heavy surface compresses by about 0.1 in., as shown in (**b**), while a 120 mph ball compresses by about 0.6 in., as shown in (**c**)

If our model ball is dropped onto a rigid surface then the spring in the model will begin to compress as soon as the ball contacts the surface, since the bottom end of the spring comes to a stop instantly while the mass at the top end continues to fall at its initial speed. As a result, the spring exerts an upward force on the mass, causing it to slow down until it comes to a complete stop. At that point the spring stops compressing and begins to expand, accelerating the ball away from the surface until the spring expands back to its original length. When the spring returns to its original length the force exerted by the spring, both on the ball and the surface, drops to zero. The ball bounces off the surface at a speed that depends on the amount of energy lost in the spring during the time it compressed and expanded back to its original length. If there is no loss of energy then the ball will bounce off the surface at the same speed that it impacted the surface. However, there is always a loss of energy when a ball bounces, and the rebound speed is always less than the original impact speed.

The time, T, taken for the spring to compress and expand depends on the mass m and the stiffness k, and is given by the formula

$$T = \pi \sqrt{\frac{m}{k}} \tag{8.1}$$

indicating that the time will be relatively long for a soft and heavy ball and relatively short for a stiff and light ball. For a baseball, $m = 0.145\,\mathrm{kg}$ and k is about $1 \times 10^6\,\mathrm{N\,m^{-1}}$, giving $T = 0.0012\,\mathrm{s}$ (1.2 ms). The stiffness of the ball depends to some extent on its initial speed, most balls becoming stiffer the more they are compressed. If a baseball impacts on a surface at low speed then it will compress by only 1 or 2 mm, and the impact time is then about 2 ms. However, if the ball impacts at high speed then it can compress by 10 mm or more, in which case the stiffness increases by a factor of four or more and the impact time is typically about 1 ms or less.

The model shown in Fig. 8.1 accounts for some of the essential features of a bouncing ball, but it does not help to explain why some balls bounce better than others or how energy is lost when a ball bounces. We will return to this problem in Chap. 9 when we consider slightly more complicated models of bouncing balls.

8.2 Contact Time and Impact Force

The very short contact time of a baseball or a softball during the bounce off a bat or off a rigid floor can be explained in terms of the large stiffness of these balls. If you try to squash a ball by hand then you might be able to squash it by 1 mm, but there is no way that you could squash it by 1 in. or even half an inch since the ball is too stiff. The force needed to a squash a ball by 1 mm is about 200 N or about 44 lb. A force of about 5,000 lb is needed to compress a ball rapidly by 1 in., although a lower force can be used to squash the ball by 1 in. if the ball is squashed slowly, as explained in Chap. 9. The only way to generate a force of 5,000 lb by hand, in baseball or softball, is to strike the ball with a bat.

An interesting question is how the force exerted by the hands on the bat gets magnified so much when the bat strikes the ball. The short answer is that it is the same effect that occurs when you hammer a nail into a block of wood. You can't push the nail into the block by hand, but you can drive the nail into the wood using a hammer. The hammer acts to magnify whatever force you exert on the handle (Fig. 8.2). In both cases, bat or hammer, you push with a small force for a relatively long time to get the bat or the hammer up to speed, then there is a very rapid change in speed of the bat or the hammer when it collides with the ball or the nail. Any change in speed of anything requires a force. The same change in speed can be generated using a small force acting for a long time or a large force acting for a

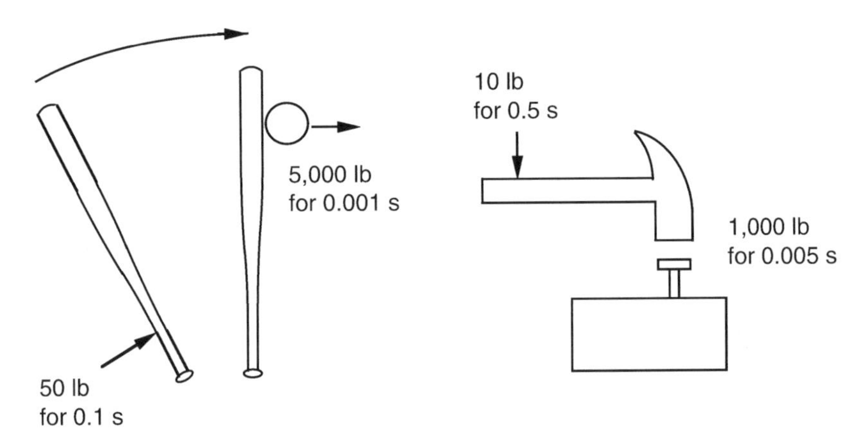

Fig. 8.2 A small force acting on the handle of a bat or a hammer for a relatively long time can generate much larger forces on impact. The impact lasts for only a very short time

short time. If it takes say 0.5 s to swing a hammer with a force of say 10 lb, and if the hammer then comes to a stop in 0.005 s, the force on the nail is magnified by a factor of 100, to 1000 lb.

The relation between impact time and impact force can be clarified by another simple calculation. Suppose that a ball is dropped onto a surface at a speed of $10\,\mathrm{ft\,s^{-1}}$. Before the ball bounces up off the surface it has to come to a complete stop and then reverse direction. Suppose it takes 1 s to come to a stop. Since the average speed during that time is $5\,\mathrm{ft\,s^{-1}}$, the ball will travel 5 ft before it comes to a stop. Obviously, we are talking here about an incredibly soft surface like a 20 ft high pile of pillows or a 50 ft square trampoline.

On a hard surface like concrete, the ball will squash by about 0.1 in. before it comes to a stop. The average speed during this time is still $5\,\mathrm{ft\,s^{-1}} = 60\,\mathrm{in\,s^{-1}}$ (starting at $10\,\mathrm{ft\,s^{-1}}$ and ending at zero) and the time to travel 0.1 in. is then $0.1/60 = 1/600 = 0.0017$ s. This is indeed what happens, since the ball is so stiff. It takes a force of about 200 lb to squash a ball by 0.1 in., so this is the force generated when the ball bounces off the concrete. Conversely, a force of 200 lb on the ball will bring it to a complete stop in 0.0017 s and then reverse its direction as the ball expands back to its original shape. A simple calculation of the force on a bouncing ball is given in Appendix 8.2.

8.3 Impact Force on a Player

A safety issue in both baseball and softball is that a player can be seriously injured or even killed if he or she is struck by a high speed ball [10–12]. Suppose that a ball of mass m strikes a person at speed v and that the impact lasts for a time T. Then the force on the person is given approximately by

$$F = \frac{mv}{T} \tag{8.2}$$

The force varies with time during the impact, and will depend on the rebound speed of the ball and the stiffness of the part of the body struck by the ball, but the formula here gives a good estimate of the average force during the impact. The important point to notice is that force is proportional to the speed of the ball, it increases with the mass of the ball, and it decreases if the collision time increases. The latter feature explains in part why fielders wear gloves. Soft gloves act to extend the collision time, reducing the impact force on the hand.

For example, if $m = 0.145$ kg (5.1 oz), $v = 44.7\,\mathrm{m\,s^{-1}}$ (100 mph) and $T = 1$ ms then $F = 6{,}481$ N (1457 lb). That is a seriously large force. It acts for only a very small time, but it is large enough to crack or break human bones. The force will be reduced if the ball strikes a soft part of the body rather than bone, since the ball then takes longer to slow down and the impact time is increased.

The impact time depends on the mass of the ball, on the stiffness of the ball and on the stiffness of the body. If the ball strikes hard bone, then the impact time will depend mainly on the stiffness, k, of the ball rather than the stiffness of the body.

The impact time is then given to a good approximation by (8.1). Since we know that T is about 1 ms in a high speed collision between a hard ball and a hard surface, the ball stiffness k is about $1.4 \times 10^6\,\mathrm{N\,m^{-1}}$ (8,140 lb in.$^{-1}$). If we combine (8.1) and (8.2) then

$$F = \frac{v}{\pi}\sqrt{mk} \tag{8.3}$$

The force can therefore be reduced, without reducing the mass or speed of the ball, by reducing the stiffness of the ball. It is for that reason that baseball is played in many countries around the world with a hollow rubber ball, similar to a tennis ball but without a cloth cover, heavier, stiffer and slightly larger in diameter. The ball is used mainly in youth leagues, especially in Japan, and is commonly known as a Kenko ball. The manufacturers in Japan claim to have sold more than 6 billion Kenko balls during the last 80 years. In the USA, Kenko balls and other similar balls are regarded primarily as training balls, but many other countries have adopted them as an official youth ball for safety reasons.

A comparison of the force exerted by various balls, when dropped onto a hard surface, is shown in Fig. 8.3. Each ball was dropped (by the author) from a height of 10 cm, at the same low impact speed, onto a 50-mm diameter piezoelectric disk to measure the impact force. The force would be proportionally larger at high impact speeds, and the impact time would be shorter, but the point of the exercise was not to measure the force at high ball speeds but to measure the relative force and impact duration of each ball at the same impact speed. In all cases, the force exerted by each ball was many times larger than the weight of the ball. The smallest force was

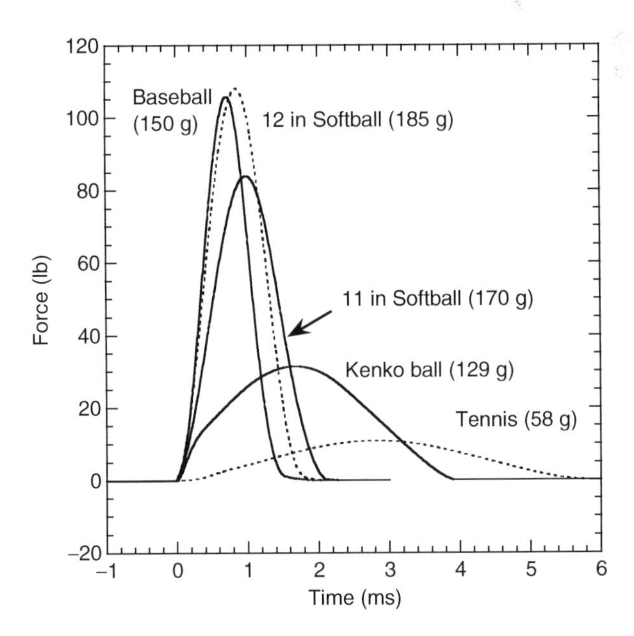

Fig. 8.3 Measured force on various balls obtained by dropping each ball from a height of 10 cm onto a hard surface

Table 8.1 Calculated stiffness of the balls shown in Fig. 8.3

Ball	m (g)	T (ms)	k (N m^{-1})	k (lb in.$^{-1}$)	F_o (lb)
Baseball	150	1.4	7.5×10^5	4,310	106
12 in. softball	185	1.7	6.3×10^5	3,610	108
11 in. softball	170	2.0	4.2×10^5	2,390	84
Kenko ball	129	3.8	8.8×10^4	504	34
Tennis ball	58	5.4	2.0×10^4	112	11

exerted by the tennis ball since it was the lightest and softest ball. The force exerted by the 12-in. softball was almost the same as the baseball. The force exerted by the Kenko ball was about three times smaller than the adult baseball, partly because it was lighter but mainly because it was about nine times softer (but not as soft as a tennis ball).

The stiffness of each ball is shown in Table 8.1, calculated from (8.1) and the measured impact time. The peak force, F_o on each ball can be calculated from the relation $(1/2)mv_o^2 = (1/2)kx_o^2$, where v_o is the impact speed (1.4 m s^{-1} for a 10 cm drop) and x_o is the maximum deformation of the ball. Hence, $F_o = kx_o = v_o\sqrt{mk}$. This calculation is also shown in Table 8.1.

A safety concern regarding aluminum bats is that the batted ball speed can be up to 5% higher than it is off a wood bat, at least for bats regulated by the NCAA. The speed off an aluminum bat could well be more than 5% higher off unregulated bats. A ball striking a person at 105 mph rather than 100 mph would exert a force that is 5% larger, assuming that the same ball strikes the person in the same spot. One way around this problem, assuming that it is indeed a real rather than a perceived problem, would be to reduce the performance of the bat by 5% in some manner. Another would be to reduce the performance of the ball, by reducing its COR. A third way would be to reduce the ball stiffness to increase the impact duration, T. The force can be decreased by 5% by increasing T by 5% which can be achieved by reducing k by 10%. In fact, baseballs and softballs can vary in stiffness by up to 50% anyway, so a decrease in ball stiffness of 10% might indeed help to reduce injuries and would not alter the traditional nature of the game in any significant way. Furthermore, a decrease in ball stiffness would have two other beneficial effects. First, it would reduce the force on the bat, leading to fewer breakages. And it would reduce the trampoline effect with hollow bats, since a greater fraction of the elastic energy would be stored (and then lost) in the ball rather than in the bat.

Given that safety is such a contentious issue, it is worth adding here an additional comment. The author has been involved in many investigations into fatal falling accidents for the police and the Coroner in NSW. The question sometimes arises as to force required to break or fracture the skull. For example, the police might allege that a victim was bashed over the head with a bat and the defence might argue that the damage to the victim's head was caused when the victim fell onto a hard surface. There is no simple answer to the force question since the required force depends on the point of impact and the individual concerned, as well as other factors. It is difficult to give a reliable estimate of the required force even within a factor of two. It is therefore unlikely that anyone will ever prove, from the injury statistics available, that a 105 mph ball is more dangerous than a 100 mph ball. However, one can say with complete confidence that it is more likely that serious injury will occur if the impact force increases. The force can increase if the batted ball speed increases, or if the ball stiffness increases, and one cannot simply attribute an increase in force to a particular type of bat. An increase in impact force can equally well be attributed to a particular type of ball or even to a particularly strong batter.

8.4 How Well Does a Ball Bounce?

Baseballs and softballs differ from most other sports balls in that they are not very bouncy. If a baseball is dropped onto a hard floor from a height of 3 ft, it will bounce to a height of only about 1 ft. A basketball or a tennis ball or a golf ball dropped from 3 ft will bounce to a height of nearly 2 ft. Baseballs are not made from rubber or from some other bouncy material, but from many layers of wool and cotton yarn wound up into a tight ball and held together with a leather cover. Softballs are made from polyurethane, or kapok, or a mixture of cork and rubber and also have a leather cover. These balls bounce better than a pair of socks rolled up into a tight ball, but not a whole lot better. There is nothing wrong with the bounce of a baseball or softball. They were simply designed that way so that batters would not be able to hit them too far or too fast. Players and officials like it that way.

The difference between a ball that bounces well and one that has a low bounce is determined by the elastic properties of the ball. Socks are stretchy but, in physics terms, socks are not very elastic. In common usage, something that is easy to stretch or compress or bend is said to be soft, elastic, flexible, springy, supple, or pliable. In physics, the word "elastic" has a specific meaning. An elastic material is one that regains its shape rapidly after it is stretched or compressed, and therefore includes steel as well as rubber. A steel ball bearing dropped from a height of 3 ft onto a hard steel plate or onto a hard, polished block of granite will bounce to a height of almost 3 ft, despite the fact that steel is very hard and does not stretch easily.

If a sock is squashed or stretched it won't spring rapidly back to its original shape. Socks spring back too slowly to bounce well. Baseballs have the same property, in that when a baseball is dropped onto a hard floor it squashes but it does not immediately return to its original shape. It "oozes" back to its original shape, meaning that it expands slowly. It expands faster than the eye can see, but still not fast enough to push itself up off the floor as fast as it hit the floor. It takes about 0.001 s for a baseball to squash when it is dropped onto a hard floor, but it takes about 0.002 s to expand back to its original shape. By the time it bounces off the floor (or off a bat), it has expanded only about half way back to its original shape, as we will show in Chap. 9.

Tennis balls and basketballs take even longer to expand back to their original shape when they are dropped on a hard floor, but that doesn't mean that they don't bounce well. The time taken to expand back needs to be compared with the time taken to squash, and that depends on the mass and stiffness of the ball. A tennis ball takes about 0.002 s to squash and about 0.0025 s to expand. A basketball takes about

0.007 s to squash and about 0.008 s to expand. When the times are about equal the ball bounces moderately well. A plasticene ball takes about 0.002 s to squash but it never expands back to its original shape, and it doesn't bounce at all.

A baseball acts like a stiff spring when it squashes, but it loses some of that stiffness as it expands and therefore expands more slowly. The behavior is analogous to a person doing pushups where the person lowers his body quickly to the floor and then pushes up slowly. If there is only a weak upwards force then there is no significant speed generated in the pushup, and there is no "bouncing" up off the floor. The end result is that when a baseball bounces up off a hard floor, some of the original kinetic energy is lost and converted to heat energy in the ball, so the ball does not bounce as high as the original drop height.

When a baseball hits a hard floor, all of the kinetic energy of the ball is used in squashing the ball. By comparison, the floor squashes and bends by a much smaller amount. The ball comes to a complete stop before it reverses direction and bounces back up off the floor. At the instant that it comes to a stop it has zero kinetic energy. Some of the original kinetic energy is stored temporarily as elastic energy in the ball, in the same way that elastic energy is stored in a compressed spring, but some of the original energy is lost as heat energy due to friction between the layers of yarn in the ball. As the ball expands back to its original shape it pushes itself up off the floor. The ball bounces up off the floor at only about half the speed that it hit the floor, and it bounces with only about 1/4 of the original kinetic energy. Three quarters of the original energy of the ball is used up in heating the ball.

Technically, when the ball expands it pushes down on the floor and the floor reacts by pushing up on the ball, so it is actually the floor that pushes the ball upwards, not the ball itself. The idea that the floor pushes up on the ball might seem strange, especially if we assume that the floor is just sitting there minding its own business and not actually doing anything. In fact, the floor compresses and bends a tiny amount, so it acts as a very heavy and very stiff spring. The floor compresses and expands like the ball, so it actually does do something. If it happens to be a floor in a physics department then it will know that it needs to push up on the ball because it will have heard about Newton's third law of action and reaction many times (an old physics joke).

8.5 Coefficient of Restitution

If a smooth spherical ball is dropped vertically onto a smooth horizontal surface, and if the ball is not spinning when it is dropped, then it will bounce vertically without spin. The bounce of a spinning ball is a more complicated process and will be considered in Chap. 15. When a spinning ball is dropped vertically onto a floor, it bounces up off the floor but it also bounces sideways. When a ball is thrown without spin at an angle onto the floor, it bounces with spin. In this chapter, we ignore the spin of the ball by considering only the case of a ball bouncing at right angles to a surface.

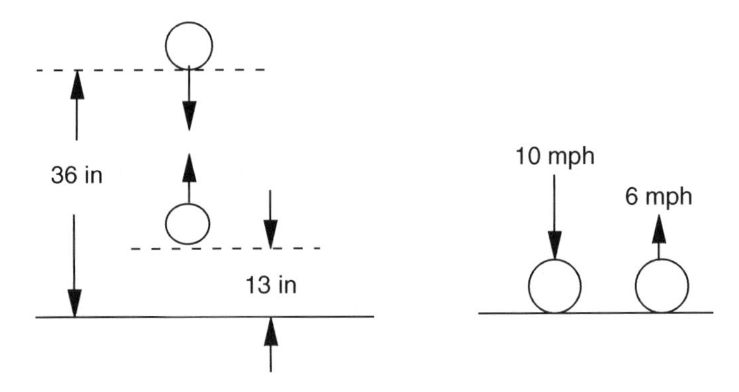

Fig. 8.4 A baseball dropped from a height of 36 in. onto a hard, heavy surface bounces to a height of about 13 in. If it hits the surface at a speed of 10 mph then it bounces at about 6 mph. Here, $\text{COR} = \sqrt{13/36} = 6/10 = 0.6$

A baseball dropped from a height of say 3 ft onto a solid floor will bounce to a height of about 13 in (Fig. 8.4). Dropped from a height of 6 ft, it will bounce to a height of about 26 in. The bounce height is proportional to the drop height but is a lot less than the drop height. The ratio of bounce height to drop height is about 13/36 = 0.36, at least when the ball is dropped from a relatively small height. At the higher ball speeds encountered in the game, this ratio drops to about 0.25.

In practice, the bounciness of a ball is specified by a number called the coefficient of restitution, or COR. Restitution is an old-fashioned word and not a particularly good one but no one has thought of a better one. It refers to the ability of a ball to return instantly back to its original spherical shape when it bounces. The COR of a ball is actually a measure of how much energy is dissipated in the ball when it bounces. A plasticene ball has a COR of zero, meaning that it loses all its kinetic energy when it bounces. A ball that doesn't lose any energy at all would bounce back to its original drop height, in which case the COR would be 1.0. Most balls used in sport have a COR between 0.5 and 0.9. Baseballs and softballs lose about 75% of their energy when they bounce.

The COR can be defined and measured either in terms of bounce height or bounce speed. A baseball incident at right angles to heavy surface at a relatively low speed, say 10 mph, will bounce at about 6 mph. The COR is defined by the relation

$$\text{COR} = \frac{\text{rebound speed}}{\text{incident speed}}$$

which in this case is 6/10 = 0.6. The ratio of the bounce speeds here is the same as the square root of the bounce height ratio ($\sqrt{0.36}$) since bounce height is proportional to the bounce speed squared. If a ball bounces twice as fast then its kinetic energy increases by a factor of four so it bounces four times higher. Intuitively, one might expect that a ball bouncing twice as fast should bounce twice as high, but it doesn't. Intuition is correct in the horizontal direction but not in the vertical direction. If you throw or hit a ball twice as fast in the horizontal direction then the ball will travel about twice as far before it hits the ground. That's because a ball projected horizontally takes the same time to fall to the ground (ignoring air resistance), regardless of its launch speed. When a ball is projected vertically into the air, it takes a certain time to come to a stop before it falls back down. The time taken to come to a stop is proportional to the initial vertical speed. If a ball bounces twice as fast then it will take twice as long to come to a stop. Since it travels twice as fast for twice the time, it bounces four times higher.

The easiest way to measure the COR of a ball is drop it onto a hard floor from a certain height, say h_1, and then measure the bounce height, h_2. The COR can then be calculated from the formula

$$\mathrm{COR} = \sqrt{\frac{h_2}{h_1}}$$

as explained in Appendix 8.1. For example, if $h_1 = 3$ ft and $h_2 = 1$ ft then $\mathrm{COR} = \sqrt{1/3} = 0.58$. That method of measuring the COR is simpler than trying to measure the ball speed just before and just after the ball hits the floor.

The COR of a baseball or a softball decreases as the incident ball speed is increased, and drops to about 0.5 at an incident ball speed of around 60 mph, since the ball then bounces at about 30 mph. If it hits the floor at 100 mph then it bounces at about 45 mph and the COR is then about 0.45. All balls, not just baseballs and softballs, have the property that the COR decreases slightly as the ball speed increases.

The COR of a ball traveling at 60 or 100 mph is not measured by dropping it on the floor. It is measured in the safety of a laboratory by firing the ball horizontally from a canon onto a vertical steel plate or a hardwood block or a metal cylinder and measuring the incident and rebound speeds with a video camera or a radar gun or with light gates. Typical results are shown in Fig. 8.5 (from [2] and [3]). The COR of a baseball or a softball is usually specified, by both ball manufacturers and sporting organizations, by quoting the value for a ball incident at speed of 60 mph on a flat, rigid wall.

8.6 COR for Two Colliding Balls

We are interested primarily in the collision of a bat and a ball, but we first need to look at the collision between two balls to understand how the COR is measured for such a collision. An example is shown in Fig. 8.6. In this situation, the coefficient of restitution is defined in a different way to that for a bounce off the floor. For a head-on collision, the COR is defined as the relative speed of the two balls after the collision, divided by the relative speed before the collision. For example, if two balls are approaching each other at a relative speed of 14 mph, and they bounce with a relative speed of 7 mph, then the COR = 7/14 = 0.5.

When the COR is less than 1, as it always is, some of the original kinetic energy of the two balls is lost. Most of the lost energy is shared by the two balls since they

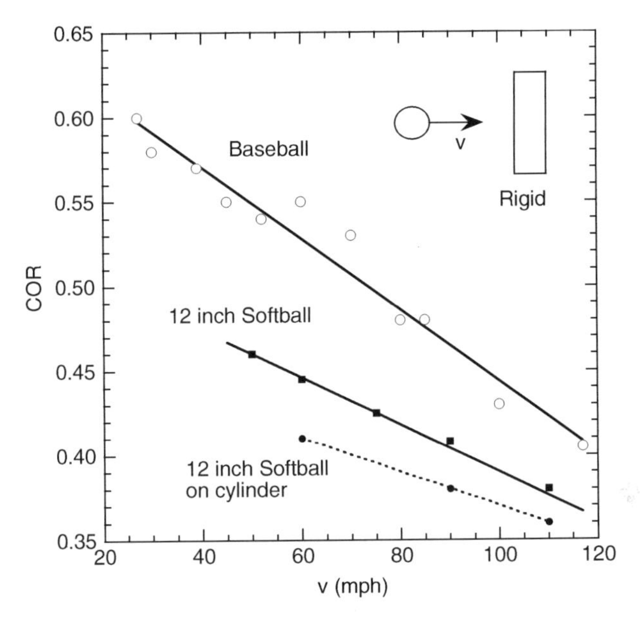

Fig. 8.5 Typical data for a baseball and a 12 in. softball incident on a heavy, flat, rigid surface showing how the COR decreases with increasing ball speed. Also shown is the COR for the same softball incident on a rigid curved surface

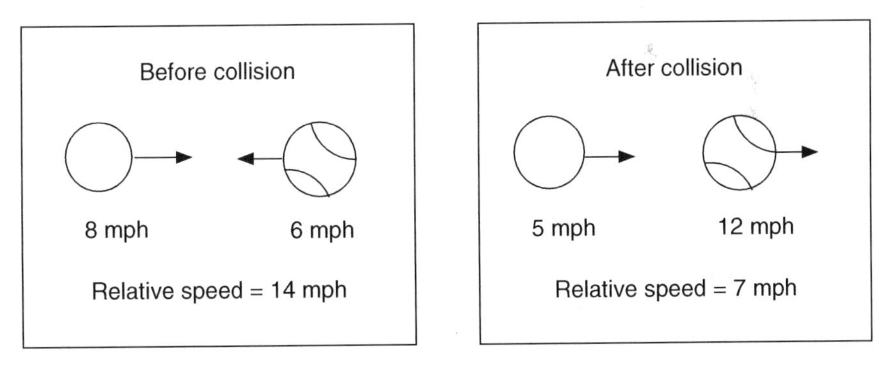

Fig. 8.6 Two balls collide head-on with a relative speed of 14 mph. After they collide, their relative speed is 7 mph. In this case, the COR is 7/14 = 0.5

both start vibrating as soon as they collide, resulting in an audible click or thud when the balls collide. The surviving energy is shared between the two balls as they each head off after the collision with a certain amount of kinetic energy.

The COR is a very useful quantity in describing such collisions because the COR for the two balls remains the same if the relative speed before the collision is the same. One ball could be at rest or they could both be moving toward each other, and the COR will still be the same. For example, the COR for the two balls in Fig. 8.6 would still be 0.5 if one ball was initially at rest at the other was approaching it at 14 mph. The relative speed after the collision would still be 7 mph, although the actual speeds after the collision would not be the same as those indicated in Fig. 8.6. The reason that the COR remains the same is that the outcome of a collision like this does not depend on the reference frame used to measure the speeds. A bug flying beside one of the balls would think that his ball was at rest and the other ball was approaching it, even though a bug sitting on the ground might see that both balls are approaching each other. The collision is the same and the COR is the same regardless of which bug is viewing the collision.

An additional feature of the COR is that the COR does not depend strongly on the relative speed before the collision. If the relative speed in Fig. 8.6 was 15 mph rather than 14 mph then the COR would still be 0.5 or very close to 0.5. However, if the relative speed was much larger, say 100 mph, then the COR would be less than 0.5. It is almost always the case that the COR for a collision between two objects decreases as the initial relative speed increases.

If a single ball bounces off a heavy floor, then the floor remains at rest. The relative speed of the ball and the floor, both before and after the bounce, is just the speed of the ball. That is why we defined the COR in that case as the speed of the ball after the collision divided by the speed of the ball before the collision.

Suppose that we measure the COR of each of the two balls by bouncing them on a heavy floor, and one is say 0.6 and the other is say 0.8. What will the COR be when they bounce off each other? The answer is given in Appendix 8.4. If ball 1 has a COR = e_1 and stiffness k_1 while ball 2 has a COR = e_2 and stiffness k_2 then when one ball collides with the other the COR is given by

$$e^2 = \frac{k_2 e_1^2 + k_1 e_2^2}{k_1 + k_2} \tag{8.4}$$

From this result, we see that:

1. If there was no energy loss in either of the balls, so that they both had a COR = 1 when bouncing off a heavy floor, then the COR would be 1 when they bounced off each other since neither of the balls would lose any energy.
2. If one of the balls is much stiffer than the other, then the softer ball will compress a lot more than the stiffer ball and the soft ball will store a lot more elastic energy during the collision than the stiff ball, as explained in Appendix 8.3. As a result, the COR will be close to the value of the softer ball.
3. If the two balls are equally stiff, then they will share the elastic energy equally. The low COR ball will lose more of its stored energy than the other ball, and the end result is that the COR is about equal to the average COR of the two balls.
4. If two identical balls collide with each other, each having the same mass, stiffness, and COR, then the COR when they collide is the same as that measured when each ball impacts a heavy floor. Sir Isaac Newton's measurement, where he obtained a COR of 5/9 = 0.55 for two colliding compressed wool balls in 1687, was way ahead of its time but a major advance in physics. Baseball wasn't even invented until the 1800s, although cricket was played in England in the 1600s.

8.7 Happy and Unhappy Balls

Arbor Scientific and Edmund Scientific sell an inexpensive pair of rubber balls known as Happy and Unhappy balls. They both look and feel the same, having almost the same stiffness. Happy has a COR of almost 1 and Unhappy has a COR of zero. Unhappy doesn't bounce at all when dropped on a hard floor. An interesting question is what will happen if Unhappy is dropped on a soft, elastic surface like a drum or the strings of a tennis racquet? Another interesting question is what will happen when Unhappy collides with Happy?

It is easy to do these experiments, and the answers agree with our conclusions in the previous section. Unhappy bounces nicely off an elastic surface. If the ball and the surface are equal in stiffness then they will share the elastic energy equally. Unhappy loses all of its stored elastic energy while the elastic surface returns almost all of its elastic energy, so only half of the stored elastic energy is lost.

A similar thing happens when Unhappy collides with Happy. Both share the total elastic energy about equally, and about half the total elastic energy is lost. The two balls collide with a COR value about 0.7. A detailed description of the behavior of these balls is given in [4].

8.8 Brick Walls and Peanuts

Suppose that a baseball strikes a brick wall at 60 mph. It will squash by about 1/4 in. before it bounces back. If it bounces back at 30 mph then the COR will be 0.5 and the ball loses 75% of its kinetic energy. Since kinetic energy is proportional to speed squared, and since the ball bounces at half the incident speed, the rebounding ball has only one quarter of its incident kinetic energy. Now consider what happens when a 60-mph baseball strikes a much lighter object. To emphasize and dramatize the physics of the situation, we will consider the collision between a baseball and a peanut (or any other suitably small object that you might prefer). In that case the ball will slow down a tiny amount, to about 59 mph. The ball will squash by no more than 1 mm, and the peanut will fly off at around 80 mph or more. Only a tiny amount of elastic energy will be stored in the baseball, and about 75% of that energy will be lost.

Given that the COR is a measure of the energy lost in the collision, and given that only a tiny amount of energy is lost, will the COR in this case be nearly 1.0 or will it still be 0.5? The surprising answer is that it depends on the stiffness of the peanut. If the peanut is much stiffer than the baseball (or if it is a walnut) then the COR will be 0.5 since the peanut will not squash or store any elastic energy but the ball still loses 75% of its stored elastic energy. If the peanut is relatively soft and elastic (meaning that it doesn't lose any of its stored elastic energy) then the COR will be greater than 0.5.

The example of a ball colliding with a peanut highlights an important point. The point is that the COR is a measure of the fraction of the elastic energy lost, not the fraction of the total energy lost. In the case of a ball colliding with a brick wall, all of the kinetic energy of the ball is converted to elastic energy when the ball temporarily comes to a stop. In that case, 75% of the original kinetic energy is lost. When a ball collides with a peanut, and slows down from 60 mph to 59 mph, only about 1% of the original kinetic energy is lost, but 75% of the stored elastic energy is lost. Even though the COR provides a measure of the energy lost in a collision, a better definition of the COR is the following:

> The COR measures the fraction of the stored elastic energy that is lost.

If a 60-mph baseball strikes a peanut, and the peanut is initially at rest, then the relative speed before the collision is 60 mph. If the peanut is only 1.6 g, and if the COR is 0.5, then the ball will slow down to 59 mph (as determined by the collision equations described in Chap. 9). The relative speed after the collision will be $0.5 \times 60 = 30$ mph, so the peanut flies off at $59 + 30 = 89$ mph. If the COR happens to be 0.6 rather than 0.5, then the relative speed after the collision will be $0.6 \times 60 = 36$ mph. In that case, the ball will slow down to 58.9 mph and the peanut will head off at $58.9 + 36 = 94.9$ mph

The COR is defined as the ratio of the relative speeds after and before the collision, but it is also a measure of the elastic energy that is lost during the collision. The fraction of the stored elastic energy that is lost is given by $1 - e^2$ where e is the COR. A derivation of this result is given in Appendix 8.3 for a collision with the floor, and in Appendix 8.4 for a collision between two balls. For example, when $e = 0.5$ then the fractional energy loss is $1 - 0.25 = 0.75$.

Now consider the collision of a baseball with a peanut having the same stiffness as a baseball. Since they have the same stiffness, and since equal and opposite forces act on the ball and the peanut, they will both compress by the same amount and store exactly the same amount of elastic energy (despite their huge difference in weight). If the peanut loses none of its elastic energy, while the ball loses 75%, then only $0.75/2 = 0.375$ of the stored elastic energy is lost. The COR will then be 0.79 since $1-e^2 = 0.375$ when $e = 0.79$. In that case, the relative speed after the collision will be $0.79 \times 60 = 47.4$ mph, and the peanut will fly off at $58.8 + 47.4 = 106.2$ mph. The COR is not just a property of the ball. It is important to note that:

> The COR depends on the elastic properties of both the ball and the object it collides with. It is only when the ball bounces off a very stiff object like the floor that the COR is determined just by the elastic properties of the ball.

If a ball squashes by 1/4 in. or more then it will lose a greater fraction of its stored elastic energy than when it squashes by only 1 mm or less. A baseball impacting on a brick wall at 60 mph will squash by about 1/4 in. and bounce off the wall with a COR about 0.5. If it impacts the wall at only 6 mph then it will squash by about 1 mm and the COR of the ball will then be about 0.6. Consequently, the COR for a 60-mph baseball colliding with a very stiff peanut will be about 0.6, while the COR for a 60-mph baseball colliding with an elastic peanut will be about 0.82 if the peanut has the same stiffness as the ball.

8.9 Bounce Off a Bat

A relatively simple experiment to measure the COR between a bat and a ball is described in Project 9. You might like to read about that project first, before reading further, to get a better idea of why the COR is so important in understanding bat and ball collisions. In the remainder of this chapter we explain how the COR is defined for a collision between a bat and a ball, and explain in more detail how energy loss in the bat and in the ball affects the COR.

Suppose that someone holds a bat in a steady horizontal position by holding onto the handle and then drops a ball onto the barrel, as shown in Fig. 8.7. If the ball is dropped from a height of say 3 ft, then the ball will bounce to a height of only about 6 in. or less depending on whether the ball bounces off the middle of the barrel or near the tip. The bounce is even worse than off a hard floor because the ball gives a large fraction of its energy to the bat, resulting in only a small amount of elastic energy being stored in the ball. As a result, the barrel rotates away from the ball and the ball bounces up at very low speed. Furthermore, the impact of the ball causes the bat to vibrate, so energy is lost because the bat vibrates. Both of these effects result in a relatively weak bounce of the ball off the bat. However, rotational energy given to the bat is not "lost" in the collision. It is simply given to the bat. As a result, the COR for the collision is much the same as it is for a collision between a ball and a hard floor, provided the bat doesn't vibrate too strongly. Bat vibrations are much reduced at the sweet spot and are enhanced near the tip of a bat, so the COR varies along the length of the barrel in a manner that we will explore in Chap. 10.

We define the COR for a collision between a bat and a ball as the relative speed of the ball and the bat after the collision, divided by the relative speed before the collision. For example, if the ball hits a stationary bat at a speed of 10 mph, then the relative speed before the collision is 10 mph. If the ball bounces up at 2 mph and the

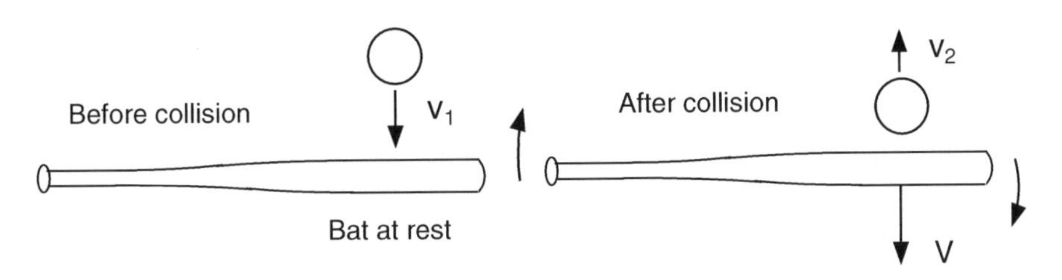

Fig. 8.7 When a ball is dropped at speed v_1 on a bat at rest, the ball bounces up at speed v_2 and the bat rotates away from the ball, at speed V at the impact point. We define the COR $= (v_2 + V)/v_1$

bat rotates downward at 4 mph then the relative speed after the collision is 6 mph. The COR is then 6/10 = 0.6. However, the COR for such a collision varies along the length of the barrel, being typically only about 0.2 or 0.3 at the far end of the barrel, due to the fact that more energy is lost in the form of bat vibrations when the ball strikes the far end of the barrel.

Different parts of the bat rotate at different speeds. If the bat rotates about an axis through the knob then the knob moves at zero speed while the tip of the barrel might rotate at 5 mph. When defining the COR in this situation, the speed of the bat after the collision is taken as the speed at the point where it is struck by the ball. It is not immediately obvious that the COR should be defined this way. For example, we could take the speed of the bat after the collision to be the speed of its center of mass. However, that would give the wrong answer for the COR, for reasons that will be explained in Chap. 10 when we consider the collision process in more detail.

Hinged and Clamped Bats

When measuring the bounce off a bat it is usually more convenient to clamp the handle in a vice or to pivot the bat about an axis through the handle, as described in Project 9 and in Chap. 10 (Sect. 10.8). In fact, most bats are tested these days by firing a ball at high speed onto the barrel of a bat that is initially at rest and free to rotate about an axis through the handle. In the latter case, we define the COR in exactly the same way as that for a completely free bat. That is, the COR is the relative speed of the bat and ball after the collision, divided by the incoming ball speed, the bat speed after the collision being the speed of the impact point (as indicated in Fig. 8.7).

In the case of a bat with a clamped handle, the velocity of the bat after the collision is zero, so the COR is just the speed of the ball after the collision divided by the speed of the ball before the collision. The bat will vibrate strongly after the collision, but the vibration speed is not counted when we measure the recoil speed of the bat. The bat doesn't recoil at all.

Even if the bat is freely suspended or pivots around an axis, it will still vibrate unless it is struck at the sweet spot. Suppose that a given point on the bat rotates through a distance of 1 in. in 0.01 s. The speed of the bat at that point is then 100 in. s^{-1}. However, if the bat is also vibrating, the surface of the bat can simultaneously move back and forth by say 0.1 in. in 0.01 s, or at a speed of 10 in. s^{-1} in one direction and then 10 in. s^{-1} in the other direction. The actual speed of the bat at that point can therefore vary rapidly between 90 and 110 in. s^{-1}. When defining the COR we need to ignore this rapid variation in speed caused by vibration of the bat, and just take the average speed, which in this case would be 100 in. s^{-1}. Energy given to the bat in the form of bat vibrations is eventually lost in the bat and in the player's arms, and is not part of the overall kinetic energy retained by the bat and the ball. Rather, it is part of the energy that is lost.

COR for a Bat and Ball Collision

If a bat is struck at its sweet spot then very little energy is lost in bat vibrations and the COR for a collision between the bat and the ball is then about the same as it is for a bounce off a heavy floor, despite the fact that the ball bounces higher when it bounces off the floor. The point here is that when a ball collides with a bat, it gives some of its energy to the bat so less energy is available for bouncing. If no energy is lost in the bat itself (that is, if it doesn't vibrate) then the COR will be essentially the same as that for a bounce off the floor. There is, however, an important exception. That is, if some of the energy given to the bat is given back to the ball, then the COR can be increased above the ball-floor value. Alternatively, if the impact occurs away from the sweet spot then the bat will vibrate, energy will be lost in the bat, and the COR will be less than ball-floor value.

The COR can be increased if a ball collides with a soft, elastic material. For example, if a ball is dropped on the strings of a tennis racquet, and if the racquet is clamped to the floor, then the ball will bounce much better off the strings than off a hard floor, and the COR for the collision will be relative large, about 0.8. The effect here is an example of the trampoline effect, whereby the bounce of a ball can be improved by allowing it to bounce off an elastic surface. Similarly, the COR for a collision between a bat and ball can be increased slightly if the bat is slightly more elastic than a hard floor. The end result, or the bottom line, is that the ball can be hit faster if the bat is slightly elastic. Aluminum and composite bats are generally more elastic than wood bats due to the hollow, spring-like wall, and therefore tend to outperform wood bats.

8.10 Wood Bats vs. Aluminum Bats

Wood bats are generally much stiffer than the ball, so the amount of elastic energy stored in the bat by virtue of its compression, when it collides with a ball, is relatively small. Nevertheless, a wood bat can bend along its length when struck by a ball, especially when it is struck near the tip. Even in that case, when the ball squashes by about 1 in., the bat bends by only about 1/4 of an inch. Most of the elastic energy stored in the collision is therefore stored in the ball but there is also some elastic energy stored in the bat. About 75% of the energy stored in the ball is lost. The elastic energy stored in a wood bat propagates away from the impact point as a bending wave and most of that energy is also lost. The net result is that the COR for a collision between a wood bat and a ball is typically about 0.3 for an impact near the tip of the bat. If the impact is 6 or 7 in. from the tip then the bat does not vibrate as much and the COR is closer to 0.5.

In the case of a hollow bat, the whole bat bends in a manner similar to a wood bat. In addition, the wall of the bat bends inwards. In this situation, the COR can be greater than 0.5 as a result of the trampoline effect, leading to enhanced performance of the bat. In this respect, a hollow bat behaves somewhat like a large and heavy elastic peanut. Of course, a bat looks nothing like a peanut but the physics of the situation is similar. The trampoline effect is examined in more detail in Chap. 12 where it is shown that there are four separate effects that together determine the COR. One is the fact that the fractional energy loss in the ball is reduced in a "soft" impact with a bat, since the ball deformation is reduced. Another is the fact that the wall stores some of the total elastic energy and returns some of that energy to the ball. A third effect is that the wall vibrates strongly after the collision, at a frequency of about 1500 Hz, and retains some of the stored elastic energy as vibrational energy in the barrel. The fourth effect is that the whole bat vibrates after the collision, at a frequency of about 170 Hz, and also retains some of the stored elastic energy as vibrational energy. The end result is that there are some locations along the bat where the COR can be greater than 0.5, and other locations where the COR can be less than 0.5, as described in Project 9.

8.11 COR vs. Bounce Speed Off a Bat

A question of interest is how the collision of a ball at 60 mph on a heavy block of wood relates to the collision of a 60 mph ball with an actual wood bat. The question relates to the fact that balls are usually tested by firing them at a speed of 60 mph onto a block of wood or a steel plate. The question is, do they behave in the same way when they impact a bat at 60 mph?

The problem here is that the force exerted on the ball is not the same so the deformation of the ball is different. In order for the force to be the same, the incident ball speed on a bat would need to be greater than 60 mph. By comparison, the force on a peanut would be miniscule. If the ball COR happened to be independent of the force on the ball or independent of the amount by which the ball compresses, and if the bat is struck at the sweet spot and doesn't vibrate, then the COR for a collision with a heavy block of wood would be the same as that with a wood bat, regardless of the speed of the ball. It would not be the same as that for a collision with an aluminum bat since aluminum bats are softer and store some of the elastic energy.

When a ball collides with a very heavy block of wood bolted to a metal frame or a brick wall, the force on the ball depends on the ball stiffness and on the deformation of the ball. To simplify the following discussion, we will assume that the ball behaves like a simple spring during its compression, in which case the force on the ball is given by $F = kx$ where k is the ball stiffness and x is its deformation (or compression, to use the physics term. In the sporting industry, the term "compression" refers to the force required to compress a ball by 1/4 in.). In that case, the elastic energy stored in the spring is $kx^2/2$. If the ball is incident at speed v_1 then its kinetic energy is $mv_1^2/2$. During the collision, the ball will compress until it comes to a stop, at which point all the initial kinetic energy is converted to elastic energy and then $mv_1^2/2 = kx_o^2/2$ where x_o is the maximum deformation of the ball.

The maximum deformation is, therefore, given by $x_o = v_1\sqrt{m/k}$ and the maximum force on the ball is given by $F_o = kx_o = v_1\sqrt{mk}$.

If the ball is incident at speed v_o on a light block of wood of mass m_2, and if the wood block is free to recoil, and if the block of wood is much stiffer than the ball, then the maximum force on the ball is given by $F_o = kx_o = v_0\sqrt{mk/(1+m/m_2)}$, as shown in Appendix 8.5. In order for the maximum force to be the same as that on a very heavy block of wood, v_o needs to be increased so that $v_o/\sqrt{(1+m/m_2)} = v_1$. For example, suppose that $v_1 = 60$ mph, $m = 145$ g and $m_2 = 600$ g. Then $v_o = 66.8$ mph. A baseball colliding with a stiff, 600 g block of wood at 66.8 mph will therefore compress by the same amount and bounce with the same COR as a 60 mph ball colliding with a very heavy block of wood bolted to a brick wall.

An additional complication in the case of a real bat is that the bat surface is curved, whereas the standard test is conducted with a block of wood having a flat surface. At high ball speeds, the ball wraps itself around the curved surface of a bat with the result that the middle part of the ball is more highly compressed than the outermost contact regions. Results obtained in Professor Lloyd Smith's softball lab at Washington State University indicate that there is a small decrease in the COR when a ball impacts on a curved cylinder rather than a flat plate, as shown in Fig. 8.5.

8.12 COR vs. Temperature and Humidity

It is well known that the COR of a ball can vary with temperature and humidity although the magnitude of these effects have only recently been determined [5, 6]. Using a standard 60 mph impact test, Drane and Sherwood [5] found that the COR of a baseball increased from 0.525 at 25°F to 0.537 at 40°F, and then increased more slowly with temperature to 0.550 at 120°F. In other words, the COR increased by about 2% as the temperature was increased from 40°F to 120°F. When the incident ball speed was increased to 100 mph, the COR decreased to 0.485 ± 0.005, but it remained constant at that value, regardless of the temperature, as the temperature was increased from 40°F to 120°F. For softballs, the COR decreases with increasing temperature.

Kagan and Atkinson [6] performed a similar 60 mph experiment with baseballs at a temperature of 23°C and found that the COR decreased from 0.55 at 0% humidity to 0.495 at 100% humidity, a decrease of 10%. Balls kept in a 100% humidity chamber gained about 12 g, while those kept in a 0% humidity chamber lost about 2 g. Home runs should therefore be easier to hit on dry days, since balls struck at 0% humidity will travel about 7% faster than those at 100% humidity according to (9.3) and (9.4) in Chap. 9. Baseball games are not played in such extremes of humidity, so the actual effect of humidity on ball speed and distance traveled will be less. In fact, Meyer and Bohn [7] calculated that changes in humidity should have an overall effect of only a few feet in batted distance since the decrease in COR at high humidity is negated to some extent by the decrease in air density as the humidity increases.

Appendix 8.1 Relation Between COR and Bounce Height

A ball of mass m at a height h has potential energy PE $= mgh$. When dropped to the floor from a height h_1 it will hit the floor at a certain speed, say v_1. As the ball falls, it loses potential energy and gains kinetic energy. By the time the ball hits the floor, it will have lost all its potential energy, so the gain in kinetic energy is given by $0.5mv_1^2 = mgh_1$ so $v_1 = \sqrt{2gh_1}$, regardless of the mass m.

If the ball then bounces at speed v_2 it will rise to a height h_2 given by $0.5mv_2^2 = mgh_2$ so $v_2 = \sqrt{2gh_2}$. The ratio of the two heights is then $h_2/h_1 = (v_2/v_1)^2 = e^2$ where $e = v_2/v_1$ is the COR. The initial energy is mgh_1, the loss in energy is $mgh_1 - mgh_2$, and the fraction of the initial energy that is lost is

$$\frac{mgh_1 - mgh_2}{mgh_1} - 1 - \frac{h_2}{h_1} = 1 \quad e^2$$

If $e = 0.6$ then 0.64 or 64% of the ball energy is lost, and if $e = 0.5$ then 0.75 or 75% of the ball energy is lost.

Appendix 8.2 Force on a Bouncing Ball

Even in a low speed bounce, the force exerted on a ball is surprisingly large. For example, if a baseball is dropped from a height of 1.0 m (39.37 in.) onto a hard wood or concrete floor then the ball will hit the floor at a speed of 4.4 m s^{-1} and bounce at a speed of about 2.6 m s^{-1}. The force on the ball has to be big enough to bring the ball to a complete stop and then push it up off the floor at 2.6 m s^{-1}. Since the force lasts for only about 0.002 s and the change in velocity is $4.4 + 2.6 = 7.0$ m s^{-1}, the average acceleration of the ball is $7/0.002 = 3{,}500$ m s^{-2}. The velocity changes from $+4.4$ to -2.6 so the change is actually 7.0 m s^{-1}. Since $F = ma$, and since $m = 0.14$ kg for a baseball, the average force on the ball is $F = 0.14 \times 3{,}500 = 490$ N (112 lb). The force is not constant during the whole time. It starts at zero, increases to about 580 N and then drops back to zero as the ball lifts off the floor. The 490 N force is just the average force during the collision.

Appendix 8.3 Sharing the Elastic Energy

When two balls collide head-on, the force acting on one ball is equal and opposite the force on the other ball. If ball 1 has a stiffness k_1 and it compresses by an amount x_1 then the force on ball 1 is k_1x_1. Similarly, the force on ball 2 is k_2x_2. Since the two forces are equal and opposite, we find that $k_1x_1 = k_2x_2$ so $x_2/x_1 = k_1/k_2$. Consequently, if k_2 is bigger than k_1 then x_2 will be smaller than x_1. In other words, if a soft ball collides with a stiff ball, then the soft ball will compress more than the stiff ball, as expected.

The amount of elastic energy stored in a ball or in a spring is given by $0.5kx^2$ where k is the stiffness and x is the compression. Let $S_1 = 0.5k_1x_1^2$ be the elastic energy stored in ball 1, and let $S_2 = 0.5k_2x_2^2$ be the elastic energy stored in ball 2. Then $S_1/S_2 = k_1x_1^2/(k_2x_2^2) = k_2/k_1$ given that $x_1/x_2 = k_2/k_1$. If a soft ball collides with a stiff ball then the energy stored in the soft ball will be greater than the energy stored in the stiff ball. If both balls are equally stiff, then they will share the energy equally, regardless of their mass.

Appendix 8.4 Relation Between e and Energy Loss

Suppose that a mass m at speed v_1 is headed toward a mass M traveling in the opposite direction at speed V_1 (Fig. 8.8). To simplify the following calculation, suppose that $mv_1 = MV_1$ so that the total momentum is zero and $V_1 = mv_1/M$. The total kinetic energy is $E_1 = 0.5mv_1^2 + 0.5MV_1^2 = 0.5mv_1^2(1 + m/M)$. During the collision, both masses will come to a complete stop since the total momentum remains zero. At that time, all of the original kinetic energy is stored as elastic energy in the two masses.

After the collision, the two masses head off in opposite directions, still with zero total momentum. If m heads off at speed v_2 and M heads off at speed V_2 then $mv_2 = MV_2$ and the remaining kinetic energy is then $E_2 = 0.5mv_2^2 + 0.5MV_2^2 = 0.5mv_2^2(1 + m/M)$. The energy loss is, therefore, $E_1 - E_2 = 0.5m(1 + m/M)(v_1^2 - v_2^2)$.

The COR for the collision is given by $e = (v_2 + V_2)/(v_1 + V_1) = v_2/v_1$. The fraction, f, of the original energy that is lost is therefore given by

$$f = \frac{E_1 - E_2}{E_1} = \frac{v_1^2 - v_2^2}{v_1^2} = 1 - e^2$$

For example, if $e = 1$ then none of the energy is lost and if $e = 0$ then all of the stored elastic energy is lost. For a bat and ball collision with $e = 0.5$, 75% of the stored elastic energy is lost.

The elastic energy stored in m is $S_1 = 0.5k_1x_1^2$, and the elastic energy stored in M is $S_2 = 0.5k_2x_2^2$. The fraction of S_1 that is lost is $1 - e_1^2$ where e_1 is the COR of

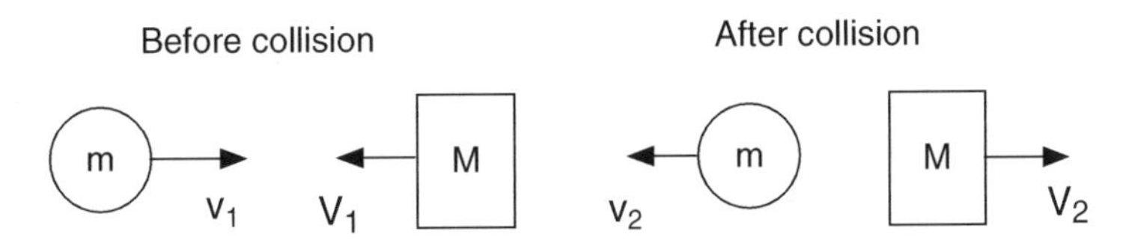

Fig. 8.8 If $mv_1 = MV_1$ then the total momentum is zero, and it remains zero during and after the collision so $mv_2 = MV_2$

m (see Appendix 8.1). Similarly, the fraction of S_2 that is lost is $1 - e_2^2$ where e_2 is the COR of M. Hence,

$$\left(1 - e^2\right)(S_1 + S_2) = \left(1 - e_1^2\right) S_1 + \left(1 - e_2^2\right) S_2$$

Using the result from Appendix 8.3 that $S_1/S_2 = k_2/k_1$, it is easy to show that

$$e^2 = \frac{k_2 e_1^2 + k_1 e_2^2}{k_1 + k_2} \tag{8.5}$$

The COR for the collision therefore depends on the relative stiffness of the two masses as well as the COR of each mass for a collision with a rigid object. According to (8.5), the COR does not depend on the mass M. However, the mass M does have an effect on the COR in a slightly subtle way, as explained in Sect. 8.8. The problem is that the e_1 and e_2 both decrease when the deformation of each mass increases. If M is small, then the force between the two masses will be relatively small, the deformation of each mass will also be relatively small, and e_1 and e_2 will both be relatively large. If the mass m collides with a large mass M, then e_1 and e_2 will both be relatively small. Furthermore, the stiffness of each mass can increase when the deformation of each mass increases. The manner in which the deformation changes with M is described in Appendix 8.5.

Appendix 8.5 Collision of a Ball with a Mass m_2

Suppose that a ball of mass m_1 and stiffness k_1 is incident at speed v_o on a stationary mass m_2 of stiffness k_2. This situation is commonly encountered when testing the performance of a bat. The ball is fired at a stationary bat and measurements are taken of the incident and rebound speeds of the ball. An important practical question is how the result depends on the mass, stiffness and COR of the ball used to test the bat. An equally important question is how the result depends on the mass and stiffness of the bat.

If m_2 is initially at rest then some of the energy of the ball is given to m_2 and not all of the initial ball energy is converted to elastic energy in the ball. To work out the deformation of the ball, it helps to consider a different situation (or a different reference frame) where the ball and m_2 are both traveling toward each other in such a way that the total momentum is zero. If the ball is traveling toward m_2 at speed v_1, then its momentum is mv_1. If m_2 is traveling at speed v_2 toward the ball, then its momentum is $m_2 v_2$ in the opposite direction. If $m_1 v_1 = m_2 v_2$, then the total momentum is zero. The deformation of the ball in this case will be the same as that when m_2 is stationary, provided the relative speeds are the same. That is, $v_1 + v_2 = v_o$, or

$$v_1(1 + m_1/m_2) = v_o \tag{8.6}$$

The two collisions are equivalent since they are the same collision viewed in two different reference frames.

During the collision, the total momentum will remain zero since any momentum lost by the ball is transferred to m_2. Consequently, when the ball comes to rest so does m_2. At that instant, the ball and the mass m_2 both have zero momentum and zero kinetic energy. All of the original kinetic energy is then converted to elastic energy in the ball and in m_2. If x_1 is the maximum deformation of the ball and x_2 is the maximum deformation of m_2, then

$$\frac{1}{2}k_1x_1^2 + \frac{1}{2}k_2x_2^2 = \frac{1}{2}m_1v_1^2 + \frac{1}{2}m_2v_2^2 \quad (8.7)$$

At maximum deformation, $k_1x_1 = k_2x_2$ since the force on each mass is equal and opposite. Since $x_2 = k_1x_1/k_2$ and since $v_2 = m_1v_1/m_2$, we find from (8.6) and (8.7) that

$$x_1^2 = \frac{m_1v_o^2}{k_1(1 + k_1/k_2)(1 + m_1/m_2)} \quad (8.8)$$

The peak force, F_o on the ball is given by $F_o = k_1x_1$. Exactly the same force would be exerted on the ball if the ball was traveling at speed v_o toward mass m_2 and if m_2 was initially at rest. If m_2 happens to be much stiffer than the ball (the usual case for a wood bat, but not an aluminum or composite bat) then

$$F_o = k_1x_1 = v_o\sqrt{m_1k_1/(1 + m_1/m_2)} \quad (8.9)$$

A slightly more elegant and more subtle derivation of (8.8) follows from the fact that we have analyzed the collision in the center of mass reference frame since the center of mass remains at rest. In that frame, it is shown in textbooks that we can define a reduced mass M and reduced stiffness K by the relations $1/M = 1/m_1 + 1/m_2$ and $1/K = 1/k_1 + 1/k_2$. The total deformation $x = x_1 + x_2$ is then given by $(1/2)Mv_o^2 = (1/2)Kx^2$, since the kinetic energy in the center of mass frame is converted entirely to elastic energy when the two masses come to rest.

Equations (8.8) or (8.9) can be used to estimate the effect of changing the ball mass, stiffness or COR on the rebound speed of a bat under test. For example, consider the softball result shown in Fig. 8.5 where COR $= 0.53 - 0.0014v$ for a ball incident on a flat, rigid surface and v is the ball speed in mph. Alternatively, COR $= 0.53 - 0.00313v$ where v is the ball speed in m s^{-1}. Suppose that this result was obtained for a ball of mass $m_1 = 0.191$ g (6.75 oz) and stiffness $k_1 = 1.23 \times 10^6$ N m^{-1} (7,000 lb in.$^{-1}$). The ball deformation on the rigid surface is $x_1 = v\sqrt{m_1/k_1}$. Substituting these values gives COR $= 0.53 - 0.00793x_1$ where x_1 is the ball deformation in mm. For example, at 60 mph (26.82 m s^{-1}), $x_1 = 10.6$ mm so COR $= 0.446$.

Equation (8.8) can be used to calculate the deformation of the ball when it impacts on a bat of effective mass m_2 (defined in Chap. 10) and stiffness k_2. Provided that there is no energy loss in the bat, then the COR for the collision can be determined from the relation between COR and x_1. Similarly, the COR for a collision with the bat can be estimated for a different ball having a different mass and stiffness, provided the COR of that ball is known from rigid wall impact experiments. This procedure allows the bat to be tested using any approved ball that is conveniently available, rather than having to rely on a standard ball with a specific mass, stiffness and COR [8, 9].

References

1. R. Cross, Differences between bouncing balls, springs and rods. Am. J. Phys. **76**, 908–915 (2008)
2. S.P. Hendee, R.M. Greenwald, J.J. Crisco, Static and dynamic properties of various baseballs. J. Appl. Biomech. **14**, 390–400 (1998)
3. L.V. Smith, A. Ison, in *Rigid Wall Effects on Softball Coefficient of Restitution Measurements*, ed. by Moritz, Haake. The Engineering of Sport 6, Developments for Sports, vol. 6, Munich, Germany (Springer, Heidelberg, 2006), pp. 29–34; More extensive data on the COR of softballs can be found in J. Duris, Experimental and numerical characterization of softballs, MSc in Mech Eng thesis, University of Washington, 2004
4. R. Cross, The coeffcient of restitution for collisions of happy balls, unhappy balls and tennis balls. Am. J. Phys. **68**, 1025–1031 (2000)
5. P.J. Drane, J.A. Sherwood, Characterization of the effect of temperature on baseball COR performance, in 5th International Conference on the Engineering of Sport, vol. 2 (UC Davis, CA, 2004), pp. 59–65
6. D. Kagan, D. Atkinson, The coeffcient of restitution of baseballs as a function of relative humidity. Phys. Teach. **42**, 330–354 (2004)
7. E.R. Meyer, J.L. Bohn, Influence of a humidor on the aerodynamics of baseballs. Am. J. Phys. **76**, 1015–1021 (2008)
8. L. Smith, Progress in measuring the performance of baseball and softball bats. Sports Tech. **1**, 291–299 (2008)
9. L. Smith, A.M. Nathan, J.G. Duris, A determination of the dynamic response of softballs. Sports Eng. **12**, 163–169 (2010)

Safety Issues

10. J.J. Crisco, S.P. Hendee, R.M. Greenwald, The influence of baseball modulus and mass on head and chest impacts: a theoretical study, Med. Sci. Sports Exerc. **29**, 26–36 (1997)
11. R.L. Nicholls, K. Miller, B.C. Elliott, A numerical model for risk of ball-impact injury to baseball pitchers, Med. Sci. Sports Exerc. **37**, 30–38 (2005)
12. M. McDowell, M.V. Ciocco, A controlled study on batted ball speed and available pitcher reaction time in slowpitch softball, Br. J. Sports Med. **39**, 223–225 (2005)

Chapter 9
Ball Hysteresis

9.1 Introduction

The most obvious response of a baseball or a softball when it is struck by a bat is that it suddenly reverses direction. The ball travels toward the batter before it is struck, and it travels away from the batter after it is struck. As far as most people are concerned that is all that happens to the ball. In fact, a lot more happens that is hidden from view since it can't be seen by eye. The advent of high speed video cameras has enabled the viewer to see what happens in slow motion, but even that does not tell the whole story. High speed video shows that the ball squashes and expands while it is in contact with the bat, and it also shows that the bat bends and vibrates. Still, there is a lot more going on during the collision that even these fast cameras can't see, the problem being that cameras can't see inside a ball.

Unless we X-ray a ball, we can't really see what is happening inside the ball, but we can get a good idea by compressing the ball to "see" what happens. It is a bit like feeling a Christmas present from the outside before we open the parcel. One of the standard tests to measure ball properties is to compress the ball by 1/4 in. and measure the required force. The NCAA rule for a 12 in. circumference softball is that the compression force must be within the range 300–400 lb. It is not a test that the average person can perform since it requires a special type of machine known as a materials testing machine. Such machines are generally computer controlled and quite expensive, costing anything from $20,000 to $200,000 depending on the size and sophistication. Most commonly, they are used by engineers to measure the stiffness and strength of materials. Big ones can exert forces of thousands of pounds, and they can stretch or compress a heavy steel or concrete structure until it breaks.

When a materials testing machine is used to squash a ball, the force exerted on the ball is commonly described by engineers as the load or the compression, and the decrease in diameter of the ball is commonly described as the displacement. The latter term is used because the ball is squashed between two parallel metal plates, one of which remains at rest and the other is moved or displaced to squash the ball. In physics, the word "compression" usually refers to a decrease in length or volume rather than a force, and displacement refers to a change in position of an object. In the following discussion, the physics term "force" is used rather than

R. Cross, *Physics of Baseball & Softball*, DOI 10.1007/978-1-4419-8113-4_9,

"compression" and the terms "compression" or "deformation" are used rather than "displacement" to denote the decrease in diameter of the ball.

Suppose that a ball is placed in one of these machines and the force on the ball is increased to 400 lb. An electronic measuring device attached to the machine will indicate that the ball has compressed by say 0.255 in. However, if the operator checks again, he will see that the ball has compressed a little more than 0.255 in. Every time he looks at the measurement he will see that the compression has increased. What happens is that the ball keeps compressing slowly over time even though the force on the ball is held constant. The effect is known as creep. It is due to the fact that baseballs and softballs, like many other materials, are viscous, meaning that they flow slowly like honey. If you pour a glass of water, the water flows freely. If you pour honey, the honey flows more slowly since it is more viscous. The flow rate of a softball or a baseball is extremely slow. Nevertheless, when a ball is subject to a force of a few hundred pounds, it can creep a fraction of an inch in only a few seconds.

There is another effect that is observed with baseballs and softballs and other viscoelastic materials, known as hysteresis (pronounced hiss-ter-eesis). Suppose that a force of 100 lb causes a ball to compress by 0.1 in., and a force of 200 lb causes the ball to compress by 0.18 in. This means that the ball is becoming stiffer since the first 100 lb compressed the ball by 0.1 in. but the next 100 lb compressed the ball only 0.08 in. Now suppose the force is reduced back to 100 lb. The compression will then be about 0.15 in. If the force is decreased to zero, the compression will be about 0.05 in. In other words, the ball remains slightly squashed even after the force is removed, although the ball will eventually return to its original shape given enough time. The same sort of thing happens, in a much more obvious way, if you compress a piece of plasticene.

Hysteresis refers to the fact that different graphs of force vs. compression are obtained depending on whether the force is increasing or decreasing. A graph of force vs. compression (or force vs. extension) for a metal spring is a straight line, and it is the same regardless of whether the force is increasing or decreasing. A graph of force vs. compression for a baseball or a softball is a curved line and its shape does depend on whether the force is increasing or decreasing. The curved shape indicates that baseballs and softballs get stiffer the more they are compressed, while hysteresis indicates that elastic energy stored in the ball cannot be fully recovered when the ball expands. The loss of energy means that baseballs and softballs don't bounce very well.

Hysteresis curves can be measured by compressing the ball slowly, using a materials testing machine, or they can be measured by compressing the ball rapidly, by bouncing the ball. A slow compression results in a "quasi static" curve, while a rapid compression results in a "dynamic" curve. The curves are not exactly the same, partly because of the effect of creep but also because of a closely related effect known as stress relaxation. If a ball is compressed slowly by say 1/4 in., the required force might be say 400 lb. If the compression is held fixed at 1/4 in. then the force drops quickly for a few seconds and then more slowly over time. Over a period of 10 s the force can drop from 400 lb to about 350 lb. The latter effect is called "stress relaxation." The same effect occurs with tennis strings. When a tennis string is stretched with a force of 60 lb and tied to a racquet frame, the string tension drops to about 55 lb within 10 min. After a few weeks, the tension will have dropped to about 50 lb.

Stress relaxation is an effect that can only be observed clearly when the deformation is held fixed and the force is allowed to decrease with time. However, the cause of that effect doesn't suddenly come into being as soon as the deformation is held fixed. It exists even when the deformation is changing. As a result, the force that is measured during a slow compression is less than the force measured during a rapid compression by the same amount. In both cases, the force increases with time while the ball is being compressed, but stress relaxation over time acts to reduce the force by a greater amount during a slow compression.

Stress relaxation in a baseball or a softball is especially rapid at large values of the applied force. Lloyd Smith at Washington State University measured this effect in 2010 by compressing a baseball by 1 in. in 1 s. During that time, the force on the ball increased from 0 to 1,800 lbs. He then left the ball in the squashed state for several seconds and recorded the force on the ball as a function of time. The result is shown in Fig. 9.1. The force on the ball immediately started dropping at a rate of about 2,500 lb s^{-1} for the first 0.2 s, the force dropping to 1,000 lb after 4 s. Stress relaxation would also have been occurring while the ball was being squashed. Had it not been for the rapid decrease in the force due to stress relaxation, the force on the ball would have risen to about 4,300 lbs rather than the observed 1,800 lbs when the ball was compressed by 1 in. That is essentially the result obtained when the ball is compressed very rapidly in a bounce test, although the results are not directly

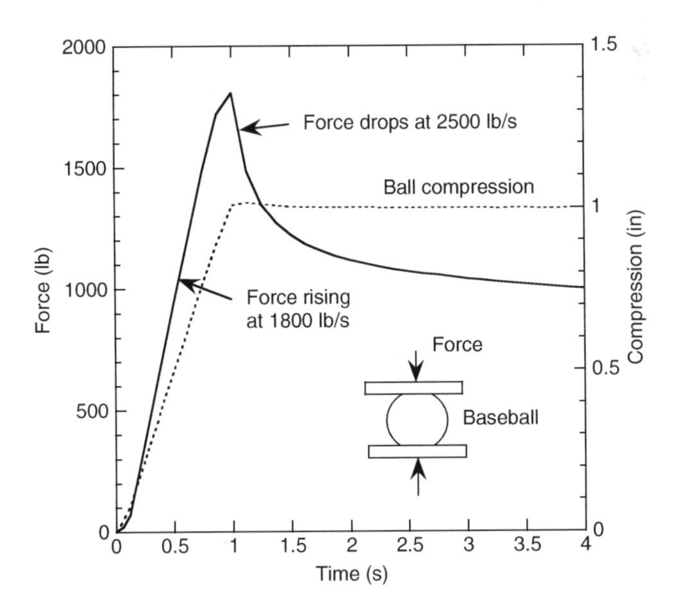

Fig. 9.1 Force and compression vs. time for a baseball compressed rapidly by 1 in. in a materials testing machine. If the ball is left in the squashed state, the force drops rapidly due to the fact that balls are viscous. The effect here is called stress relaxation

comparable since the ball squashes on only one side in a bounce test and on both sides in a quasi static test. During a 0.001-s bounce test, stress relaxation causes the force on the ball to decrease by only a few lbs. A valid force vs. compression curve, for compressions of more than 1/4 in., can therefore only be obtained by compressing the ball very rapidly so that stress relaxation does not interfere with the measurement of the force on the ball. Even for a 1/4 in. compression, the force on the ball is underestimated in a quasi-static test, but the result does provide an indication as to whether a particular ball is hard or soft compared with other balls.

9.2 Static Hysteresis Curves

Force vs. compression curves for a baseball are shown in Fig. 9.2. These results were obtained in a small materials testing machine by increasing the load force to 150 lb and then decreasing the force back to zero. Similar results are obtained at larger forces, although the ball deformation is correspondingly larger. At any given point on a hysteresis curve, the ratio F/x at that point represents the stiffness of the ball. The ball stiffness increases along the compression curve, and is smaller along the expansion curve. A convenient measure of ball stiffness is the ratio of F/x at the maximum compression point. For example, the stiffness of the ball in Fig. 9.1 is 150 lb/0.095 in. = 1579 lb in.$^{-1}$.

The area under the increasing F vs. x curve (between the curve and the horizontal axis) represents the work done to compress the ball (since work = force times distance), most of which is stored as elastic energy in the ball. The area under the decreasing F vs. x curve represents the energy that is recovered when the ball expands. The area enclosed by the two curves therefore represents the energy lost. The loss of energy during the expansion of the ball is responsible for the fact that a ball dropped onto a hard surface does not bounce back to the original drop height.

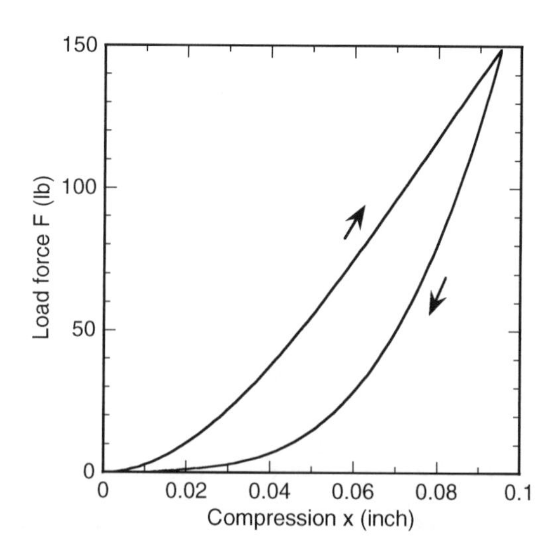

Fig. 9.2 Force vs. compression curves for a baseball measured in a materials testing machine. The resulting curve depends on whether the force is increasing or decreasing. Softball curves are very similar in shape and magnitude

In Fig. 9.2, the area of the whole graph box is 150 lb $\times$ 0.1 in. = 15 lb-in. The area under the increasing F curve is 5.56 lb-in., and the area under the decreasing F curve is 3.13 lb-in. If the force and compression for a bouncing ball were the same as that in Fig. 9.2, then the incident kinetic energy would be 5.56 in.-lb and the kinetic energy after the bounce would be 3.13 in.-lb, giving a COR of 0.75. In fact, the COR for a ball bouncing at low speed is typically about 0.6 since the ball bounces before the ball fully expands. The curves in Fig. 9.2 were obtained by compressing and expanding the ball slowly over a total period of about 15 s, allowing the ball to fully expand back to its original size, so the actual recovery of energy is overestimated by this method. Furthermore, the force on the ball is underestimated when the ball is compressed slowly. The force on the ball increases as it is being compressed, but it drops below the value that would be obtained in a very fast compression of the ball, due to stress relaxation while the ball is compressing slowly.

In 1998, a study [1] was published where the authors measured the properties of 11 traditional baseballs with an average mass of 143 g. Each ball was compressed by 1 cm (0.39 in.) in a materials testing machine to measure its static stiffness, defined as the peak force divided by the compression (1 cm). Each ball was compressed slowly over a period of 10 s using a force up to about 560 lb and then the force was reduced slowly to zero over another 10 s. The balls varied in stiffness from 1,940 to 3,310 N cm^{-1} (1,108–1,890 lb in.$^{-1}$), consistent with the result shown in Fig. 9.2. From the area enclosed by the hysteresis curve, it could be estimated that the average COR of these balls should be about 0.68. However, the average COR found by firing the balls at a rigid aluminum plate, at 60 mph, was 0.56. A slow compression therefore underestimates the amount of energy lost in the ball when it bounces at high speed.

Each ball was also fired at a speed of 90 mph onto the aluminum plate, giving a measured peak force on each ball of about 27 kN (6,100 lb), and an average COR of 0.50. The ball compression was not measured, but if we assume that each ball compressed by about 0.5 in., then the ball stiffness during this rapid compression was about 12,000 lb in.$^{-1}$. Balls are much stiffer during a rapid, high speed compression than when measured in a materials testing machine at low compression values, partly because the compression is much larger when a ball impacts a surface at speeds around 90 mph. The 0.5-in. compression here is not just a wild guess. The compression can be estimated from the measured force on the ball, as described in Appendix 9.1. Similar data for both baseballs and softballs has been obtained in Professor Lloyd Smith's lab at Washington State University [2]. Some of his data are shown in Fig. 9.6.

9.3 Dynamic Hysteresis Curves

The compression of a ball in a materials testing machine or in a vice is a relatively slow process compared with the compression occurring during a bounce off the floor or during the collision with a bat. In a machine, it can take anything from 1 s to a minute to compress a ball by 1/4 in., depending on the speed setting chosen by the operator. When a ball bounces at moderately high speed, it takes about 0.0004 s to compress and about 0.0008 s to expand. A machine compression is, therefore, about 2,000 times too slow to capture the actual force vs. compression behavior of a ball when it bounces.

A better test is to allow the ball to impact directly on a piezo disk or on a small metal plate with several piezo disks or load cells attached to the rear side of the plate. That way, the force on the ball can be measured while it bounces off the disk or off the plate. It is difficult to measure the actual ball compression while it is bouncing, although it can be done by filming the bounce with a high speed camera operating at 10,000 fps or more. However, it is easy to calculate the position of the center of mass of the ball while it bounces, using the measured force signal. The acceleration, a, of the ball during the bounce is given by $a = F/m$ where F is the force on the ball and m is its mass. If x is the position of the center of mass and v is the ball velocity, then $v = \mathrm{d}x/\mathrm{d}t$ and $a = \mathrm{d}v/\mathrm{d}t$. The velocity can be calculated by integrating the acceleration waveform, and then x can be calculated by integrating the velocity waveform. The elastic behavior of the ball can then be determined by plotting an F vs. x hysteresis curve or by calculating the ball stiffness from the ratio F/x.

A simple arrangement used by the author to measure F vs. x in this way is shown in Fig. 9.3. The measurements were done at low ball speeds, partly to avoid damaging the piezo disk used to measure the force, and partly to improve the accuracy of the measurement. If a ball is fired at 60 mph or more onto a metal plate, the force on the plate is much larger and can cause both the plate and the structure holding the plate to vibrate. As a result, the force signal might include not only the force of the ball but also the vibrations of the plate or the ball itself. A typical high speed result is

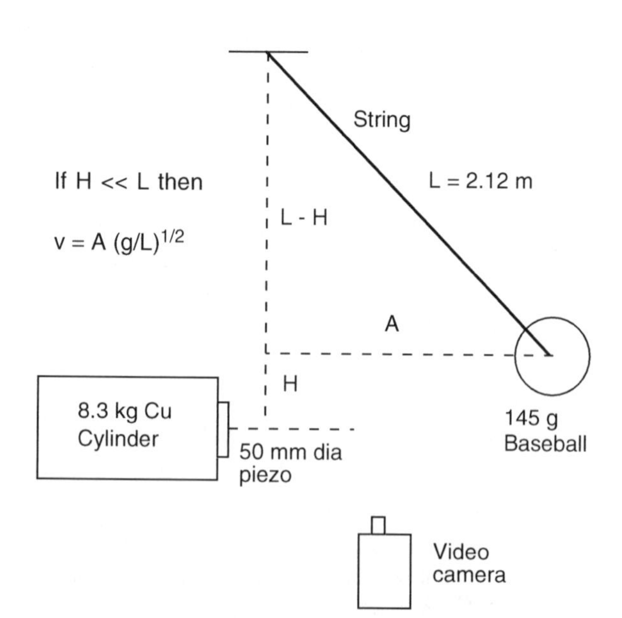

Fig. 9.3 Apparatus to measure the impact force on a baseball and the amplitude, A, of the pendulum swing. For small A, COR = ratio of the rebound to initial values of A

shown in Fig. 9.6 where vibrations are clearly present. A similar high speed result is shown in [1]. An alternative method of measuring the force would be to impact the ball on one end of a long, cylindrical rod so that the force can be measured before the pressure wave in the rod, reflected off the far end, gets back to the impact point. This method has been used to measure the force on a high speed golf ball [3] and should also work well with baseballs and softballs.

Experimental results for the same baseball tested in Fig. 9.2, but obtained with the apparatus in Fig. 9.3, are shown in Figs. 9.4 and 9.5. Results were obtained for ball speeds in the range from about 1 m s^{-1} to about 5 m s^{-1}. Figure 9.4 shows

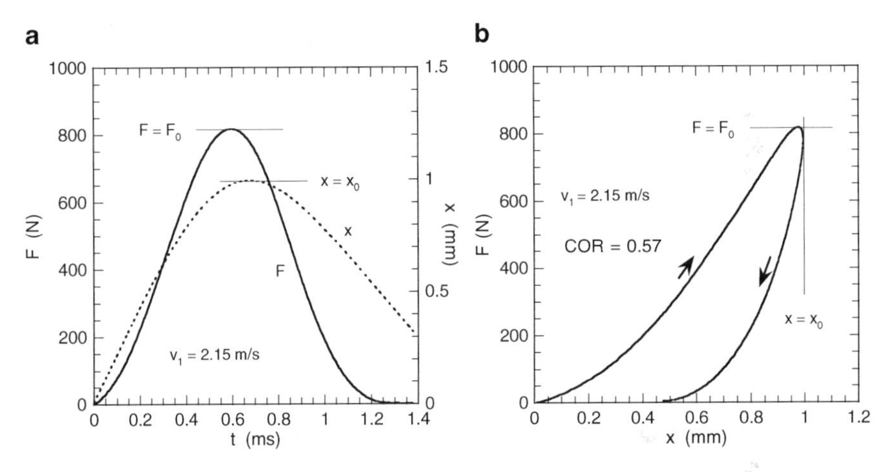

Fig. 9.4 Experimental data at $v_1 = 2.15$ m s^{-1} showing (**a**) F and x vs. t and (**b**) the corresponding F vs. x hysteresis curve, x being the displacement of the ball CM. 1,000 N = 225 lb

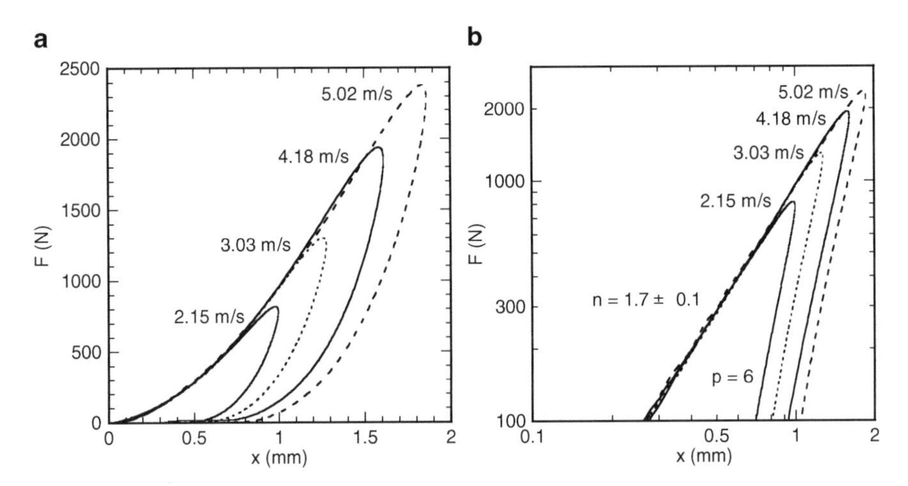

Fig. 9.5 Hysteresis curves at different incident ball speeds plotted on (**a**) linear and (**b**) log scales. During the compression phase, $F = (930 \pm 20)x^{1.7\pm0.1}$ regardless of the incident speed. During the expansion phase, F is proportional to x^6

the time variation of the force on the ball during the bounce, and also shows the displacement, x, of the center of mass. The force increased to a maximum of 810 N (182 lb) and then decreased back to zero as the ball bounced clear. The force is approximately symmetric in time. That is, the shape of the decreasing F vs. t curve is almost a mirror image of the increasing F vs. t curve. However, the x vs. t curve is not symmetric in time. x does not decrease to zero when F decreases to zero, and reaches a maximum slightly after F reaches a maximum. Regardless of the incident ball speed, it was found that x decreases to about half its maximum value when F dropped back to zero. In other words, the ball bounced while it was still compressed, and resumed its normal spherical shape some time after it bounced.

Figure 9.5 shows the corresponding hysteresis curves (F vs. x) at four different ball speeds. The ball follows the same F vs. x path as it compresses, regardless of the incident ball speed, but compresses by a greater amount as the ball speed increases. However, the return or expansion path varies with ball speed, since the expansion starts at a different point along the compression curve. The area enclosed by each curve represents the energy lost during a complete compression and expansion cycle, giving the result that the COR is about 0.57 for these low speed bounces.

The ball is about three times stiffer during a rapid compression than during a slow compression. For example, in Fig. 9.4, the peak force is 182 lb and the maximum value of x is 1.0 mm = 0.039 in. so $F/x = 4{,}667$ lb in.$^{-1}$. In Fig. 9.2, $F/x = 1{,}579$ lb in.$^{-1}$. Part of the difference is due to the fact that x is the actual compression in Fig. 9.2, while x in Fig. 9.4 is the displacement of the center of mass. Furthermore, the ball was compressed on both sides in Fig. 9.2, but when a ball strikes a surface it compresses on only one side. If we assume that the ball compressed by 1 mm when the center of mass moved 1 mm, as suggested by the result in Fig. 8.1b, then a better comparison with the slow compression result is to double the slow compression stiffness result to allow for compression on only one side of the ball. The dynamic stiffness is then 50% larger than that for a slow compression, an effect that can be attributed mainly to stress relaxation during the slow compression.

9.4 High Speed Measurements

Measurements of the force on a high speed softball incident on a solid cylindrical surface were used to obtain the hysteresis curves shown in Fig. 9.6. The front surface of the cylinder had the same shape as a softball bat, and the back surface was flat so that it could be firmly attached to a rigid wall. Load cells located between the cylinder and the wall were used to measure the force on the cylinder. These results were obtained at Washington State University in 2008 [2]. During the compression phase, the force F increases with the displacement of the CM, x, and can be described by the relation $F = kx^n$. The results in Fig. 9.6 can be described with $n = 1$ or with $n = 1.25$, both being similarly good, given the small fluctuations in the force signal.

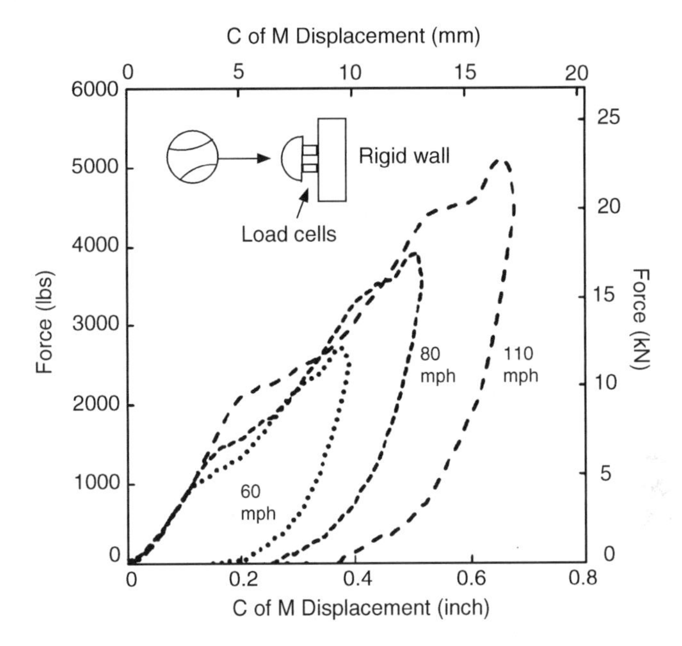

Fig. 9.6 Hysteresis curves measured for a high speed softball impacting on a cylindrical surface. The force signal was integrated twice to measure the displacement of the center of mass (C of M) of the ball [2]

When compressed by 1/4 in., softballs are required to have a stiffness of about 1,400 lb in.$^{-1}$, as measured in a materials testing machine. However, the ball stiffness increases dramatically when tested dynamically. For a 1/2 in. displacement of the ball CM, the force in Fig. 9.6 is about 4,000 lb, giving a stiffness of 8,000 lb in.$^{-1}$. Part of the increase is simply due to the factor of two difference between the actual compression and the displacement of the ball CM. Lloyd Smith and Joseph Duris [2] tested 150 different softballs and found that the dynamic stiffness was typically two or three times larger than the static stiffness for all balls. The effect was not due to the cylindrical impact surface. Essentially, the same result was obtained when impacting a flat plate, although the COR is slightly higher when impacting on a flat plate.

An interesting aspect of the results in Fig. 9.6 is that they are similar to those in Fig. 9.5a, in that the balls compress along the same F vs. x curve regardless of the ball speed. The baseball curve in Fig. 9.5a is not the same as the softball curve in Fig. 9.6, but both sets of results show that the compression curve is independent of ball speed. The same effect has also been observed with cricket balls [4]. In each of these experiments it was found that the force law for compression of a ball does not depend on the speed of the ball. These results contrast sharply with those obtained during quasi-static tests where it is found that F vs. x compression curves depend on the speed at which the ball is compressed. In a quasi-static test, a larger force is needed to compress the ball by 1/4 in. if the ball is compressed more quickly.

All of these results can be explained by the fact that stress relaxation occurs rapidly over a time scale of a few seconds and more slowly over longer time periods. During a very rapid compression, lasting only 1 or 2 ms, the time scale is too short to allow for any significant decrease in force due to stress relaxation. That is why the F vs. x curves in Figs. 9.5 and 9.6 are independent of the rate at which the ball compresses. Over a longer time scale, stress relaxation results in a significant drop in the force on the ball, as indicated in Fig. 9.1. The latter result explains why balls are much softer when compressed slowly.

Another interesting feature of dynamic hysteresis results is that force vs. time curves, such as the one in Fig. 9.4a, are quite symmetrical. This result suggests that the impact region of a baseball or a softball (or a cricket ball) is quite elastic, meaning that there is a relatively small loss of energy in this region, and that the impact region behaves like a nonlinear, metal spring. A model of the bounce process, based on these observations, is given in the following section.

From a practical point of view, the most important result from study in [2] was the fact that the COR increased significantly when high speed softballs were fired at two composite bats, rather than at a rigid surface. The bats, including a wood bat, were allowed to pivot about an axis through the handle. Balls fired at high speed onto the wood bat, or onto the rigid cylinder, bounced with a COR of about 0.38, regardless of the dynamic stiffness of the ball. However, balls fired onto the composite bats bounced with a COR between 0.50 and 0.60, the COR being largest when using balls with a large dynamic ball stiffness (about 9,000 lb in.$^{-1}$). This result demonstrates an important feature of the trampoline effect, noted in the previous chapter. That is,

> The elastic energy stored in a ball during the compression phase decreases as the ball stiffness increases, while the elastic energy stored in the wall of the bat increases.

In the limit where the ball is infinitely stiff, all of the elastic energy of the impact would be stored in the bat and none in the ball. Since the wall returns most of its elastic energy, the COR increases as the ball stiffness increases. The moral is, if a batter wants to hit more home runs, then he or she should use a stiffer ball. Alternatively, if the regulating body wants to reduce the number of home runs, then a solution is to use softer balls.

9.5 Bounce Models

The remainder of this chapter may be a bit too mathematical for some readers. It is aimed at those with an undergraduate level background.

Apart from the bounce model originally developed in 1882 by Hertz (see [5]), there have been very few attempts to model the bounce of a ball from first principles. The simplest model of a bouncing ball, described in Sect. 8.1, treats the ball as a rigid mass m attached to a linear, massless spring of spring constant k. In that case, the ball undergoes simple harmonic motion for one half period and then bounces, without loss of energy, after a time $\tau = \pi\sqrt{m/k}$. If the spring is nonlinear, with a force law $F = k_1 x^n$, then the displacement of the ball CM as a function of time can be found by solving the equation $md^2x/dt^2 = -k_1 x^n$. The peak force, F_o, on the ball can be found by assuming that the ball comes to rest at maximum compression, in which case $\int_0^{x_o} F\,dx = mv_1^2/2$, where v_1 is the incident ball speed. Simple integration indicates that

$$F_o = C_o v_1^{2n/(n+1)}, \tag{9.1}$$

where C_o is a constant that depends on n, m and k_1. In the original Hertz model, $n = 1.5$ in which case $F_o = C_o\, v_1^{1.2}$. For the low speed impacts of a baseball in the range $1 < v_1 < 5\,\mathrm{ms}^{-1}$, shown in Fig. 9.5, it was found that $n = 1.7$, in which case (9.1) indicates that $F_o = C_o v_1^{1.26}$. The latter result is consistent with the experimental data in this speed range, but it provides no information on the magnitude or the mechanism of energy loss during the bounce.

Energy loss in the ball can be described in terms of separate power laws for compression and expansion, as indicated by the results in Fig. 9.5b. If the expansion phase is described by $F = k_2 x^p$, then simple integration to obtain the area enclosed by the hysteresis curve indicates that $e^2 = (n+1)/(p+1)$ where e is the COR. For example, if $n = 1.7$ and $p = 6.76$ then $e = 0.59$, the measured value of the COR at the higher incident ball speeds found in this experiment. However, the two separate (x^n and x^p) power laws do not provide an accurate description of the force on the ball. If one attempts to reconstruct the original F vs. t waveform by integrating $k_1 x^n$ while x increases and then integrating $k_2 x^p$ while x decreases, the result is a strongly asymmetrical F vs. t waveform, in disagreement with the data in Fig. 9.4a.

An alternative method of accounting for energy loss is to model the ball as a rigid mass connected to a nonlinear spring and a series (Maxwell) or parallel (Kelvin–Voight) dashpot [4, 6]. The force on the spring is given by $F = kx^n$ and the force on the dashpot is given by $F = k_D dy/dt$, where x is the compression of the spring, y is the compression of the dashpot and k_D is a damping constant. Dashpots are not commonly encountered in physics, but they exist in the real world in door closers to stop the door slamming shut. The main feature of a dashpot is that it exerts a damping force that is proportional to speed, so it drops to zero when the speed drops to zero, allowing a door to shut in a controlled manner. In the same way, a dashpot would allow a ball to bounce in a controlled manner but with a loss of energy during the bounce.

In the dashpot models, the ball undergoes a compression and expansion cycle with $e < 1$ and with an impact force that is asymmetric in time, unlike actual data. The Kelvin–Voight version of the model causes the force on the ball to jump to a finite value as soon as the ball impacts a surface, whereas experimental data shows that the force is zero at that time. Some of the features of the experimental data can be reproduced with these models, but a much better description is obtained by incorporating the dashpot into a two-part model of the ball, as described in the following section.

9.6 Two-Part Ball Model

A feature of interest observed for all baseball and softball bounces is that the force waveform is approximately symmetrical either side of the peak. The impulse, $\int F\,dt$, after the peak is typically about $5 \pm 1\%$ smaller than the impulse before the peak. The change in ball speed during the first half of the bounce is therefore approximately equal to the change in ball speed during the second half of the bounce. Given that the ball is incident at speed v_1 and exits at speed v_2, F reaches its peak value F_o at a time when the ball speed is approximately $(v_1 - v_2)/2$. Since the ball is still moving toward the surface at this time, the displacement of the CM is a maximum shortly after F reaches its maximum value, not at the same instant. This effect can be seen in the hysteresis curves shown in Figs. 9.4b and 9.5. Such a result is not consistent with the $F = 930x^{1.7}$ relation observed during most of the compression phase. When F is close to its peak value there is a departure from this simple relation. There is a major departure from this force law during the expansion phase.

The symmetry of the F vs. t curves contrasts sharply with the asymmetry of the F vs. x curves. These results can be explained if it is assumed that the contact region compresses and expands, with only minor energy loss, according to the relation $F = k_1 x^n$, where x is the actual ball compression in the contact region and not the displacement of the CM of the whole ball. Such a result is consistent with the fact that the impulse after the peak force is approximately equal to the impulse before the peak force. It is also consistent with the fact that the compression phase of the ball obeys the same force law, regardless of the incident ball speed.

The region of maximum stress in a bouncing ball is known to be confined to a volume of radius approximately a, where a is the radius of the circular contact area between the ball and the surface with which it contacts [5]. Nevertheless, the rest of the ball is connected elastically to the contact volume and will compress and expand via this elastic connection. To model this situation, and to distinguish between actual ball compression and displacement of the ball CM, we can assume that the ball consists of two connected parts, one of mass m_1 in contact with the surface and one of mass m_2 which acts on m_1 via a damped spring S_2. The situation is shown in Fig. 9.7. The elastic properties of m_1 are represented by a nonlinear

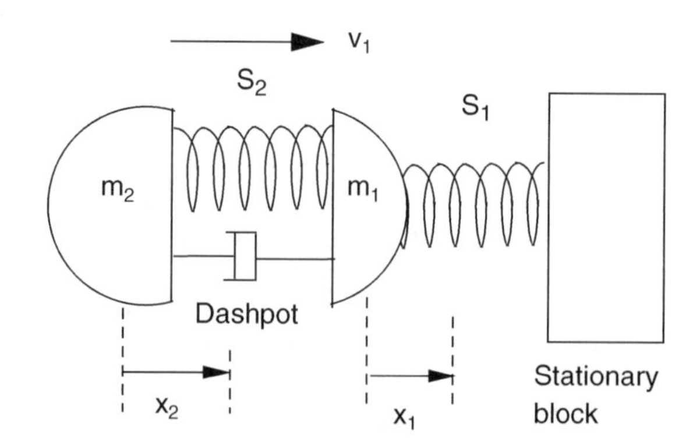

Fig. 9.7 Model where ball consists of two masses m_1 and m_2 and two springs, S_1 and S_2. The dashpot in parallel with S_2 accounts for energy losses in the ball. The ball is incident normally at speed v_1 and rebounds at speed $v_2 < v_1$. During a time t, m_1 moves a distance x_1 and m_2 moves through a distance x_2

spring, S_1, obeying a force law $F = k_1 x_1^n$ where x_1 is the compression of S_1, taken to be equal to the displacement of the CM of m_1. To account for possible energy loss in S_1 we can allow the spring to expand according to the relation $F = k_3 x_1^p$ where $p > n$ and where $k_3 x_1^p = k_1 x_1^n$ at maximum compression.

To account for energy loss in spring S_2, one could assume that this spring also obeys a different power law during its expansion, but such an approach does not yield any information on the energy loss mechanism or its time history. The area under the hysteresis curve is simply a measure of the total energy dissipated on completion of the hysteresis cycle. A better model of the energy loss process is obtained by assuming that S_2 consists of a linear spring in parallel with a linear dashpot and hence obeys a force law $F = k_2 y + k_D \mathrm{d}y/\mathrm{d}t$ where $y = x_2 - x_1$ is the compression of the spring and k_D is a damping constant. Inclusion of a specific damping term allows for an estimate of the energy dissipated as a function of time during both the compression and expansion phases. The integral $\int_0^y F\,\mathrm{d}y$ is the total work done on the spring to arrive at a compression y, regardless of whether the spring is compressing or expanding at that point. The component $k_2 y^2/2$ represents the stored elastic energy, and the component $\int k_D(\mathrm{d}y/\mathrm{d}t)\,\mathrm{d}y$ represents the energy dissipated in S_2. In the present context, the physical mechanism responsible for damping within the ball is unknown, but the damping model assumed here provides very good agreement with the experimental data, as shown in Fig. 9.8.

It is relatively easy to set up and solve the equations describing the model in Fig. 9.7, as described in Appendix 9.2. Typical numerical solutions are shown in Fig. 9.8 for a case where the ball is incident at speed $v_1 = 5.02\,\mathrm{ms}^{-1}$. An excellent fit to the experimental F vs. t waveform is obtained with $m_2/m_1 = 10$, $n = 1.8$, $p = 2.8$, $k_1 = 2.6 \times 10^8$, $k_2 = 0$ and $k_D = 2{,}080\,\mathrm{Nsm}^{-1}$. The k values are quoted here in SI units. The ball bounces, when $F = 0$ and $x_1 = 0$, at a time when it is still compressed. The ball remains compressed well after the bounce in this case since the numerical solution was obtained with $k_2 = 0$. A finite value of k_2 allows the ball to expand back to its initial diameter after the ball bounces, but there is no significant

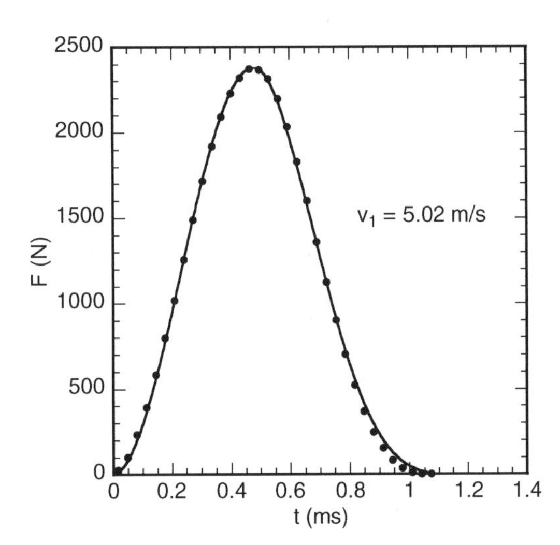

Fig. 9.8 Experimental F vs. t waveform, shown by *dots*, and the best fit solution of (9.2) and (9.3) in Appendix 9.2 when $m_2/m_1 = 10$ and $k_2 = 0$

improvement in the fit to the experimental F vs. t waveform if a finite value of k_2 is assumed or if m_2/m_1 is reduced to a value as low as 1 or 2.

One of the nice features of the ball model described here is that it predicts that the COR will decrease as the ball speed increases, as found experimentally. The actual decrease depends on the chosen parameters, but there is no need to adjust the parameters as the ball speed increases. A good fit to experimental data can be obtained by adjusting the various parameters to suit any particular ball.

9.7 What the Model Tells Us

Based on results obtained with the two-part ball model, a physical description of a bouncing baseball or softball can be given in the following terms. A real ball is three-dimensional and consists of many connected parts, but can be treated in a simplified manner as consisting of only two connected parts. When the ball first makes contact with a rigid surface, both parts of the ball continue to move toward the surface for a short time at about their initial speed, resulting in significant compression of spring S_1 but not S_2. Compression of S_1 acts to decelerate m_1 and hence S_2 also starts to compress, but with a time lag. By the time m_1 comes to a complete stop, m_2 has slowed but is still approaching m_1 and hence S_2 is still compressing. As S_1 expands, m_1 reverses its direction of motion, m_1 and m_2 approach each other and S_2 completes its compression phase. The result is a time lag between compression of S_1 and motion of the ball CM, as shown in Fig. 9.4a.

The elastic energy stored in S_2 is relatively small since the compression of S_2 is smaller than the compression of S_1 and since elastic energy stored in S_2 is largely dissipated during the compression phase. The subsequent behavior of the ball does not depend significantly on the stiffness of S_2, and can be modeled even with $k_2 = 0$. Without any damping, the two parts would oscillate strongly after the ball bounces. In fact, baseballs and softballs are strongly damped and the two parts of the ball remain compressed after the bounce since there is insufficient elastic energy stored in S_2 to cause separation, at least on the short timescale of the bounce. Most of the stored elastic energy resides in the more highly stressed contact region, represented by spring S_1. This part of the ball expands rapidly and accounts for about 40% of the energy loss in the ball, resulting in an impact force that is approximately symmetrical in time.

The model solution is consistent with the Hertz collision model for solid spheres in that most of the elastic energy is stored in the contact region of the ball. A question of interest is why this region should be more elastic than the rest of the ball, given that only about 40% of the total energy loss occurs in S_1. The experimental results do not provide an answer to this question. A plausible explanation for a baseball is that the more highly compressed region of the ball behaves more like an ideal solid while the rest of the ball, composed of many layers of wound yarn, contains voids that allows threads to slide past each other, leading to enhanced frictional losses and a slow recovery rate. The volume of the weakly compressed region is also larger than that of the compressed region, resulting in a proportionally larger energy loss.

Appendix 9.1 Estimating Dynamic Ball Compression

When a ball bounces, it squashes and expands too rapidly to measure its compression easily, unless it is filmed with a very fast and very expensive camera operating at around 10,000 frames s^{-1} or more. A simple alternative is to measure the force and impact duration with a piezo device, as described in Project 10. Suppose that the force rises from zero to a maximum value F_o after a time t_o, and suppose also that the ball comes to rest temporarily at time t_o. Assuming that we also know the incident ball speed v_o, then we can estimate the compression of the ball as follows.

The force, F, will vary with time in an approximately sinusoidal manner up until maximum compression with $F = F_o \sin(\pi t/2t_o)$. The total impulse during the compression is $\int F\,dt = 2F_o t_o/\pi = mv_o$, where m is the ball mass. Hence, $F_o = \pi m v_o/(2t_o)$, indicating that the peak force is proportional to the ball speed. In fact, the time t_o tends to decrease slightly as the ball speed increases, but we can measure t_o if we wish.

If $v = dx/dt$ is the velocity of the ball center of mass during the compression, where x is the displacement of the center of mass, then $F = -m\,dv/dt$ so

$$v = \frac{dx}{dt} = v_o \cos\left(\frac{\pi t}{2t_o}\right)$$

and

$$x = \frac{2v_o t_o}{\pi} \sin\left(\frac{\pi t}{2t_o}\right)$$

The value of the maximum displacement of the center of mass, x_o is, therefore given, by

$$x_o = \frac{2v_o t_o}{\pi} = 0.637 v_o t_o$$

For example, if $v_o = 40\,\mathrm{m\,s^{-1}}$ (90 mph) and $t_o = 0.5$ ms, then $x_o = 0.0127$ m $= 1.27$ cm (0.50 in.) and $F_o = 18.22$ kN (4,096 lb) for a 145 g (5 oz) baseball, and $F_o = 21.36$ kN (4,802 lb) for a 170 g (6 oz) softball. Defining the dynamic stiffness as $k = F_o/x_o$ gives $k = 8{,}192\,\mathrm{lb\,in.^{-1}}$ for the baseball and $k = 9{,}04\,\mathrm{lb\,in.^{-1}}$ for the softball.

A bell-shape force waveform, like that in Fig. 9.4, can be described by

$$F = \left(1 + \frac{4t}{t_o} - \frac{3t^2}{t_o^2}\right)\frac{F_o t}{2t_o},$$

which also has the property that $F = F_o$ and $dF/dt = 0$ at $t = t_o$. For this waveform, $\int F\,dt = 13F_o t_o/24 = mv_o$ in which case the peak force is given by $F_o = 24mv_o/(13t_o)$, and $x_o = 0.677v_o t_o$, both being slightly larger than the values obtained for a sinusoidal force waveform.

Appendix 9.2 Equations Describing the Two-Part Ball in Fig. 9.7

The two-part ball model shown in Fig. 9.7 requires some damping to prevent the ball vibrating during and after the bounce. If $k_D = 0$ it is found that m_1 and m_2 undergo undamped, large amplitude oscillations after the ball bounces, which is not consistent with experimental observations. Rather, baseballs and softballs behave as strongly damped springs, consistent with the fact that they don't bounce very well.

The results of the model calculations do not depend significantly on the value of k_2 or whether S_2 is linear, as assumed, or nonlinear. For most of the bounce period, the dominant term in the force law for spring S_2 is the damping term. A finite value of k_2 allows the ball to expand back to its original shape toward the end of the bounce period and after the ball bounces.

If x_2 is the displacement of the CM of m_2, then $y = x_2 - x_1$, in which case the equations of motion while S_1 compresses have the form

$$m_1 \mathrm{d}^2 x_1/\mathrm{d}t^2 = -F_1 = F_2 - F \tag{9.2}$$

$$m_2 \mathrm{d}^2 x_2/\mathrm{d}t^2 = -F_2 \tag{9.3}$$

where F_1 is the net force acting to the left on m_1, $F = k_1 x_1^n$ is the component of the force on m_1 arising from compression of S_1 and $F_2 = k_2(x_2 - x_1) + k_D \mathrm{d}(x_2 - x_1)/\mathrm{d}t$ is the net force acting to the left on m_2 due to compression of S_2. Expressed in this form, we see that $F = F_1 + F_2$ is the net force acting on the block. When S_1 is expanding, the term $k_1 x_1^n$ can be replaced by $k_3 x_1^p$, both terms being equal at maximum compression. The ball CM has coordinate $x_{CM} = (m_1 x_1 + m_2 x_2)/(m_1 + m_2)$. Addition of (9.2) and (9.3) yields the equation of motion for the ball CM during the compression phase

$$(m_1 + m_2)\mathrm{d}^2 x_{CM}/\mathrm{d}t^2 = -k_1 x_1^n \tag{9.4}$$

During the expansion phase, the exponent n in (9.4) is replaced by p. If the displacement of the ball CM is assumed to be equal to or proportional to the ball compression, then (9.4) alone would be sufficient to determine the bounce parameters. Such a solution does not provide a good fit for a baseball since x_{CM} and x_1 are different functions of t.

Solutions of (9.2) and (9.3) can be found by assuming that $x_1 = x_2 = 0$ and $\mathrm{d}x_1/\mathrm{d}t = \mathrm{d}x_2/\mathrm{d}t = v_1$ at $t = 0$, and by choosing the constants to obtain best fits to the data. Despite the relatively large number of arbitrary constants at our disposal, it turns out that solutions of interest are not particularly sensitive to the ratio m_2/m_1 or to the value of k_2. Useful solutions can be obtained when $m_2/m_1 = 2$ or 10 and $k_2 = 0$ or $1 \times 10^6\,\mathrm{N\,m^{-1}}$. All four solutions provide good fits to the F vs. t waveforms. The impact duration and the time at which F reaches its peak value are determined primarily by k_1. The slope of the F vs. x_{CM} compression curve is determined primarily by n. The peak value of F is determined primarily by k_D once n is fixed. The value of the COR obtained by assuming that $p = n$ is typically about 15%

larger than the experimental value. A better fit to the observed COR is obtained with $p > n$, the choice of p having very little effect on any other fitted parameter. Any particular bounce can therefore be fitted by a unique set of parameters (apart from m_1/m_2 and k_2) each of which is easily determined from the experimental data.

References

1. S.P. Hendee, R.M. Greenwald, J.J. Crisco, Static and dynamic properties of various baseballs. J. Appl. Biomech. **14**, 390–400 (1998)
2. L. Smith, A. Nathan, J.G. Duris, A determination of the dynamic response of softballs. Sports Eng. **12**, 163–169 (2010)
3. I.R. Jones, in *Is the Impact of a Golf Ball Hertzian?*. Science and Golf 4, chap. 44 (Routledge, London, 2002), pp. 501–513
4. M.J. Carre, D.M. James, S.J. Haake, "Impact of a non-homogeneous sphere on a rigid surface". J. Mech. Eng. Sci. **218**(3), 273–281 (2004)
5. B. Leroy, "Collision between two balls accompanied by deformation: A qualitative approach to Hertz's theory". Am. J. Phys. **53**, 346–349 (1985)
6. W.J. Stronge, in *Impact Mechanics* (Cambridge University Press, London, 2000), pp. 86–93

Chapter 10
Collisions

10.1 The Top Two Rules of Baseball and Softball

One of the things that physicists like to do is to study collisions. Indeed, modern physics had its origins in the 1600s with attempts by Sir Isaac Newton and others to understand collisions between various objects such as hard or soft balls. These days, most collision experiments in physics are concerned with tiny subatomic particles colliding at extraordinary high speeds. That is where the frontiers of physics are being pushed forward, but there are still a lot of small discoveries being made about collisions between ordinary objects such as bats and balls. In fact, a large part of this book is based on theoretical and experimental work done during the last 10 years to understand bats and balls.

The problem faced by physicists in the 1600s was that even the simplest and most basic elements of physics were not properly understood. People before them had spent a few thousand years thinking about the subject, which they called natural philosophy, without making significant progress. The breakthrough came when people realized that thinking about these things wasn't enough. In the 1600s, scientists started taking careful measurements to determine what happened when two objects collide. They knew how to measure the weight and the speed of the colliding objects, but that was all. They had no idea how forces operated and they did not understand that energy was a fundamental concept underlying everything that happens in the physical world. The idea that energy was involved in all physical phenomena was not discovered until much later. The problem was that energy comes in many different forms. When two objects collide, some of the energy is transformed into heat and vibration. Since the total kinetic energy of the colliding objects was not retained after the collision, the early scientists did not even recognize the concept of energy as being a useful quantity. Instead, they invented terms like "life force" and "quantity of motion" to describe moving objects. These days, when people use or invent terms like that, we tend to regard them as crackpots.

Eventually, Newton and his contemporaries came up with two very simple and useful laws or rules to describe head-on collisions. Laws are always valid whereas rules can sometimes be broken or modified. We will describe the following two findings as rules since the wording is slightly imprecise in the interests of simplicity.

R. Cross, *Physics of Baseball & Softball*, DOI 10.1007/978-1-4419-8113-4_10,

The two rules survive to this day and are still absolutely essential if we want to understand the collision of a bat and a ball or any other two objects. The two rules are:

1. The total momentum before a collision is equal to the total momentum after the collision.
2. The relative speed of two objects after a collision is a fixed fraction of the relative speed before the collision, regardless of whether one of the objects is initially at rest, or the other object is initially at rest or both objects are approaching each other.

These rules happen to be two of the most important rules in baseball and softball since they determine the speed of the ball as it exits the bat. They were cast when the universe was created and cannot be altered by anyone, including the NCAA and the NBA and the President of the United States. However, the two rules apply only under certain conditions. The first rule applies to the collision between any two objects provided that the only force on each object is the force exerted by the other object, as is usually the case. In the case of a bat and ball collision, the force between the bat and the ball is so much larger than the force exerted on the bat by the batter that the addition force exerted by the batter can be ignored during the collision itself. The "fixed fraction" in the second rule refers to the fact that the relative speed after the collision does not depend on the reference frame in which the collision is observed. However, the actual fraction decreases slightly as the relative speed before the collision is increased.

The two rules can be described in a more meaningful way using specific examples from baseball or softball. The implications for bat and ball sports are as follows:

1. When a bat collides with a ball, the momentum lost by the bat during the collision is given to the ball. The bat itself acquires momentum when the batter swings it, the object of the exercise being to transfer some of that momentum to the incoming ball during the split second when they collide. At first sight it might seem like a better idea to transfer *all* of the bat's momentum to the ball, rather than just some of it, but the bat would then come to a complete stop when it struck the ball. In order for that to happen, the bat would need to be as light as the ball or the ball would need to be as heavy as the bat. Either way, the ball speed off the bat would then be much too slow for anyone to enjoy the game.

2. Bats and balls have the joint property that the relative speed of the bat and ball after the collision is about half of that before the collision. If the ball is pitched at say 80 mph and the bat is swung at say 70 mph, then the relative speed of the bat and the ball before they collide is 150 mph. After the collision, the relative speed will be about half that, or 75 mph. If the bat slows down to say 30 mph immediately after the collision, then the ball will head off at $30 + 75 = 105$ mph. The second rule described above ensures that this speed fraction will be essentially the same regardless of the speed of the pitched ball and regardless of the speed at which the bat is swung. For example, the relative speed after the collision will still be 75 mph even if the ball is pitched at 90 mph and the bat is swung at 60 mph, since the relative speed before the collision is unaltered. If the ball is pitched at 90 mph and the bat is swung at 70 mph, then the relative speed

before the collision increases slightly to 160 mph, but the relative speed after the collision will be still be very close to half of that or 80 mph. Bat manufacturers can tinker with the value of the relative speed fraction to boost it slightly, but they cannot alter the basic physics of the situation, and neither can anyone else. The NCAA passed a regulation for baseball that from January 2011, the relative speed fraction must not be greater than 0.50 at any impact point along the barrel of a baseball bat when a ball is fired at 136 mph to impact a stationary bat.

The momentum of each colliding object is defined as its mass multiplied by its speed. If an object traveling to the right has positive momentum then an object traveling to the left has negative momentum. The point of multiplying mass and speed in this way can be illustrated by a simple example. If you were struck in the head by a mosquito at 20 mph or by a baseball at 20 mph then you would notice a big difference. Similarly, a baseball at 1 mph would not hurt as much as a 20 mph baseball. The force on your head is determined not by the mass of the object striking you, or by its speed but by both quantities multiplied together.

The early scientists also tried to understand collisions by multiplying mass by speed squared, without much success. We now know that kinetic energy is $mv^2/2$, but it is possible and even useful to study collisions without any reference at all to the kinetic energy of the colliding objects. In fact, we will adopt Newton's approach in this Chapter since it is simpler and more useful than the modern energy approach. There are two problems with the kinetic energy in collisions between bats and balls. One is that the total kinetic energy after a collision is always less than the kinetic energy before the collision. The other problem is that the fraction of the kinetic energy that is lost depends on the actual speeds of the bat and the ball. On the other hand, Newton's original approach is much more generally useful since the relative speed of colliding objects after a collision can be calculated from the relative speed before the collision.

Newton himself discovered the second collision rule by taking careful measurements of various balls colliding together. The technique he used was to swing each ball through the air as a pendulum bob. He calculated the speeds of the colliding balls by measuring the height of the bob before the collision, the maximum height after the collision, and the lengths of the arcs traveled by the bob before and after the collision. Newton described the ratio of the relative speeds, after and before the collision, as the coefficient of restitution or COR, a term that we still use today. The COR for any two colliding objects is the same regardless of the speed of each object before the collision, provided the relative speed before the collision is the same. It doesn't matter if one of the balls is initially at rest when the collision occurs or if both balls are approaching each other. It is the relative speed of the two balls that is the important quantity. It was found much later that the COR decreases slightly as the initial relative speed is increased, but the effect was too small for Newton to measure. It was also discovered much later that the COR is a direct measure of the amount of elastic energy lost or converted to other forms during the collision.

The COR is a number that is typically about 0.5 for bat and ball collisions. Other objects collide with a COR value that is always between 0 and 1. If the COR is 0 then the relative speed after the collision is zero. That is, the two objects stick together after the collision and the collision is said to be totally inelastic. A common example is a head-on collision between two vehicles. If the COR is 1 then the relative speed after the collision is the same as it was before the collision. The collision is then said to be completely elastic and there is no loss of energy at all during the collision. The collision of a bat and ball is described as being inelastic, meaning that some of the initial kinetic energy is converted to heat energy in the ball and to vibrational energy of the bat during the collision.

10.2 Collision Equations

The remainder of this chapter includes the equations that are needed to fully understand the collision of a bat and a ball. These equations are taught in high school and in freshman University physics courses, but readers who are unfamiliar with the mathematical language of physics might prefer to skip the equations. The results are explained in words and with graphs that we hope most readers will understand. The basic physics concepts are explained in more detail in Chap. 1.

In Fig. 10.1, we show a bat of mass M colliding with a ball of mass m. The bat is traveling to the right at speed V_1 and the ball is traveling to the left at speed v_1. Subscript 1 is used to denote the speeds before the collision and subscript 2 denotes the speeds after the collision. Later we will show that only part of the mass of the bat is "involved" in the collision, and hence M is not the mass of the whole bat. For the moment it is convenient to distinguish the two colliding objects simply as a bat and a ball. In terms of what follows, they could equally well be a fly and a mosquito, a Ferrari and a Porsche, two opposing football players or a tennis racquet and a tennis ball. The mathematics of the collision is given in Appendix 10.1, where it is shown that the outgoing ball speed, v_2, is given by

$$v_2 = \left(\frac{1+e}{1+m/M}\right) V_1 + \left(\frac{e-m/M}{1+m/M}\right) v_1, \tag{10.1}$$

where e is the COR for the collision. This equation looks quite formidable, as indeed it is. To make things easier, we will shortly derive a much simpler and more useful version of (10.1). Before we do, it will help to explain the physical significance of the various terms in (10.1).

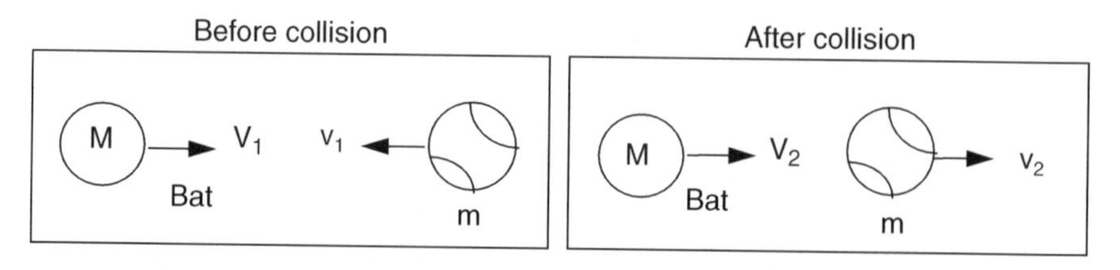

Fig. 10.1 Collision between a bat and a ball, showing the speeds before and after the collision

The outgoing ball speed v_2 consists of two parts. The first part is proportional to the incoming speed V_1 of the bat and the second part is proportional to the incoming speed v_1 of the ball. If the ball was initially at rest, with $v_1 = 0$, then the outgoing ball speed would just be proportional to the speed of the bat. If the bat was initially at rest, with $V_1 = 0$, then the outgoing ball speed would just be proportional to v_1.

The speed of the bat before the collision is determined mainly by the strength and technique of the batter. The speed of the bat also depends on its weight, since light bats can usually be swung faster than heavy bats. Equation (10.1) tells us nothing about how fast a bat can be swung. That is a separate issue. In (10.1), we just assume that it is swung at speed V_1, and that depends to some extent not only on how fast the batter wants to swing the bat but also on where the bat strikes the ball. The tip of the bat travels faster than other points along the bat. Later on, when we get to the nitty gritty of this issue, V_1 in (10.1) will be the speed of that part of the bat that actually makes contact with the ball, and M will be the effective mass of that part of the bat.

The second part of the outgoing ball speed, when $V_1 = 0$, represents the speed at which the ball bounces off the bat when the bat is not swung at the ball. The bounce speed is then proportional to the incoming ball speed. This is such an important part of the performance of a bat that it deserves a special name of its own. A powerful bat is one that allows the ball to bounce off the bat at high speed even when the bat is not being swung at the ball. Lots of names have been suggested for the inbuilt power of a bat, and several manufacturing techniques have been developed to make sure the ball bounces well. One technique is to make the bat slightly springy so the ball springs off the bat as a result of the trampoline effect. However, the main way to make sure that the ball bounces well is to use a heavy bat. A ball will bounce much faster off a heavy bat than off a long, thin pencil, even if the pencil is slightly springy. Players sometimes attempt to cheat by corking their bat, either to make the bat lighter and easier to swing or to make it behave more like a spring, but this technique has been shown to have very little effect on the outgoing ball speed. A hollow wood bat with a thick wall does not behave like a hollow aluminum bat with a thin wall since wood is much stiffer and since it is also less elastic. The main practical effect of drilling a hole in a wood barrel is to reduce its strength so it will be more likely to break.

The intrinsic or "inbuilt" power of a bat is represented by the second term in brackets in (10.1) since this is the term that tells us how well the ball bounces off the bat. In the interests of giving this term a simple and meaningful name, we will call it the Bounce Factor and describe it using the symbol q. Other authors have called it the apparent coefficient of restitution or ACOR, or the collision efficiency or the rebound power, all referring to the same thing, namely the ratio of v_2 to v_1 when the bat is initially at rest. That is,

$$\text{Bounce Factor} = q = \frac{v_2}{v_1} = \frac{e - m/M}{1 + m/M} \tag{10.2}$$

The bounce factor is related to the COR but it is not the same thing since the formula for q ignores the recoil speed of the bat and it is just the ratio of the outgoing ball speed to the incoming ball speed for a situation where the bat is initially at rest. Equation (10.2) indicates that q is always less than e.

Using the definition of q, and the relation $1+q=(1+e)/(1+m/M)$, we find that (10.1) simplifies to

$$v_2 = (1+q)V_1 + qv_1 \tag{10.3}$$

This is the primary physics equation that describes the outgoing speed of a struck ball, regardless of whether the ball is struck by a bat or a racquet or a club. The performance of any given striking implement depends mainly on the value of q for that implement, but there are two complicating factors. One is that q and V_1 both vary with the impact point. The tip of a bat travels faster than any other point on the bat, but it is the point with the lowest value of q. The second complicating factor is that light bats can be swung faster than heavy bats, but q is relatively small for light bats. Consequently, the variation of outgoing ball speed with bat weight is not as large as one might expect, and neither is the variation of the outgoing ball speed with the impact point. Nevertheless, there are small variations, as we will see shortly.

We can also calculate the speed of the bat after the collision. From (10.3) and (10.17) (in Appendix 10.1), we find that

$$V_2 = (1+p)V_1 + pv_1, \tag{10.4}$$

where $p = q - e$ is the recoil factor. If the bat is initially at rest (with $V_1 = 0$) and is struck by the ball, then p is the ratio V_2/v_1, in the same way that q is the ratio v_2/v_1. Since the bat recoils to the left in this situation, p is a negative number and V_2 is negative. Normally, the bat is swung at the incoming ball with $V_1 > 0$ and follows through after the collision, with positive V_2, as shown in Fig. 10.1. However, the bat will slow down according to (10.4). Since p is negative, V_2 is less than V_1.

Equation (10.3) was derived from (10.1) and (10.2) via the slightly laborious calculation given in Appendix 10.1. There is a much easier way to arrive at (10.3), but it is more subtle. A ball incident on a stationary bat at speed v_1 bounces at speed $v_2 = qv_1$. The same collision can be viewed in a different reference frame where the bat approaches the ball at speed V_1, the ball is incident at speed $v_{\text{in}} = v_1 - V_1$ and the ball rebounds at speed $v_{\text{out}} = v_2 + V_1 = qv_1 + V_1 = q(v_{\text{in}} + V_1) + V_1$. Hence, $v_{\text{out}} = (1+q)V_1 + qv_{\text{in}}$ which is equivalent to (10.3).

10.3 Examples of Collisions

Some simple examples will illustrate the significance of (10.2)–(10.4).

(a) Impact at Sweet Spot vs. Impact at the Tip

Suppose that a ball is pitched at speed $v_1 = 80$ mph and the impact point on the bat approaches the ball at $V_1 = 70$ mph. The outgoing speed of the ball will then be $v_2 = 70(1 + q) + 80q$. Near the sweet spot of a baseball bat, q is about 0.2, so $v_2 = 70 \times 1.2 + 80 \times 0.2 = 100$ mph. Struck at a point near the tip of the bat where $V_1 = 80$ mph and $q = 0.1$, $v_2 = 80 \times 1.1 + 80 \times 0.1 = 96$ mph. This is a typical result, where the outgoing ball speed off the tip of a bat is less than that near the sweet spot, despite the fact that the tip travels faster.

(b) Trampoline Effect

The trampoline effect with a hollow bat can increase the COR from about 0.50 to about 0.55, depending on the stiffness of the bat and the ball. The reasons are discussed later, but how will that affect the batted ball speed? Suppose that the ball mass is 5 oz and the effective mass of the bat at the impact point is 20 oz (as described in the following section). If $e = 0.50$ then $q = 0.20$ from (10.2), as we assumed in the previous example. If e is increased to 0.55 then $q = 0.24$. In the above example, $v_2 = 100$ mph at the sweet spot when $q = 0.2$. If q increases to 0.24 as a result of the trampoline effect, then $v_2 = 70 \times 1.24 + 80 \times 0.24 = 106$ mph. That is an important effect, and it is why aluminum and composite bats generally outperform wood bats.

(c) Billiard Ball Collisions

One of the simplest and best known examples of a collision is the head-on collision of two billiard or pool balls of the same mass. If the target ball is at rest then the incident ball comes to a dead stop and the target ball takes off with the speed of the incident ball. The incident ball does not bounce at all, so the bounce factor q is zero even though the COR is 1. The relative speed after the collision is the same as the relative speed before the collision. This result is correctly described by (10.2) and (10.3). Since $e = 1$ for billiard balls and $m = M$ we find from (10.3) that $q = 0$ so $v_2 = V_1$ when $v_1 = 0$. Similarly, the recoil factor $p = q - e = -1$ so from (10.5) we find that $V_2 = 0$ when $v_1 = 0$.

(d) Heavy Bat or Ball Colliding with a Light Ball

In general, a ball will bounce faster off a heavy object, with a larger bounce factor, than off a light object. If a heavy ball collides with a light ball, then the heavy ball won't bounce backward at all. Rather, it will follow through in the same direction with reduced speed, in which case the bounce factor is negative. This is the usual situation when a heavy bat collides with a light ball. The bounce factor of the ball is positive, but the bounce factor of the bat is negative. For example, if $m/M = 2$ then $q = (e - 2)/3$. If $e = 1$ then $q = -1/3$.

(e) Dead Spot on a Bat

It is possible for a light ball to come to a dead stop, without bouncing backward, even if it collides with a heavy object. An example is a ball of plasticene dropped onto the floor. The bounce factor in that case is zero because the COR is zero and because the mass of the floor is effectively infinite.

A baseball bounces off a bat with q typically about 0.1 or 0.2. However, q depends on the impact point and can even be zero or less than zero. Consider a collision between a bat and a ball with $m/M = 0.5$. That is, we assume that the bat is twice as heavy as the ball. Real bats are about six times heavier than the ball but the effective mass of the bat near the fat end is only about twice the ball mass. The COR between a bat and a ball is also about 0.5. In that case, we find from (10.2) that $q = 0$. A ball incident on a stationary bat near its far end does not bounce at all in this case since the bounce factor is zero. If you try this with your own bat and ball you might find that the ball bounces slightly, but the bounce is very weak and may even be zero or slightly negative [1].

Nevertheless, the ball will still come off the bat at high speed if the bat is swung at the ball. According to (10.3), the outgoing speed of the ball will be $v_2 = V_1$ when $q = 0$, regardless of the incoming speed of the ball. That is almost the same situation as a billiard ball collision. The only difference is that the bat does not necessarily come to a complete stop. Since $p = q - e = -0.5$, we find from (5) that $V_2 = 0.5V_1 - 0.5v_1$. The impact point on the bat will therefore come to a complete stop if $V_1 = v_1$, although the handle end will continue to move at whatever speed the batter is swinging it.

(f) High and Low COR Balls

Large changes in batted ball speed can result from large variations in the COR. To illustrate, suppose that $m/M = 0.5$. The minimum possible value of the COR is zero, in which case we find from (10.2) that $q = -1/3$. The maximum possible value of the COR is 1.0, in which case $q = +1/3$. The two extremes correspond to perfectly inelastic and perfectly elastic collisions respectively, as one might find when using a heavy plasticene ball or a heavy superball rather than a baseball or a softball. For a pitch speed $v_1 = 80$ mph and a bat speed $V_1 = 70$ mph, the outgoing speed of the ball is $v_2 = 70(1+q) + 80q$, giving $v_2 = 20$ mph when $e = 0$, or $v_2 = 120$ mph when $e = 1$. If $e = 0.5$ and $m/M = 0.5$, then $q = 0$ and $v_2 = 70$ mph. It is easy to see from these examples why the COR has such a big effect on batted ball speed and why the COR needs to be closely specified by the rules of the game.

(g) Light Balls

Suppose that the ball mass is reduced to such an extent that m/M is almost zero. Suppose also that $e = 1$ so that the ball bounces as fast as possible. Then $q = 1$

and the batted ball speed in the previous example increases to $70 \times 2 + 80 \times 1 = 220$ mph. That is why baseballs and softballs are relatively heavy and why the COR is relatively low.

10.4 Effective Mass of a Bat

We mentioned earlier that only part of the total mass of a bat was involved in a collision with the ball. To find an appropriate value for the "effective" mass of the bat, consider the situation shown in Fig. 10.2. A ball of mass m collides with a bat at a distance b from the bat center of mass (CM). Let the mass of the whole bat be M and suppose that the bat is initially at rest and freely supported. That is, no-one is holding onto the handle, although the bat could be suspended using a long length of string. In that case, the ball will bounce off the bat and the bat will be set in motion. The bat CM recoils at speed V_{CM} and the impact point on the bat recoils at speed V. Because the bat rotates when it is struck by the ball, V will be greater than V_{CM}. The impact point therefore accelerates faster than the CM, as if it was an isolated mass separate from the rest of the bat and lighter than the bat.

The whole bat is involved in the collision, but the effect on the ball is equivalent to a collision with an isolated mass M_e that is less than the mass of the whole bat. Furthermore, the impact point recoils as if it was a mass M_e. In other words, we can treat the collision as being equivalent to one between a ball of mass m and an object of mass M_e where M_e is less than the mass of the whole bat [2, 3].

The effective mass of the impact point, M_e, is derived in Appendix 10.2, and the result is

$$M_e = \frac{M}{1 + Mb^2/I_{CM}}, \tag{10.5}$$

where I_{CM} is the moment of inertia (MOI) of the bat for rotation about its center of mass.

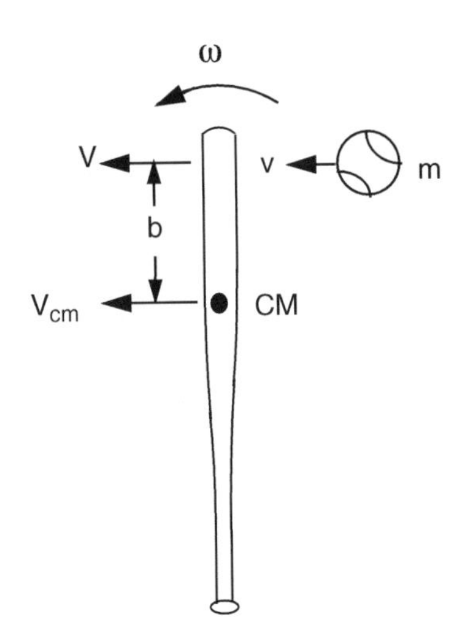

Fig. 10.2 A ball impacting on a stationary bat, at distance b from the bat CM, will cause the bat to rotate. The speed and acceleration of the impact point is greater than that for an impact at the CM, so the effective mass at the impact point is less than the mass of the whole bat

The effective mass of the bat at the impact point varies along the bat according to (10.5). M_e is a maximum at the bat CM where $b = 0$ and then $M_e = M$. The ball therefore "sees" the whole mass of the bat when it collides at the CM. The bat does not rotate at all in that case and all points on the bat will recoil at the same speed as the CM. The value of I_{CM} for a baseball bat is approximately $ML^2/12$ where L is the length of the bat. That is, it is about the same as that for a uniform cylindrical rod of mass M and length L. In practice, I_{CM} for a bat is about 12% smaller, but if we take the uniform rod as a guide, then (10.5) becomes

$$M_e \approx \frac{M}{1 + 12(b/L)^2}$$

M_e is a minimum at the very tip of the bat, where b is about $L/3$, in which case M_e is about $M/2$ (or about $M/2.3$). For an impact at the tip, the tip will recoil at about twice the speed of the CM due to rotation of the bat, and the ball will behave as if it collided with only half the mass of the bat. The ball will not bounce as well. Furthermore, an impact at the tip causes the bat to vibrate strongly, so the COR at the tip is less than it is further along the barrel. The end result is that the bounce factor near the tip is close to zero for most bats, and is sometimes even negative. If the bounce factor is negative and if the bat is initially at rest then the ball will continue moving forward at reduced speed after it strikes the bat, rather than bouncing backward.

Figure 10.3 shows M_e as a function of b for two different 33-in. bats, one being a 31 oz wood bat (Slugger R161) and one being a 30-oz aluminum bat (Easton BK7).

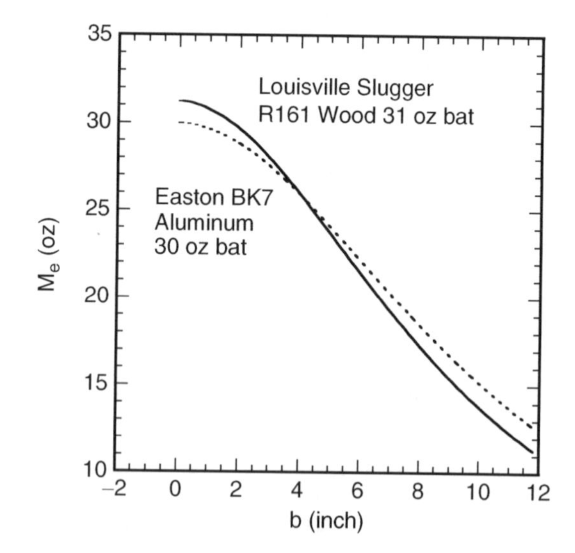

Fig. 10.3 The effective mass M_e for two different baseball bats vs. the impact distance from the center of mass of each bat. The tip of the wood bat was 10.9 in. from its CM. The tip of the aluminum bat (*dashed line*) was 12.3 in. from its CM

M_e was calculated using (10.5). For the R161, $I_{CM} = 2{,}460\,\text{oz in.}^2$. For the BK7, $I_{CM} = 3{,}060\,\text{oz in.}^2$. The CM of the wood bat was located 10.9 in. from the barrel end. The CM of the aluminum bat was located 12.3 in. from the barrel end, or 1.4 in. further back than the wood bat. It might appear in Fig. 10.3 that the slightly lighter aluminum bat is heavier at its the tip than the wood bat, but in fact it was about the same, given the 1.4 in. difference in the location of the CM.

10.5 Bat and Ball Collisions

The collision of a bat and ball is shown in Fig. 10.4. It appears at first sight to be a much more complicated situation than the simple head-on collision between two balls shown in Fig. 10.1. The collision in Fig. 10.4 is not a head-on collision since the ball can impact anywhere along the barrel. An added complication in Fig. 10.4 is that the bat is rotating before the collision and the rotation rate changes during the collision. Nevertheless, the collision can be described by exactly the same equations as those for two colliding balls. For the situation shown in Fig. 10.4, we show in Appendix 10.3 that the outgoing ball speed v_2 is related to the bat speed and the incoming ball speed by

$$M_e V_1 - m v_1 = M_e V_2 + m v_2, \tag{10.6}$$

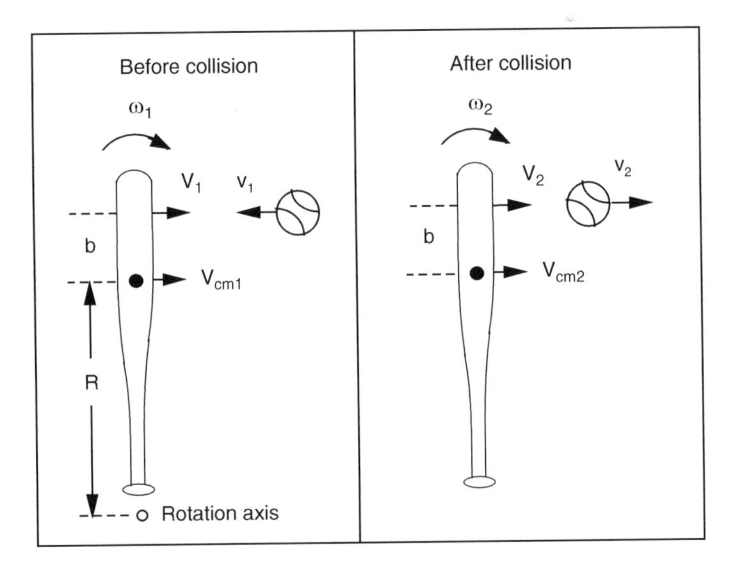

Fig. 10.4 A bat swung toward an incoming ball at angular velocity ω_1. After the collision, the bat rotates at angular velocity ω_2. The *black dot* indicates the position of the center of mass of the bat

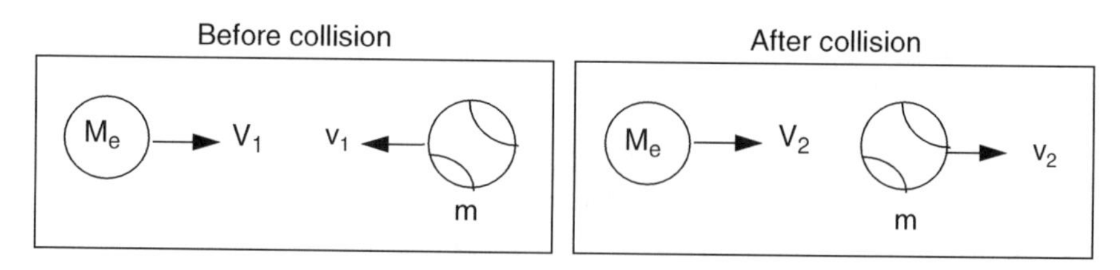

Fig. 10.5 The collision between a bat and a ball is equivalent to one between an object of mass M_e and a ball of mass m

where the effective mass M_e is given by

$$\frac{1}{M_e} = \frac{1}{M} + \frac{b^2}{I_{\mathrm{CM}}} \tag{10.7}$$

which is just a re-arranged version of (10.5). The collision between a bat and a ball is therefore equivalent to one between an object of mass M_e and a ball of mass m, as indicated in Fig. 10.5. Equation (10.6) indicates simply that the total momentum before the collision is equal to the total momentum after the collision. The added complications of the bat and ball collision, including bat rotation and translation (but not vibration), are conveniently collected together in (10.7). Energy losses due to bat vibrations affect the coefficient of restitution but have no effect on the equivalent mass of the bat or on conservation of linear or angular momentum.

The coefficient of restitution, e, for the collision in Fig. 10.4 or 10.5 is given by

$$e = \frac{(v_2 - V_2)}{(v_1 + V_1)}$$

Using this expression for e to eliminate V_2 in (10.6) gives the result that

$$v_2 = \left(\frac{1+e}{1+m/M_e}\right) V_1 + \left(\frac{e - m/M_e}{1+m/M_e}\right) v_1, \tag{10.8}$$

which is essentially the same as (10.1) corrected for the fact that the effective mass of the bat at the impact point is M_e rather than the actual mass of the bat, M. As before, we can write this much more simply as

$$v_2 = (1+q)V_1 + qv_1, \tag{10.9}$$

where the bounce factor q is given by

$$q = \frac{e - m/M_e}{1+m/M_e} \tag{10.10}$$

The outgoing ball speed v_2 therefore depends on (a) the speed v_1 of the incoming ball, (b) the speed V_1 at which the bat approaches the ball, and (c) the bounce factor q. The pitcher determines v_1, while V_1 is determined by the strength of the batter, the swing weight of the bat, and by the speed at which the batter wants to swing the bat. The mass (or "inertial") and elastic properties of the bat and the ball together determine q.

The quantity q is a measure of the intrinsic power of a bat, and it depends on both its effective mass, and the coefficient of restitution for the bat–ball collision. We describe in the following sections the factors that are involved in determining the effective mass and the COR. Effective mass depends mainly on the swing weight of the bat (a property of the bat) and the impact point along the barrel (a matter of skill or luck) while the COR depends, amongst other things, on the relative stiffness of the bat and the ball.

10.6 Ball Speed Calculations

A few examples will illustrate the significance of the factors that determine the outgoing ball speed. We would like to know what happens to the ball speed if we increase each of the factors in (10.9) and (10.10) by say 10%. Will the result be an increase in ball speed by 10% or will the effect be bigger or smaller than 10%? To find out, consider a typical example from baseball where the ball approaches the batter at a speed $v_1 = 80$ mph, the batter swings the bat at $V_1 = 60$ mph, the ball mass $m = 5$ oz, the effective mass of the bat at the impact point is $M_e = 20$ oz and the COR is $e = 0.5$. Then $q = 0.20$ so the batted ball speed $v_2 = 1.2 \times 60 + 0.2 \times 80 = 88$ mph. Increasing each of these factors by 10%, one at a time, gives the results shown in Table 10.1:

Table 10.1 Typical batted ball speeds in baseball

m (oz)	M_e (oz)	e	v_1 (mph)	V_1 (mph)	q	v_2 (mph)	Increase (%)
5	20	0.5	80	60	0.200	88.0	0
5.5	20	0.5	80	60	0.176	84.7	−3.7
5	**22**	0.5	80	60	0.222	91.1	+3.5
5	20	**0.55**	80	60	0.240	93.6	+6.4
5	20	0.5	**88**	60	0.200	89.6	+1.8
5	20	0.5	80	**66**	0.200	95.2	+8.2
5	**22**	**0.55**	**88**	**66**	0.263	106.5	+21

Table 10.2 Typical batted ball speeds in men's slow pitch softball

m (oz)	M_e (oz)	e	v_1 (mph)	V_1 (mph)	q	v_2 (mph)	Increase (%)
6.75	18	0.5	25	80	0.091	89.5	0
7.42	18	0.5	25	80	0.062	86.5	−3.4
6.75	**19.8**	0.5	25	80	0.119	92.4	+3.2
6.75	18	**0.55**	25	80	0.127	93.4	+4.3
6.75	18	0.5	**27.5**	80	0.091	89.8	+0.2
6.75	18	0.5	25	**88**	0.091	98.3	+9.7
6.75	**19.8**	**0.55**	**27.5**	**88**	0.156	106.0	+18.4

Table 10.3 Typical batted ball speeds in women's fast pitch softball

m (oz)	M_e (oz)	e	v_1 (mph)	V_1 (mph)	q	v_2 (mph)	Increase (%)
6.75	15	0.5	60	60	0.034	64.1	0
7.42	15	0.5	60	60	0.004	60.4	−5.8
6.75	**16.5**	0.5	60	60	0.064	67.7	+5.6
6.75	15	**0.55**	60	60	0.069	68.3	+6.4
6.75	15	0.5	**66**	60	0.034	64.3	+0.3
6.75	15	0.5	60	**66**	0.034	70.3	+9.7
6.75	**16.5**	**0.55**	**66**	**66**	0.100	79.2	+23.5

Each change is shown by a bold number, and the percentage increase in batted ball speed is shown in the last column. For example, increasing the ball mass by 10% from 5 to 5.5 oz changes v_2 from 88 to 84.7 mph, a decrease of 3.7%. The biggest single effect is to increase the swing speed of the bat. The next most important effect is to increase the COR. The last row of numbers shows that the batted ball speed can be increased by 21% by increasing all of the various factors (apart from the ball mass) by 10%.

Similar calculations are shown in Table 10.2 for slow pitch softball and in Table 10.3 for fast pitch softball. A 12 in. softball has a mass about 6.75 oz. Slow pitch bats weigh about 28 oz but the effective mass of the bat at the impact point is about 18 oz, depending on exactly where the batter strikes the ball (as well as on other factors such as the swing weight of the bat). Fast pitch bats weigh about 23 oz and the effective mass of the bat at the impact point is typically about 15 oz. The typical bat and ball speeds in the tables were measured by Lloyd Smith [4]. In all three tables, the biggest effect results from increasing the swing speed of the bat, and the smallest effect arises from the 10% increase in pitch speed.

10.7 Coefficient of Restitution, e

The COR for a baseball or a softball bouncing vertically off a solid horizontal surface is easy to measure. It is simply a matter of measuring the incident speed v_1 and the bounce speed v_2 and taking the ratio $e = v_2/v_1$. If a ball is dropped onto a solid surface such as a concrete driveway or a solid wood floor, then $e = v_2/v_1 = \sqrt{h_2/h_1}$, where h_2 is the bounce height and h_1 is the drop height. The answer will be about 0.6 at low ball speeds or low drop heights, up to a few feet. If the ball is incident on the driveway at 100 mph then the answer will be about 0.45 since a larger fraction of the initial kinetic energy is lost when the ball squashes more.

It is more difficult to measure the COR for a baseball bouncing off a bat, since it is then necessary to measure or estimate the speed of the bat as well as the speed of the ball, and it is also necessary to measure the impact point on the bat. If the speed of the bat is not measured, then some reliable estimate of its speed is needed to calculate the COR. It doesn't matter whether the ball is initially at rest or the bat is initially at rest or whether the bat and the ball are both moving toward each other. The answer will be the same if the initial relative speed is the same. If the bat is initially at rest, and if the impact point on the bat recoils at speed V_2 then $e = (v_2 + V_2)/v_1$, where v_1 is the incident speed of the ball and v_2 is the exit speed of the ball. The answer will depend on the impact point on the bat, and it won't necessarily be the same as when the ball impacts on a slab of concrete or wood floor. When a ball impacts on a bat, some of the kinetic energy will be lost to bat vibrations, unless the ball impacts at the so-called vibration node (the sweet spot). In addition, the trampoline effect in hollow bats will affect the result.

Low speed results are easy to obtain, by suspending a bat horizontally using two lengths of string, by suspending the ball as a pendulum bob, and by filming the bounce off the bat with a video camera. That way, the impact point can be selected and measured quite accurately and the experiment can safely be done indoors. High speed measurements are more difficult, since it needs a high speed ball launcher, safety precautions need to be taken to avoid possible damage, and the ball can easily hit the edge of the bat rather than a point along the axis. If the ball does strike the edge of the bat then the ball will bounce off the bat at a large angle.

The most interesting result is that the bounce speed of a ball off a bat does not depend on whether the bat is suspended freely by a length of string or whether the bat is hand held or whether the handle is clamped in a vice. By clamping the handle in a solid vice, the bat is hard to bend or rotate by hand and feels quite solid. If the bat is suspended by string then it is easy to rotate the bat with one finger. Consequently, it would be reasonable to expect that a ball will bounce much better off a clamped bat than off a freely supported bat. This is not the case. The bounce is the same, at least for impacts along most of the length of the barrel. However, it is found that with some bats, the ball bounces best off the tip of the barrel when the handle is clamped in a vice. The reason is that the ball can bounce twice if the handle is clamped. After the first bounce, the ball comes almost to a stop, setting the bat into a strong vibration. The vibrating bat then strikes the ball and the ball bounces clear. Double bounces can be avoided by striking the bat 3 in or more from the tip, in which case the ball bounces clear after the first bounce.

When a ball bounces off the barrel of a bat, the barrel bends slightly and a bending wave travels toward both ends of the bat. The wave reaches the handle after a delay of about 1 ms, then reflects off the handle and heads back to the impact point. However, the ball bounces clear before the bending wave gets back to the impact

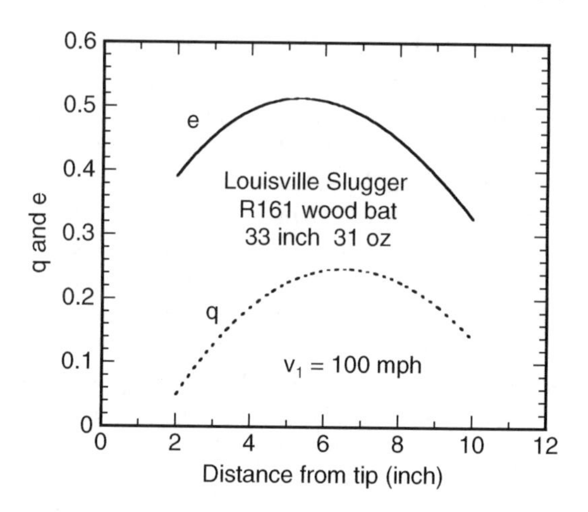

Fig. 10.6 Typical profiles of the bounce factor q and the COR, e, versus impact distance from the tip of the barrel for a wood bat

point, so the ball has no way of knowing whether the handle was held in a vice or by a human hand or whether it was completely free. The bounce speed is therefore unaffected by the clamping method. The bounce would be affected if the ball bounced off a point near the handle or if the ball was a basketball and spent a long time in contact with the bat, then the reflected wave would influence the bounce speed. After the ball bounces clear, the bending wave travels at high speed up and down the bat about 170 times each second, causing the bat to vibrate at about 170 Hz. The wave travels at a speed of about $2 \times 33 \times 170 = 11{,}220\,\text{in. s}^{-1} = 638\,\text{mph}$, but this is too slow to affect the bounce of the ball. Higher frequency components of the bending wave travel faster along the bat but most of the bending energy ends up in the low frequency, 170 Hz component and in rotation and translation of the bat if the bat is free, or hand-held, or is free to pivot about an axis in the handle.

Measurements of the COR for a wood bat are shown in Fig. 10.6 for a ball impacting at 100 mph at various points along the barrel. To obtain this data, the bat was pivoted at a point 6 in. from the knob and struck at various points along the barrel with a ball incident at $v_1 = 100$ mph. The bounce speed v_2 of the ball was measured to calculate $q = v_2/v_1$, and then e was calculated using (10.7) and (10.10). The results in this case show that e at a point 6 in. from the tip is essentially the same as the COR measured for an impact on a solid surface, but e decreases away from this point due to the extra energy lost in bat vibrations. The COR between a bat and ball is not a fixed number, but varies along the bat.

10.8 What Determines the Bounce Factor?

It is interesting to calculate the bounce factor from (10.10) to see how it depends on e and on the effective mass of a bat. For wood bats the value of e near the sweet spot is typically in the range 0.47–0.50, and for hollow bats e can vary from about 0.48 to about 0.55 or even higher near the sweet spot. Near the tip of the barrel, e is smaller, and so is the effective mass of the bat. We present calculations below using $e = 0.4$ or $e = 0.5$ to show the difference between an impact near the sweet spot and a few inches away from the sweet spot. Taking a value of ball mass $m = 0.145\,\text{kg}$ (or 5.1 oz) then gives a value for q of

$$q = \frac{e - 0.145/M_e}{1 + 0.145/M_e} \quad \text{or} \quad q = \frac{e - 5.1/M_e}{1 + 5.1/M_e} \tag{10.11}$$

depending on whether we measure M_e in kg or oz, where

$$M_e = \frac{M}{1 + Mb^2/I_{\text{CM}}} \tag{10.12}$$

Note that b can be measured in m or inch, and I_{CM} can be measured in kg m^2 or in oz in.^{2}, but we need to be consistent here and can't mix the units when using these formulas.

Figure 10.7 shows q as a function of M_e for a collision where e is either 0.5 or 0.4. As one would expect, light balls bounce better than heavy balls and balls bounce best off a bat with a large value of M_e and with a large value of e. If energy losses increase due to increased vibration of the bat then e decreases and the ball does not bounce as well. If a bat has a low value of M_e, or if M_e is low because the ball bounces near the tip of the bat, then the incoming ball gives more of its energy to the bat and therefore, bounces off the bat at a lower speed. The bounce factor can be negative if M_e is small enough, meaning that a ball incident on a stationary bat will continue in the same direction after the bounce, rather than bouncing back off the bat. Such a result is obtained for impacts near the tip of light bats commonly used in fast pitch softball and also with relatively light baseball bats.

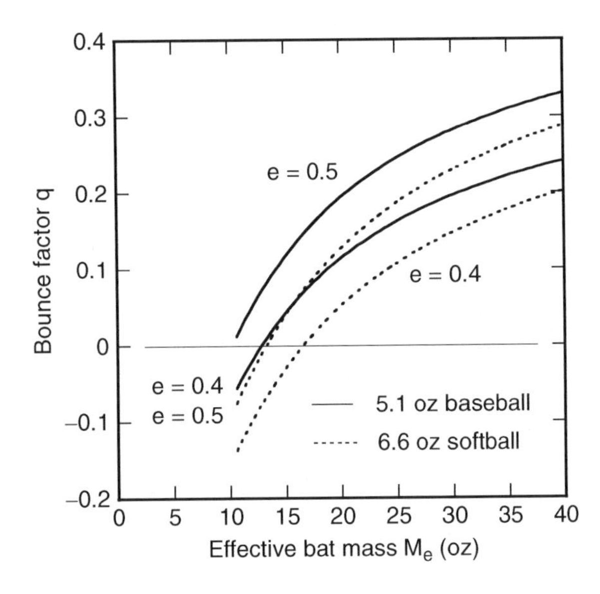

Fig. 10.7 The bounce factor q depends on the mass of the ball, the effective mass, M_e, of the bat and the COR, e. For a collision in the sweet spot region, e is typically about 0.5, but for an impact closer to the tip of a bat e can drop to about 0.4 or less due to vibration energy losses in the bat. Some hollow bats can have a value of e as high as 0.55 near the sweet spot, for a well struck ball, due to the trampoline effect

Baseballs and softballs do not bounce very well off any bat. A typical bounce off a baseball bat is one with q about 0.2, meaning that a ball incident at say 100 mph on a stationary bat will bounce off the bat at only about 20 mph. An obvious consequence is that the batter must do most of the work himself if he wants to return the ball at high speed. He cannot simply dangle the bat in front of the ball and expect the ball to bounce off the bat at 100 mph. A batter can use a fungo bat or hit a ball off a tee (with zero pitch speed) at relatively high speed, but cannot bunt a ball (at zero bat speed) at anything like the same high speed.

10.9 Effective Mass vs. Swing Weight

Equation (10.12) shows that the effective mass of a bat depends on three separate factors, namely its actual mass M, the impact distance b, and I_{CM}. In fact, there is a fourth quantity involved here, namely the location of the CM, since we need to know where it is to calculate the impact distance. Since we are interested mostly in the performance of a bat where it performs best, we will look at this region in more detail. It turns out that a remarkable simplification arises in this case, where the effective mass of the bat at any given impact point depends on only one factor rather than the several factors just listed. Furthermore, that factor is exactly the same factor that determines how fast a batter can swing the bat. In both cases, the factor involved is the swing weight or MOI of the bat [5].

The quantity I_{CM} in our calculations thus far refers to the MOI of the bat when it rotates about an axis through its center of mass (CM). When people measure the MOI of a bat, they don't rotate the bat through its CM. They rotate it about an axis 6 in. from the end of the knob. There is nothing special about this 6 in. distance. It just happens to be a convenient location, and it is the location specified in the ASTM standards document F2398 so that everyone can measure the MOI using the same method and the same rotation axis. In theory, the actual rotation axis used to measure the MOI is irrelevant since there is a simple formula, given by the parallel axis theorem, that says

$$I_A = I_{\text{CM}} + Mh^2, \tag{10.13}$$

where I_A is the MOI of a bat of mass M about an axis located a distance h from the CM of the bat, and where I_{CM} is the MOI about a parallel axis passing through the CM. If we measure I_A for rotation about any axis, and if we measure M and h, then we can calculate the MOI about any other axis we like, including the axis 6 in. from the knob. If I_6 is the value of the MOI for an axis 6 in. from the knob, then we show in Appendix 10.4 that the effective mass of the bat at any distance d from the tip of the barrel is given to a very good approximation by

$$M_e = \frac{I_6}{(L-6-d)^2}, \tag{10.14}$$

where L is the length of the bat in inches, d is also measured in inches, M_e is measured in oz and I_6 is measured in oz in.^{2}. The quantity $(L - 6 - d)$ is the distance from the 6-in. axis to the impact point.

To see how good an approximation this is, the effective mass and the bounce factor at a point 6 in from the end of the barrel was calculated for 320 different bats all 33 in. long, using (10.11) and (10.12) and the known values of M, B (balance point) and I_{CM} for each bat. The bats ranged in weight from 21 oz to 34 oz and had a very wide range of weight distributions, as described in Appendix 10.5. Some straight, uniform rods were also included, where the weight of the "barrel" was the same as the weight of the handle. The results are shown in Figs. 10.8 and 10.9. Figure 10.8 shows the effective mass vs. the weight of the bat, and Fig. 10.9 shows the effective mass and bounce factor vs. the swing weight, I_6 (for an axis 6 in. from the knob).

In general, a ball bounces better off a heavy bat than a light bat but this is not always the case since the bounce also depends on the weight distribution. The ball bounces best if the handle is very light and most of the weight is in the barrel. Figure 10.9a shows clearly that the effective mass depends only on the swing weight, at least for an impact 6 in. from the tip of the barrel. M_e is very close to the value predicted by (10.14) for all of the bats, even the uniform rods. Our approximation is not as reliable at impact points near the tip of the barrel or near the CM of the bat, but is very good at impact points in the range $4 < d < 8$ in. Figure 10.9b shows the resulting bounce factor, q for all 320 bats, as calculated from (10.11), for cases where e is either 0.5 or 0.45. Since q is a measure of the inbuilt power of a bat, it is clear that bats with a large swing weight have more inbuilt power than bats with a low swing weight. The only problem is, bats with a large swing weight can't be swung as fast as bats with a low swing weight, so we can't yet say whether bats with a large swing weight are more powerful or not. We will address that issue in the next chapter.

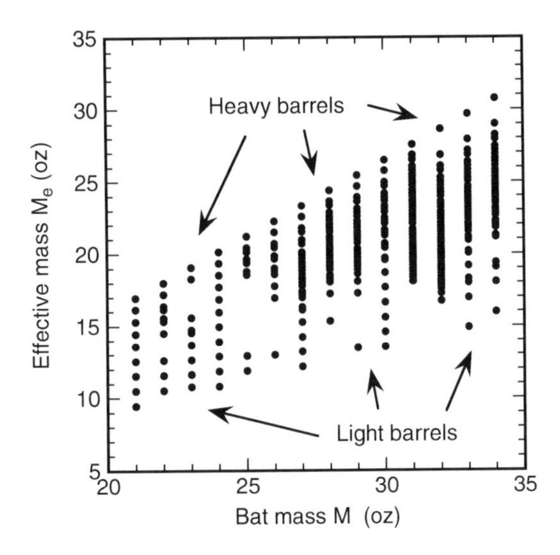

Fig. 10.8 Effective mass M_e vs. bat mass M for 320 different bats, all 33 in., for an impact 6 in. from the tip of the barrel. For any given bat mass M, M_e is largest when the barrel is much heavier than the handle, and smallest when the handle is as heavy as the barrel (i.e., a uniform rod)

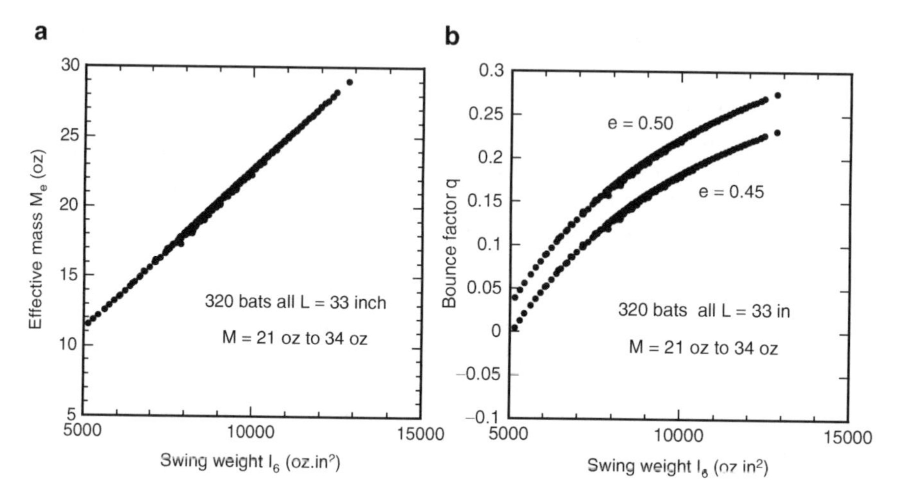

Fig. 10.9 (**a**) Effective mass M_e and (**b**) Bounce factor q vs. swing weight I_6 for 320 different bats, all 33 in., for an impact 6 in. from the tip of the barrel. In (**b**), $e = 0.45$ or 0.50. For wood bats, e varies typically from 0.47 to 0.50. Aluminum and composite bats have a wider range of e values due to the trampoline effect, with e typically in the range 0.48–0.55 at a point 6 in from the tip of the barrel

Equation (10.14) is valid for bats of any length, not just 33 in. For example, consider a bat of length 34 in. If a ball impacts at a distance of 6 in. from the end of the barrel, and if the bat rotates about an axis 6 in. from the knob, then the distance between these two points is still $L - 12$ in. and the formula is exactly correct. Similarly, if the actual rotation axis is an inch or two away from the 6 in. axis then the formula is still a very good approximation even though it is not exact. However, if a 33-in. bat and a 34-in. bat have the same swing weight, I_6, then the inbuilt power of the 33 in. bat would seem to be slightly larger since it has a slightly larger M_e. In both cases, the ball impacts 6 in. from the tip of the bat, but it impacts the 34-in. bat 1 in. further away from the knob. We have a problem here in comparing the inbuilt power of the two bats since the ball is effectively impacting in different spots. To resolve this problem properly we would need to consider the variation of M_e along the barrel of each bat, at different impact points, to see which bat had the largest bounce factor.

10.10 Summary

The main conclusions of this chapter are summarized in Fig. 10.10.

The effective mass of a bat, M_e, is less than its actual mass since the ball collides with only part of the bat. M_e can be described as the "hitting weight" of the bat, to distinguish it from the actual weight and the swing weight. The hitting weight

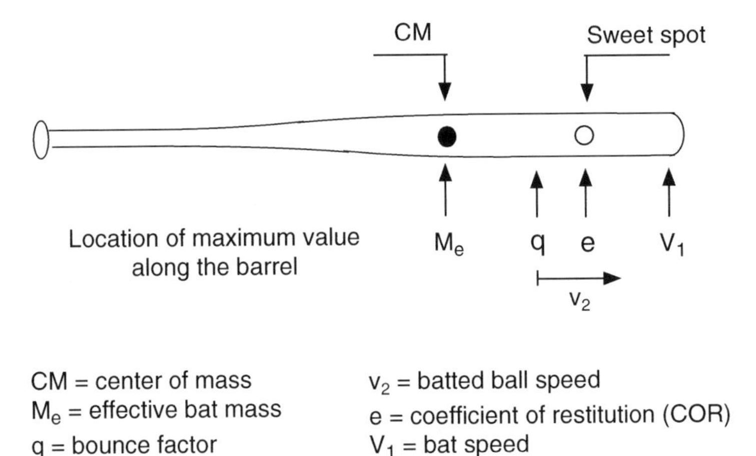

Fig. 10.10 Location along the barrel where the batted ball speed factors are a maximum

depends on the impact point, being a maximum at the bat center of mass and a minimum at the tip of the bat. For an object such as a hammer or a golf club, the hitting weight is a maximum at the tip since that is where almost all of the mass is located. A bat is different since the useful hitting area extends over the whole length of the barrel and since the center of mass is located well back from the tip.

The coefficient of restitution, e, is a maximum at the sweet spot where bat vibrations are a minimum, since e is a measure of the energy loss in the bat and the ball. The maximum value of e is typically about 0.5 for most wood bats. e is a minimum at the tip of the bat and is typically about 0.3 near the tip of a bat.

The bounce factor, q, describes how well the ball bounces off the bat. The bounce factor is a measure of the inbuilt power of the bat and it depends primarily on the swing weight of the bat and the actual point of impact on the bat. q is maximum between the bat CM and the sweet spot, since the ball bounces best off heavy parts of the bat but the bounce also depends strongly on the local value of the COR (e). The maximum value of q is typically about 0.25. q is a minimum at the tip of the bat and can even be zero or slightly negative at the tip of a light bat.

The bat speed, V_1 is a maximum at the tip of the bat, and decreases linearly to a minimum at the knob end.

The location on the bat where the batted ball speed is a maximum depends on the bat speed, and it moves toward the tip of the bat as the bat speed increases. If the bat is not swung at all, and is struck by a ball incident at speed v_1, then the outgoing ball speed is given by $v_2 = qv_1$. In that case, v_2 is a maximum at the point where q is a maximum. At high bat speeds, v_2 is a maximum closer to the tip of the bat since the bat travels fastest at the tip. However, the tip itself is the spot where M_e is a minimum, and so is e, since an impact at the tip results in strong vibrations. Consequently, v_2 is a maximum near the sweet spot when the bat is swung at high speed.

Appendix 10.1 Derivation of (10.1)

In Fig. 10.1, the total momentum before the collision is $MV_1 - mv_1$ and the relative speed is $v_1 + V_1$. If the bat and ball collide head-on, then they will continue to move along the same straight line after the collision. As shown in Fig. 10.1, the ball has speed v_2 after the collision and the bat has speed V_2. The total momentum after the collision is $MV_2 + mv_2$ and the relative speed after the collision is $v_2 - V_2$. Applying the two collision rules to this collision, we have

$$MV_1 - mv_1 = MV_2 + mv_2 \tag{10.15}$$

and

$$e = \frac{(v_2 - V_2)}{(v_1 + V_1)}, \tag{10.16}$$

where e is the symbol that is always used for the COR. If we know the masses of the bat and the ball, their initial speeds and the value of e then we can solve these two equations to find the two speeds after the collision. If the ball or the bat was at rest before the collision then the algebra would be a bit easier but as all baseball and softball fans know, the bat and the ball are headed toward each other before the collision. To calculate the outgoing ball speed v_2 we can eliminate V_2 using (10.15) which gives

$$V_2 = V_1 - \frac{m(v_1 + v_2)}{M} \tag{10.17}$$

If we substitute (10.17) into (10.16), then we find that

$$e(v_1 + V_1) = v_2 - V_1 + \frac{m(v_1 + v_2)}{M}$$

which can be rearranged, by moving the two v_2 terms to the left side of the equation, and collecting the v_1 and V_1 terms on the right side of the equation, to give (10.1).

Appendix 10.2 Derivation of (10.5)

Let M_e be the effective or equivalent mass of the bat at the impact point. A force F acting at the impact point will cause that point to accelerate according to the relation $F = M_e \mathrm{d}V/\mathrm{d}t$ and it will cause the CM of the bat to accelerate according to the relation $F = M\mathrm{d}V_{\mathrm{CM}}/\mathrm{d}t$. The whole bat rotates as a result of the torque Fb acting about the CM. If ω is the angular speed of the bat then $Fb = I_{\mathrm{CM}}\mathrm{d}\omega/\mathrm{d}t$, where I_{CM} is the MOI of the bat about an axis through the CM. Consequently,

$$F = M\frac{\mathrm{d}V_{\mathrm{CM}}}{\mathrm{d}t} = M_e\frac{\mathrm{d}V}{\mathrm{d}t} = \frac{I_{\mathrm{CM}}}{b}\frac{\mathrm{d}\omega}{\mathrm{d}t} \tag{10.18}$$

The impact point rotates at a speed $b\omega$ with respect to the bat CM, so

$$V = V_{CM} + b\omega \tag{10.19}$$

indicating that

$$\frac{dV}{dt} = \frac{dV_{CM}}{dt} + b\frac{d\omega}{dt} \tag{10.20}$$

From (10.18) and (10.20) we find that

$$\frac{F}{M_e} = \frac{F}{M} + \frac{b^2 F}{I_{CM}}$$

so

$$\frac{1}{M_e} = \frac{1}{M} + \frac{b^2}{I_{CM}}$$

which can be rearranged to give

$$M_e = \frac{M}{1 + Mb^2/I_{CM}}$$

Appendix 10.3 Derivation of (10.6)

A bat and ball collision is shown in Fig. 10.4. The bat is swung about an axis near the end of the handle at angular velocity ω_1, with the result that the center of mass (CM) of the bat approaches the ball at speed V_{CM1} and the impact point on the bat approaches the ball at speed V_1. If the distance from the axis to the CM is R, then $V_{CM1} = R\omega_1$. If the distance from the axis to the impact point is $R + b$ then $V_1 = (R + b)\omega_1 = V_{CM1} + b\omega_1$. Similarly, $V_2 = V_{CM2} + b\omega_2$.

If v is the velocity of the ball at any given time, V_{CM} is the velocity of the CM, and V is the velocity of the impact point, then the force F on the ball and the bat is given by $F = m\mathrm{d}v/\mathrm{d}t = M\mathrm{d}V_{CM}/\mathrm{d}t$ where m is the mass of the ball and M is the mass of the bat. These two forces are actually equal and opposite but both the forces can be regarded as being positive if the bat and ball are traveling in opposite directions before the collision, as shown in Fig. 10.4, and if we assume that v and V are both positive quantities. The force F acting at the impact point causes the velocity V to change according to the relation $F = M_e\mathrm{d}V/\mathrm{d}t$, where M_e is the effective mass of the bat at the impact point. The force F also causes the bat to change its angular velocity according to the relation $Fb = I_{CM}\mathrm{d}\omega/\mathrm{d}t$. The force F here is much larger than the force of the hands on the bat during the collision, although the hands do affect the motion of the bat before and after the collision.

By integrating over the collision time, we find that

$$\int F\,\mathrm{d}t = m(v_1 + v_2) = M(V_{\mathrm{CM1}} - V_{\mathrm{CM2}}) = M_e(V_1 - V_2) \tag{10.21}$$

and

$$b \int F\,\mathrm{d}t = I_{\mathrm{CM}}(\omega_1 - \omega_2) \tag{10.22}$$

Note that the ball reverses direction after the collision, but we have assumed for convenience that v_2 is a positive number, so the change in ball speed is $v_1 + v_2$. If it was incident at 80 mph and exited at 80 mph then the change is not zero but 160 mph. Note also that (10.21) is an expression showing that momentum is conserved during the collision since the impulse $\int F\,\mathrm{d}t$ given to the ball is the same as the impulse given to the bat.

Given that $V_1 = V_{\mathrm{CM1}} + b\omega_1$ and $V_2 = V_{\mathrm{CM2}} + b\omega_2$, we can also write (10.21) as

$$\int F\,\mathrm{d}t = M_e(V_{\mathrm{CM1}} + b\omega_1 - V_{\mathrm{CM2}} - b\omega_2) \tag{10.23}$$

By combining the results in (10.21)–(10.23) we find that

$$\int F\,\mathrm{d}t = M_e\left(\frac{\int F\,\mathrm{d}t}{M} + \frac{b^2 \int F\,\mathrm{d}t}{I_{\mathrm{CM}}}\right) \tag{10.24}$$

giving

$$\frac{1}{M_e} = \frac{1}{M} + \frac{b^2}{I_{\mathrm{CM}}} \tag{10.25}$$

as we found previously when deriving (10.5). A bat and ball collision is therefore equivalent to one between an object of mass M_e and a ball of mass m. We can rearrange (10.21) as

$$M_e V_1 - m v_1 = M_e V_2 + m v_2 \tag{10.26}$$

which describes conservation of momentum in the situation shown in Fig. 10.5.

Appendix 10.4 Derivation of (10.14)

Consider the geometry shown in Fig. 10.11, and suppose that a stationary bat is struck at some point along the barrel at a distance d from the tip. If the bat is freely supported (not hand-held) then it will rotate about an axis near the other end, at some point in the handle. Technically, the impact point is known as the center of percussion for that particular rotation axis. It is easy to calculate the distance h between the rotation axis and the CM. From (10.18),

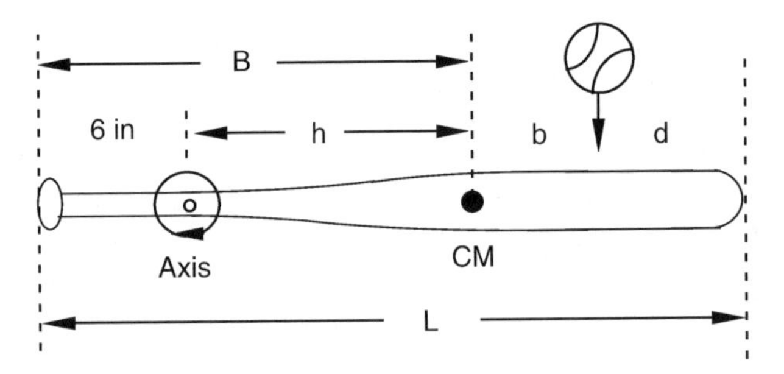

Fig. 10.11 A bat struck at a distance b from its CM rotates through an axis at a distance h on the other side of the CM. The location of the axis depends on the impact point but is typically about 6 in. from the knob. For that reason, the swing weight of a bat is always measured about an axis 6 in. from the knob

$$F = M\frac{\mathrm{d}V_{\mathrm{CM}}}{\mathrm{d}t} = \frac{I_{\mathrm{CM}}}{b}\frac{\mathrm{d}\omega}{\mathrm{d}t} \tag{10.27}$$

If the bat rotates at angular speed ω about the axis, then $V_{\mathrm{CM}} = h\omega$, so

$$\frac{\mathrm{d}V_{\mathrm{CM}}}{\mathrm{d}t} = h\frac{\mathrm{d}\omega}{\mathrm{d}t} \tag{10.28}$$

Substituting (10.28) into (10.27) then gives

$$h = \frac{I_{\mathrm{CM}}}{Mb}, \tag{10.29}$$

where b is the distance from the impact point to the CM. The MOI of the bat for rotation about this axis is $I_A = I_{\mathrm{CM}} + Mh^2$. The axis here is not necessarily 6 in. from the knob, but it will be reasonably close for most cases of interest, perhaps 1 or 2 in. away.

The effective mass of the bat at the impact point is derived in Appendix 10.2 and is given by

$$M_e = \frac{M}{1 + Mb^2/I_{\mathrm{CM}}} = \frac{MI_{\mathrm{CM}}}{I_{\mathrm{CM}} + Mb^2} \tag{10.30}$$

From (10.29), $I_{\mathrm{CM}} = Mbh$. Substituting this in (10.30), we find that $M_e = Mh/(b+h)$. We also find that $I_A = Mbh + Mh^2 = Mh(b+h)$, so

$$M_e = \frac{I_A}{(b+h)^2} \tag{10.31}$$

We see here that the effective mass of the bat depends on the swing weight I_A, and it also depends on $b + h$.

An alternative and slightly simpler derivation of (10.31) is as follows. Suppose we support the bat like a pendulum so it can swing freely about an axis located at distance h from the CM. If we then strike the bat with a ball somewhere along the barrel, the ball will exert a torque $F(h + b) = I_A \mathrm{d}\omega/\mathrm{d}t$ about this axis and the impact point will rotate at speed $V = (h + b)\omega$. Since $F = M_e \mathrm{d}V/\mathrm{d}t = M_e (h + b)\mathrm{d}\omega/\mathrm{d}t$, we find that $M_e = I_A/(h + b)^2$, as we found in (10.31).

If the rotation axis happens to be 6 in. from the knob then $b + h = L - 6 - d$, and we arrive at the result that

$$M_e = \frac{I_6}{(L - 6 - d)^2}, \tag{10.32}$$

where I_6 is the MOI for rotation about an axis 6 in. from the knob. The effective mass of a bat therefore depends on the impact distance d, the bat length L and the swing weight I_6. However, for bats of any given length, and for a given impact distance d, the effective mass of a bat depends only on its swing weight, as shown in Fig. 10.9a.

Suppose that the ball impacts somewhere along the barrel, and the bat actually rotates about an axis 7 in. from the knob rather than 6 in from the knob. The actual value of M_e is then given by $M_e = I_7/(L - 7 - d)^2$, where I_7 is the MOI for rotation about the axis 7 in from the knob. Since I_7 is slightly smaller that I_6 and $L - 7$ is slightly smaller than $L - 6$, the actual value of M_e is essentially the same as that given by (10.32). For example, consider a typical 33 in., 30 oz bat with the CM located 20 in. from the knob and with $I_{\mathrm{CM}} = 3{,}000$ oz in.^{2}. For this bat, $I_6 = 8{,}880$ oz in.^{2} and $I_7 = 8{,}070$ oz in.^{2}. At an impact point with $d = 6$ in., (10.32) gives $M_e = 20.14$ oz, while $I_7/(L - 7 - d)^2 = 20.17$ oz, which is only 0.2% larger. Equation (10.32), therefore, gives a very good estimate of the effective mass of a bat, regardless of the actual location of the rotation axis and regardless of the actual impact point.

Appendix 10.5 Three Section Bat

A very large range of bats can be modeled as shown in Fig. 10.12, assuming that the bat consists of three cylinders each of length $L = L_{\mathrm{bat}}/3$ where L_{bat} is the overall length of the bat. The mass of each cylinder is shown in Fig. 10.12 as M_1, M_2 and M_3, the total mass of the bat being $M = M_1 + M_2 + M_3$. The knob of the bat is located at $x = 0$, the tip is located a $x = 3L$ and the center of mass is located at

$$x = B = \frac{L}{2M}(M_1 + 3M_2 + 5M_3)$$

The MOI of a cylinder of mass M and length L, about a transverse axis through the center of the cylinder is given by $I_c = ML^2/12$. The MOI about a parallel axis

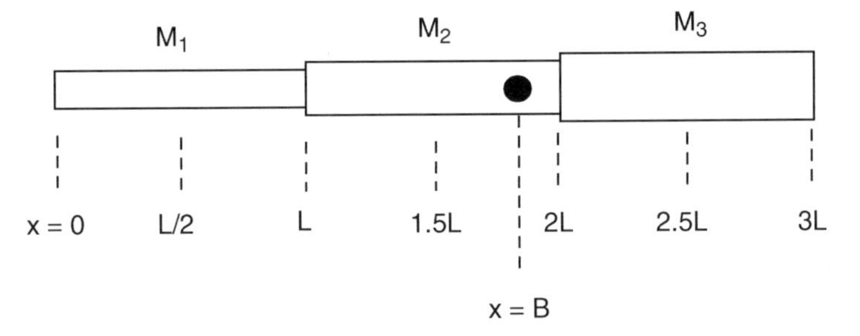

Fig. 10.12 A simple bat model consisting of three sections each of length L. The mass of each section increases from the handle end to the barrel end. The *black dot* at $x = B$ denotes the center of mass of the bat

located at a distance d from the center of the cylinder is given by $I_d = I_c + Md^2$. Using these expressions, the total MOI of the bat, about an axis at $x = 0$ is given by

$$I_o = L^2 \left(\frac{M_1}{3} + \frac{M_2}{12} + 2.25M_2 + \frac{M_3}{12} + 6.25M_3 \right)$$

and the MOI about the centre of mass is given by $I_{\text{CM}} = I_o - MB^2$. The bats in Figs. 10.8 and 10.9 were modeled by increasing M_1, M_2, and M_3 in one oz steps to determine the properties of 320 different bats ranging in weight from 21 oz to 34 oz.

References

1. R. Cross, The dead spot of a tennis racket. Am. J. Phys. **65**, 754–764 (1998)
2. R. Cross, Impact of a ball with a bat or racket. Am. J. Phys. **67**, 692–702 (1999)
3. A.M. Nathan, Dynamics of the baseball bat collision. Am. J. Phys. **68**, 979–990 (2000)
4. L. Smith, Progress in measuring the performance of baseball and softball bats. Sports Technol. **1**, 291–299 (2008)
5. R. Cross, A. Nathan, Performance vs moment of inertia of sporting implements. Sports Technol. **2**, 7–15 (2009)

Further Reading

L.L. Van Zandt, The dynamical theory of the baseball bat, Am. J. Phys. **60**, 172–181 (1992)

S. Redner, A billiard-theoretic approach to elementary one-dimensional elastic collisions, Am. J. Phys. **72**, 1492–1498 (2004)

R.L. Nicholls, B.C. Elliott, K. Miller, M. Koh, Bat kinematics in baseball: implications for ball exit velocity and player safety, J. Appl. Biomechan. **19**, 283–294 (2003)

J.M. Penrose, D.R. Hose, An impact analysis of a flexible bat using an iterative solver, J. Sports Sci. **17**(8), 677–682 (1999)

A.S. Weyrich, S.P. Messier, B.S. Ruhmann, Effects of bat composition, grip firmness and impact location on postimpact ball velocity, M.J. Berry, Med. Sci. Sports Exerc. **21**(2), 199–205 (1989)
S.S. Hughes, B.C. Lyons, J.J. Mayo, Effect of grip strength and grip strengthening exercises on instantaneous bat velocity of collegiate baseball players, J. Strength Condit. Res. **18**, 298–301 (2004)

Chapter 11
Bat Performance

The 2004 bat standard has a maximum batted ball speed (BBS) limit of 98 mph when tested according to the ASTM F2219 test method. Amateur Softball Association of America (ASA) Bat and ball certification program, current in 2010.

The BBCOR, as determined from an average of six (6) consecutive valid hits at the maximum BBCOR location described above, must not exceed 0.500.

– NCAA Standard for testing baseball bat performance, effective January 1, 2011

11.1 Introduction

Of all the technical issues that concern bat and ball sports, the one that is of most interest to most players concerns bat performance. Players would like to hit the ball as far as possible, using the most powerful bat that they can afford and that they are allowed to use. They would also like the bat to be as light as possible so that it is easy to swing and connect with the ball. Therein lies a dilemma. Heavy bats are more powerful, but they are harder to swing. Light bats are easy to swing but they lack the punch of a heavy bat.

Choosing a bat for slow pitch softball is relatively easy since the batter has plenty of time to swing the bat. The batter just needs to choose the heaviest bat that he or she feels comfortable with. Baseball and fast pitch softball players have a different problem since they have very little time to get the bat on the ball. As a result, most players prefer to use light bats since it makes it easier to hit the ball. Players are prepared to lose a few mph in batted ball speed if it helps them to swing the bat faster. For that reason, most of the bats made by manufacturers for NCAA baseball are minus 3 bats, since they are the lightest that players are allowed to use. That is, the bat weight is 3 oz less than the bat length in inches. A typical NCAA baseball bat is 34 in. and 31 oz. Fast pitch softball bats are even lighter, −10 and −12 bats being relatively common. That is, the bat weight is typically 10 or 12 oz less than the bat length in inches. A typical fast pitch bat is 34 in. and 24 oz. The difference Weight (oz) – Length (inch) is commonly known as the bat drop. A 30 oz/33 in. bat has a bat drop of −3, and a 31 oz/33 in. bat has a bat drop of −2.

Aluminum bats first became popular in the mid-1970s when it was discovered that they didn't break as easily as wood bats. Soon after, aluminum bats began to outperform wood bats, not only in terms of durability but also in terms of swing speed and batted ball speed. Almost all players, other than those in major and minor baseball leagues, switched to aluminum bats in the 1980s. Officials have been fighting a battle ever since to place reasonable, practical limits on bat performance.

R. Cross, *Physics of Baseball & Softball*, DOI 10.1007/978-1-4419-8113-4_11,

The word "performance" here generally refers to the power of a bat, as measured by the batted ball speed, but it can also refer to the speed of the bat itself. The two different ways of viewing "performance" are actually diametrically opposite. Maximum batted ball speed results when the batter uses a bat with a *large* swing weight. Maximum swing speed or bat speed results when the batter uses a bat with a *small* swing weight. Generally, players like to think in terms of actual weight rather than swing weight, and would say that light bats can be swung faster while heavy bats are more powerful. However, it is the swing weight rather than the actual weight that determines swing speed and power. The trampoline effect in hollow bats adds to that power, with the result that a low swing weight bat with a strong trampoline effect can outperform a higher swing weight bat with a weak trampoline effect.

"Swing weight" is a user-friendly term for the moment of inertia (MOI) of a bat, and is measured in oz in.^{2} or kg m^2. In this book, the notation I_6 is used to denote the swing weight since the standard physics symbol for MOI is I and since the swing weight of a bat is always quoted for an axis through the handle located 6 in. from the knob.

Manufacturers rarely advertise the swing weight of their bats, with the result that players are not generally familiar with the term swing weight or its significance. Some manufacturers might mention that their bats have a low swing weight for faster swing speed, or that their bats are evenly balanced for maximum swing speed or that a bat has a swing weight rating of 70, but none of that information is particularly useful. In fact, the statement implying maximum speed from an evenly balanced bat is not even close to being technically correct. For a start, a bat can't be evenly balanced. It has a definite balance point, at a certain distance from the knob, but it can't be balanced "evenly." The term has no meaning in a scientific sense, unless the manufacturer means that half the weight is on one side of the balance point and the other half is on the other side, but even that is not true. Even if the weight of the barrel is evenly distributed along the barrel, that doesn't mean that the bat will be easy to swing.

If a bat is advertised as being light or heavy, a player would definitely like to know the actual weight in oz. In the same way, if a bat has a low swing weight, the manufacturer should be quoting the actual value and the player should know what it is before he or she buys it. The balance point of a bat is also important but it just one of several factors that determine the swing weight of a bat.

In the late 1990s, most baseball and softball organizations began introducing regulations concerning the performance of aluminum bats in an attempt to control their swing speed and power or to make them more "wood like." For example, the National Collegiate Athletic Association (NCAA) baseball rules committee recommended in 1999 that three new rules be introduced to help limit the performance of aluminum baseball bats. Their objective was threefold: (1) to preserve the integrity of the game, (2) to maintain a balance between offense and defense, and (3) to minimize the risk of injury, especially to the pitcher. The committee decided that these three objectives could best be met by the introduction of three new rules: (1) The Minus 3 Rule: the mass of a bat (in oz) should be at least equal to the length of the bat (in inches) minus 3, (2) the bat diameter should be no larger than 2 5/8 in., and (3) the batted ball speed should not be greater than 94 mph when measured with a Baum hitting machine.

At the time, almost all 33 in. wood bats were 2 5/8 in. in diameter and weighed between 30 oz and 33 oz depending on the density of the wood and the exact shape of the bat. However, 33 in. aluminum bats were being made that were 2 3/4 in. in diameter and that were generally lighter, as low as 28 oz. Before 1999, a "Minus-5" rule ensured that 33-in. aluminum bats had to be at least 28 oz. Being lighter, and having a center of mass closer to the handle, these bats had a lower swing weight and were easier to swing. Combined with the larger diameter, it was easier for batters to connect with the ball and get a big hit. The first two rules were adopted (but not the 94 mph rule) to ensure that aluminum bats would be more like wood bats in having the same maximum diameter and that the weight of a 33-in. aluminum bat would be at least 30 oz, the same as the lighter versions of 33-in. wood bats.

The Baum hitting machine was used to swing a bat at 66 mph toward a ball pitched at 70-mph ball, giving batted ball speeds greater than 90 mph. The NCAA eventually settled on a maximum batted ball speed of 97 mph for non-wood bats so that their performance would be no better than the best wood bats, at least in these laboratory tests. The 97 mph limit corresponded to a bounce factor $q = 0.228$ for the best wood bats. This particular test was later refined [1] and led to the Maximum BESR rule, standing for Ball Exit Speed Ratio, where BESR $= q + 0.5$, q being the bounce factor defined in Chap. 9.

In a similar way, the ASA modified their 2000 standard for softball bats in 2004 so that the maximum batted ball speed would not exceed 98 mph when swung at a speed calculated from the swing weight of the bat. The performance of softball bats is measured in the laboratory by firing a ball at a speed of 110 mph onto a stationary bat to measure the bounce factor [2]. By combining the measured bounce factor with the calculated swing speed, the batted ball speed can be calculated using (11.3) in Chap. 10. The NCAA introduced an additional rule in 2002 to set a minimum value of the swing weight of bats of different length, so that aluminum bats would be even more like wood bats, both in terms of actual weight and swing weight. The minimum MOI set by the NCAA depended on bat length, varying up to 11,767 oz in.^{2} for a 36-in. bat and down to 5,407 oz in.^{2} for a 29-in. bat. For a 33-in. bat, the minimum MOI set by the NCAA was 8,538 oz in.^{2}, and for a 34-in. bat the minimum MOI was set at 9,530 oz in.^{2}. These values are shown by the curved lines in Fig. 11.1, together with typical values of 500 different wood and aluminum bats measured for the NCAA in 2006–2007. The minimum allowed MOI was set to a value less than the MOI of typical wood bats, to allow for the fact that the center of mass of an aluminum bat is closer to the handle and therefore has a lower MOI than a wood bat of the same length and weight. Aluminum bats were therefore allowed by the minimum MOI rule to have a slightly lower MOI than wood bats, allowing low MOI aluminum bats to be swung slightly faster. However, as shown in Fig. 11.1, many aluminum bats continued to be manufactured with a higher MOI than the allowed minimum, because that is what players preferred to use.

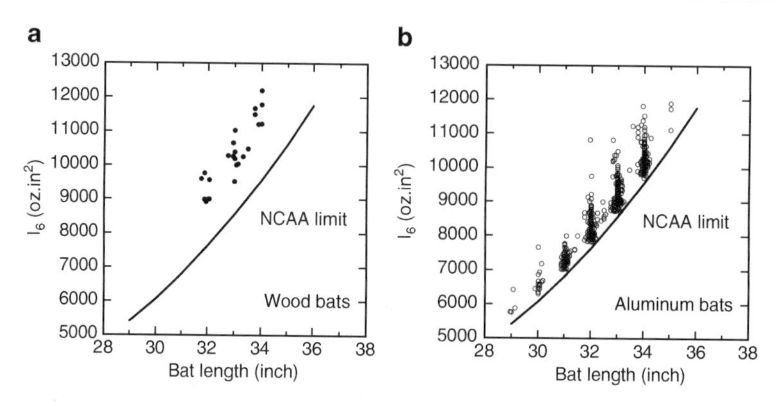

Fig. 11.1 Swingweight I_6 vs. bat length for a sample of 500 NCAA (**a**) wood and (**b**) aluminum bats approved in 2006-2007. The curved line shows the minimum value of I_6 allowed, vs. bat length

Bat performance continues to be a contentious issue, especially when it concerns player safety. Regulated non-wood bats now outperform wood bats by only a small margin. Nevertheless, some organizations have banned the use of aluminum bats due to the perceived danger of high speed batted balls acting as missiles and causing serious injury to players. New York City and North Dakota banned their use in high schools in 2007. The NCAA passed an indefinite ban on baseball bats with composite barrels in 2009, not because they were deemed to be unsafe but because a significant fraction of bats with composite barrels failed the official BESR test during the 2009 baseball season. The NCAA subsequently adopted an accelerated break-in procedure to allow manufacturers to submit composite bats for approval during the 2010 season. The NCAA web site in May 2010 listed 152 bats with composite barrels that had been declared illegal, together with about 600 other bats of various descriptions that were approved for use.

11.2 Issues Regarding Bat Performance

The main physics issues regarding bat performance are that (a) bats with a small swing weight can be swung faster, (b) bats with a small swing weight do not have the same intrinsic or "inbuilt" power as bats with a large swing weight, (c) hollow bats tend to outperform wood bats due to the trampoline effect, and (d) batted ball speed depends on the properties of the ball as well as the properties of the bat. Effects (a) and (b) tend to cancel in the sense that the batted ball speed does not depend as strongly on the swing weight of the bat as one might otherwise expect. Nevertheless, wood bats with a large swing weight are generally more powerful than wood bats with a small swing weight, while small swing weight bats are easier to control since they are more manouverable.

The focus of bat performance issues is usually on the properties of bats, since most balls used in baseball or softball are assumed to be similar in terms of mass, stiffness and COR. However, wide variations in ball properties have been reported, as described in Sect. 2.2 and in [3] and [4]. In particular, variations in ball stiffness will have a strong effect not only on batted ball speed with non-wood bats, but also on the impact force on a player if he or she is struck by a ball.

The performance of a bat is not as easy to measure as one might expect. High performance equates to high batted ball speed, so in principle it should be relatively easy to measure the batted ball speed using various bats to see which bats perform best. Given that the batted ball speed depends on the swing speed of the bat, and that low swing weight bats can be swung faster, that immediately raises questions as to whether all bats should be tested at the same swing speed, whether bats should be tested by allowing players to swing the bats or whether the bats should be swung by a machine at an accurately controlled speed. If players swing the bats, then they could get tired after swinging 20 bats, so the next 20 bats might appear to perform poorly. If a machine swings the bats, then what speed should each bat be swung at? Should the machine be programmed to swing light bats faster than heavy bats? How much faster? And where should the ball be struck? At the tip of the bat or at some other point along the barrel? What if one bat is 1 in. longer than the others? Where then should the ball be struck? And what if some bats are tested with a ball that bounces better than others? These are the sorts of questions that bat testing organizations have been grappling with ever since bat testing got underway in ernest around 2000. In the remainder of this chapter, we look at the physics of the problem to learn how bat performance can best be defined and measured.

11.3 Swing Speed vs. Swing Weight

Players don't always swing bats as fast as they can, but if we want to determine the maximum performance of a bat then we need to assume that players will swing it as fast as possible. It is obvious that light bats can be swung faster than heavy bats. However, several studies have shown that the swing speed is determined not so much by the weight of a bat but by its swing weight. This is a slightly subtle point, since for any given bat length and weight distribution, swing weight is proportional to actual weight. That is, heavy bats tends to have a large swing weight and light bats tend to have a small swing weight. However, the swing weight of a bat also depends on its length and its weight distribution, so it is possible to manufacture a range of bats all of the same weight but with different swing weights, and vice versa.

An extensive study of swing speeds using baseball bats was undertaken in 1999 by Greenwald, Penna, and Crisco [5]. They found that good batters swung bats in such a way that, just before impact with the ball, the bat swings in a circular arc centered within an inch or two of the knob. They also found that the angular velocity of a bat, averaged over several strong batters, was given by

$$\omega = \frac{659}{(I_0)^{0.277}}, \tag{11.1}$$

where ω is the angular velocity of the bat (in rad/s) just before impact and I_0 (in oz in.^{2}) is the MOI of the bat about an axis through the knob. The fact that I_0 is at the bottom of this equation shows that the swing speed decreases as the MOI increases, as expected. The factor 0.277 shows that the swing speed does not decrease by very much. For example, if I_0 is decreased by a factor of 2 then ω increases by only 20%. In practice, most bats vary in swing weight by less than a factor of 2, since their actual weights and lengths also differ by factors less than 2, so a batter might be able to increase his swing speed by 10% or so by changing to a low swing weight bat.

The speed of the bat at the impact point can be calculated from its angular velocity. For example, suppose the bat is 33 in. long, the impact point is 6 in. from the end of the barrel, and the bat is swung about an axis 1 in. beyond the knob. The impact point then swings around in an arc of radius 28 in. at a speed $V_1 = 28\,\omega$ in. s^{-1} for a 33-in. bat or at $V_1 = 29\,\omega$ in. s^{-1} for a 34-in. bat. We can convert those speeds to mph using 1 in. s^{-1} = 0.05682 mph. The result for a 33-in. bat is $V_1 = 1048/(I_0)^{0.277}$, giving $V_1 = 70.6$ mph when $I_0 = 17{,}000$ oz in.^{2}, this being a typical value of I_0 for a bat with a swing weight value $I_6 = 10{,}000$ oz in.^{2} (see Fig. 11.2).

Another very nice experiment of this type was conducted in 2002 in Alabama at a slow pitch softball tournament, by Lloyd Smith and colleagues [6], using 20 different aluminum softball bats all 34 in. long. Ten of the bats were 28 oz in weight but their swing weight varied from 7,000 to 11,000 oz in.^{2}, measured with an axis 6 in. from the knob (toward the barrel). The other ten bats all had a swing weight of 8,400 oz in.^{2} but their weights varied from 24.5 oz to 31 oz. Fourteen good hitters swung at pitched softballs using each bat and were filmed with a high speed video

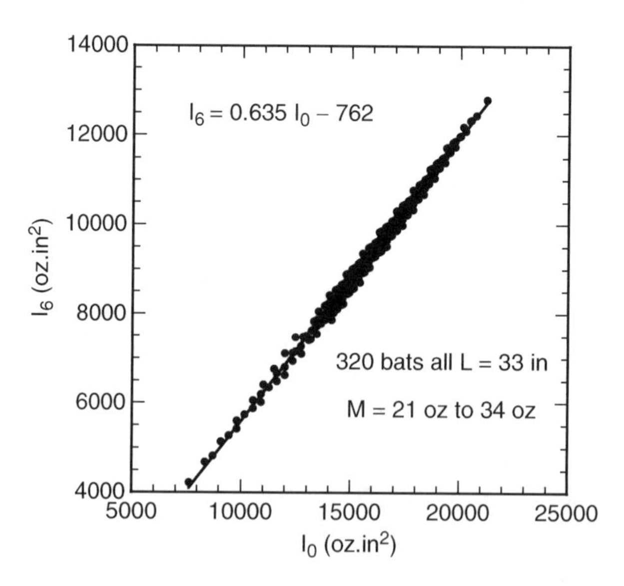

Fig. 11.2 I_6 vs. I_0 for 320 different bats, all 33 in. I_6 is the conventional swing weight measured about an axis 6 in. from the knob and I_0 is the swing weight measured about an axis through the knob

camera to measure their swing speed. It was found that the bats of the same swing weight were all swung at the same speed on average, regardless of their actual weight. On the other hand, the swing speed of the ten 28 oz bats decreased as the swing weight I_6 increased, and was proportional to $1/(I_6)^{0.25}$.

In 2004, the ASA modified their bat performance standards in the light of Smith et al's study, by assuming that a softball bat can be swung at a speed

$$V_{\text{bat}} = 85\left(\frac{D+2.5}{30.5}\right)\left(\frac{9000}{I_6}\right)^{0.25}, \tag{11.2}$$

where V_{bat} (in mph) is the speed of the bat at the impact point and D is the distance (inches) from the knob to the impact point. For example, if the swing weight of the bat is 9,000 oz in.^{2} and if $D = 28$ in then $V_{\text{bat}} = 85$ mph, which is slightly larger than that deduced from (11.1). Softball players often wrap both hands around the knob to increase bat speed, but the difference in bat speed here might just reflect the different group of batters included in each study.

The numbers in (11.2) are based on average values determined from the 2002 field study. The average swing weight of the bats was 9,000 oz in.^{2}, the average bat speed was 85 mph when measured at a point 6 in. from the end of the bat, and the bats were swung about an axis located 2.5 in. from the end of the knob just before impact. The average bat speed was therefore given by $85 = R\omega$ where $R = 30.5$ in. was the radial distance from the axis to the measurement point and ω is the angular velocity of the bat. The bat velocity at any other distance D from the knob was therefore given by $V_{\text{bat}} = (D + 2.5)\omega = 85(D + 2.5)/30.5$, at least for the 34 in., 9,000 oz in.^{2} bats. Equation (11.2) gives the bat speed at any given impact point for any other bat having a swing weight I_6.

In a similar study by Cross and Bower [7] using various rods with a much wider range of swing weights than in the [6] study, it was found that the swing speed was proportional to $1/(I_6)^{0.27}$. Many years ago, another similar experiment was undertaken by Daisch [8] using golf clubs. He found that the swing speed of the clubs was proportional to 1/(swing weight)$^{0.26}$. Measurements of swing speed have also been reported by other authors [9, 10]. All of these experiments tell us essentially the same thing, which gives us confidence in the results. That is,

> It is the swing weight, not the actual weight of the bat, that determines the swing speed of a bat

Intuitively, one might expect that a factor of two decrease in bat swing weight might give a factor of two increase in swing speed, not just a 20% increase. However, a large part of the effort exerted by a batter is needed just to swing his own arms. Adding a 2-lb bat to the batter's 15-lb arms is not going to change his swing speed by much, as noted previously in Sect. 5.3. Similarly, a fielder can throw a 6-oz ball almost as fast as he can throw a 4-oz ball. There is possibly a 5% difference, but not a 50% difference in throw speed.

To use (11.1) to estimate swing speeds, we need a relation between I_0 and the standard I_6 value of the MOI for a bat. I_0 is larger than I_6 since the swing axis is 6 in. further away from the bat center of mass when we measure I_0. We can calculate I_0 using the Parallel Axis Theorem described in Chap. 1. If the balance point is a distance B from the knob then $I_0 = I_{CM} + MB^2$ and $I_6 = I_{CM} + M(B-6)^2$, where I_{CM} is the MOI for an axis through the center of mass. Hence, $I_6 = I_0 - 12M(B-3)$. The difference between the two swing weights is different for every bat since it depends on both the weight and the balance point for each bat. However, I_0 is typically about 70% larger than I_6 for most bats, and a reasonably good approximation is that $I_6 = 0.58 I_0$. A better approximation is given by

$$I_6 = 0.635 I_0 - 762 \tag{11.3}$$

as can be seen in Fig. 11.2 where we have plotted I_6 vs. I_0 for the same 320 bats described in Chap. 9. The bats were all 33 in. long.

The calculations were repeated for two more sets of 320 bats, one 32 in. set and one 34 in. set. The results were

$$I_6 = 0.625 I_0 - 722 \quad L = 32\,\text{in} \tag{11.4}$$

$$I_6 = 0.645 I_0 - 801 \quad L = 34\,\text{in} \tag{11.5}$$

11.4 Batted Ball Speed vs. Swing Weight

We now have sufficient information to calculate the batted ball speed for a maximum effort swing by an average, relatively strong player. For a baseball bat, the swing speed is given by (11.1), I_0 is given by (11.3)–(11.5) and the batted ball speed is given by

$$v_2 = (1+q)V_1 + q v_1, \tag{11.6}$$

where

$$q = \frac{e - m/M_e}{1 + m/M_e} \quad \text{Bounce factor}$$

and

$$M_e = \frac{I_6}{(L-6-d)^2} \quad \text{Effective bat mass}$$

L is the bat length in inches and d is the distance in inches between the impact point and the tip of the bat. The bat can have any weight, balance point and swing weight that the batter likes, provided that it is with legal limits. The question is, which bat will do the best job?

Figure 11.3 shows the batted ball speed for several cases where the bat is swung about an axis 1 in. beyond the end of the knob and the ball impacts 6 in. from the

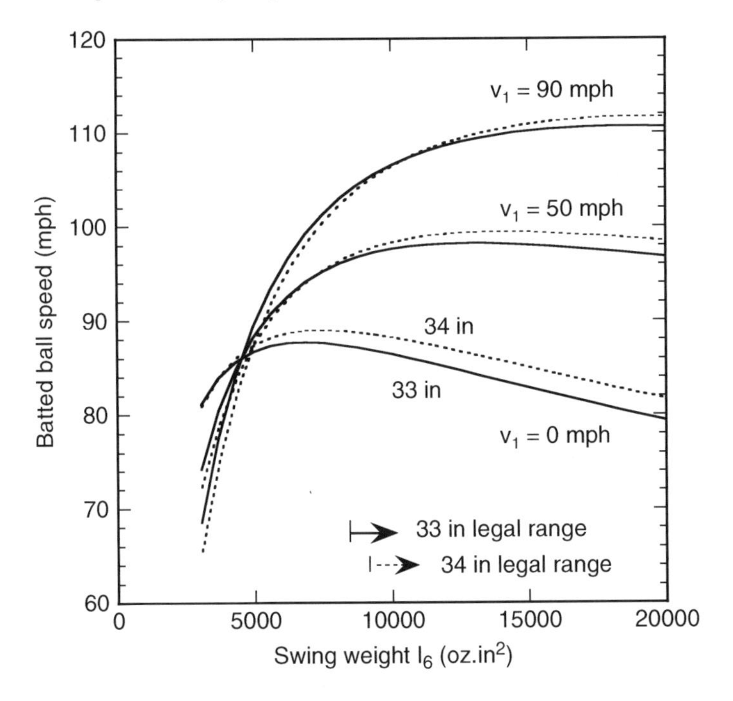

Fig. 11.3 Batted ball speed v_2 vs. swing weight I_6 for 33-in. and 34-in. bats when $e = 0.5$, when the ball impacts 6 in. from the end of the barrel and when the ball mass is 5.25 oz. The allowed range of bats is shown for NCAA baseball

end of the barrel. These are the results for a good batter, being based on the average swing speed of several good, male batters. Results are shown for bats of two different lengths, assuming that $e = 0.5$, and for a wide range of different bat weights and weight distributions, slightly wider than the normal range of bats encountered in practice. The results apply equally well to wood and aluminum bats since the only properties that we need to consider, apart from length, is the swing weight of the bat and the values of e and the impact distance, d. Results for softball are qualitatively similar, apart from the fact that batted ball speeds are about 5 or 6 mph lower because of the larger ball mass, and it is lower still for women since they usually swing at a lower speed than that given by (11.1).

The swing weight determines the speed at which the bat can be swung, it determines the effective mass of the bat and hence it determines the bounce factor, q, or the intrinsic power for any given value of e. A very strong batter would be able to swing the bats faster and generate proportionally higher batted ball speeds, but the manner in which batted ball speed varies with swing weight and bat length would be unchanged.

The results in Fig. 11.3 show that the batted ball speed increases rapidly with swing weight at first, but at high values of swing weight, the batted ball speed starts to decrease. These results are consistent with our expectations. If a batter used a light broom stick, with a very small swing weight, the batted ball speed would be quite low, and possibly even negative. If a batter used a 10-lb steel bar, the batted ball speed would also be low since the batter would have trouble swinging it. In fast pitch softball, where the swing weight of bats is typically about 6,500–8,500 oz in.^{2}, and the pitched ball speed is typically about 50–70 mph, increasing the swing weight has the advantage of increasing the batted ball speed slightly and the disadvantage of being more difficult to swing. The same situation applies in NCAA baseball, even though the bat swing weights are larger (as shown in Fig. 11.1), since the pitched ball speed is also larger. There is no special advantage in using a high swing weight bat in slow pitch softball since the pitched ball speeds are relatively low.

In theory, there is an optimum swing weight for a bat, around 10,000 oz in.^{2}, which will generate high batted ball speeds and that will be relatively easy to swing at the ball. The optimum swing weight varies with the incoming ball speed, but 10,000 oz in.^{2} is about the best compromise. In fact, 10,000 oz in.^{2} is close to the allowed limits set by the NCAA (for 33- and 34-in. baseball bats) so amateur players have very little choice in the matter anyway. The legal limits set by the NCAA are shown in Fig. 11.3. There is a lower limit but no official upper limit. Major League players facing high speed pitched balls would benefit by using bats with a higher swinger weight, at least in terms of bat performance, but at the expense of missing the ball more often. For that reason, most players use bats that have a lower than optimum swing weight so that the bats will be easier to swing.

Figure 11.3 does not show the effect of the COR on the batted ball speed, since a fixed value $e = 0.50$ was chosen to work out the batted ball speeds. In practice, the COR can vary over a relatively wide range, from about 0.45 to about 0.6, depending on the type of bat, the type of ball and the relative speed of the bat and the ball before impact. The COR is not just a property of the ball, since it also depends on the elastic properties of the bat. For that reason, the COR is commonly known as the BBCOR in the case of a bat–ball collision, standing for Ball–Bat Coefficient of Restitution. The calculations in Fig. 11.3 for $L = 33$ in. bats are repeated in Fig. 11.4 with $e = 0.45$ and $e = 0.55$.

It is clear from Fig. 11.4 that bats with a high COR easily outperform bats with a low COR, which is why the ASA, the NCAA and other organizations began to introduce limits on bat performance in the late 1990s. Over the legal range of swing weight values, the performance of a non-wood bat depends more strongly on the COR than it does on the swing weight. The best wood bats have a BBCOR value of about 0.50 at incident ball speeds above about 50 mph, so the NCAA formulated a new rule from January 2011 that the BBCOR of non-wood bats must not exceed 0.50.

One of the main reasons for the new NCAA COR limit rule is that the batted ball speed correlates closely with the COR, at least for bats of any given length, whereas the correlation with q or I_6 is not as strong, despite the obvious dependence of v_2 on I_6 shown in Figs. 11.3 and 11.4. In theory, the batted ball speed depends on both the COR and on the swing weight, as shown in Fig. 11.4. Nevertheless, almost all non-wood bats of any given length are manufactured with the same weight. For example, almost all 34 in. NCAA bats are close to 31 oz, and almost all 33 in. bats are close to 30 oz, due to the Minus 3 rule and because batters much prefer to use

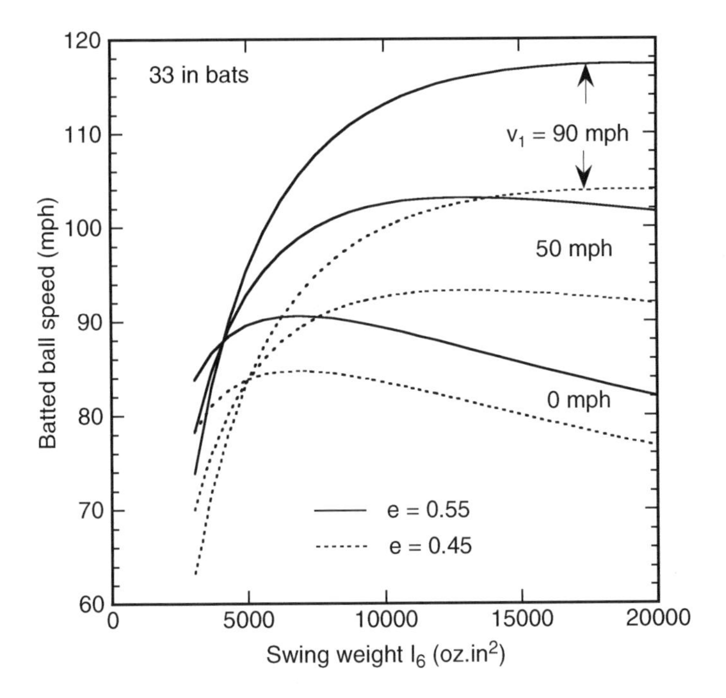

Fig. 11.4 Batted ball speed v_2 vs. swing weight I_6 for 33 in bats when $e = 0.45$ or $e = 0.55$ and when the ball impacts 6 in. from the end of the barrel. v_1 is the incident ball speed. The bat is swung at a speed given by (11.1)

the lightest bats possible. Consequently, all non-wood bats of the same length, having essentially the same weight and only small differences in weight distribution, have very similar values of swing weight. For example, most non-wood 33 in. bats have a swing weight of about 9,000 oz in.^{2} and most non-wood 34 in. bats have a swing weight of about 10,000 oz in.^{2}, as shown in Fig. 11.2. The corresponding values for commonly available wood bats are 10,200 oz in.^{2} for 33-in. wood bats and 11,400 oz in.^{2} for 34-in. wood bats (also shown in Fig. 11.1).

11.5 Ball Speed vs. Sweet Spot Location

The impact point on the bat affects the batted ball speed. In Figs. 11.3 and 11.4 it was assumed that the impact point was located at a distance $d = 6$ in. from the tip of the barrel since that is a typical location for the sweet spot. It is the spot where bat vibrations are minimized and hence where the COR is a maximum along the barrel. However, the sweet spot is not always at the 6 in. location. On many bats, the sweet spot is located closer to $d = 7$ in.. If additional mass is deliberately added at the tip of the barrel, then the sweet spot shifts closer to the tip of the barrel and could even shift to $d = 5$ in.

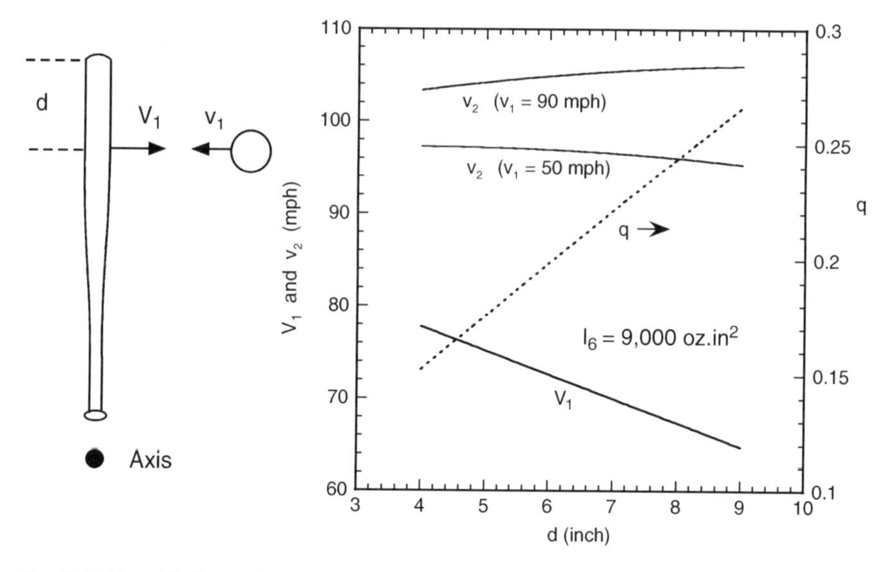

Fig. 11.5 Batted ball speed v_2 vs. impact distance d for 33 in. bats when $e = 0.5$ and when the ball is incident at 50 or 90 mph. The bat is swung at a constant angular speed about a fixed axis 1 in. from the knob so the speed V_1 decreases as d increases. The bounce factor q (*right side scale*) increases as d increases

Figure 11.5 shows a graph of the batted ball speed vs. the impact distance d, for $L = 33$ in., $I_6 = 9,000\,\text{oz in.}^2$ bats, assuming that $e = 0.50$ at the impact point. For any given bat, the sweet spot is located at only one point along the barrel, but it could be anywhere in the region from about $d = 5$ in. to about $d = 8$ in. depending on the design of the bat. If the sweet spot is located at $d = 7$ in. for example, and if $e = 0.50$ at the sweet spot, then e will actually be less than 0.50 at points either side of the sweet spot (as shown in Project 9 and in Fig. 11.6). Figure 11.5 does not represent the variation of batted ball speed with d for a given bat. Rather, it represents the variation of batted ball speed with d for a range of different bats, all having the same swing weight, when the sweet spot happens to be located at the distance d plotted in the graph and when $e = 0.5$ at the sweet spot.

The question of interest is where the sweet spot would be best located, given that it is the point where the COR is a maximum. Figure 11.5 shows that if the ball is incident at 50 mph, then the batted ball speed will be maximized if the sweet spot is located as close to the tip of the barrel as possible. It is not a big effect, so adding a large weight to the tip of the barrel, in an attempt to shift the sweet spot in that direction, could be counter-productive. The bat would then be harder to swing, unless mass was removed from somewhere else to keep I_6 constant. In fact, if the ball is incident at high speed, the best location for the sweet spot is as far as possible from the tip of the bat, as shown by the result in Fig. 11.5 when $v_1 = 90$ mph. We conclude from this that the location of the sweet spot is not particularly important in terms of batted ball speed.

There are two opposing effects when the impact point moves further away from the tip of the barrel. One is that the bat speed, V_1 decreases, and the other is that bounce factor, q increases. The two effects tend to cancel, with the result that the batted ball speed varies by only a small amount in Fig. 11.5 as the sweet spot location is varied. A larger variation in batted ball speed could be achieved by increasing I_6 when the sweet spot location is shifted, but the increase in batted ball speed in that case would be due primarily to the increase in I_6 rather than the shift in the location of the sweet spot. Alternatively, a larger variation in batted ball speed could be achieved if the shift in the sweet spot incidentally enhances the trampoline effect and therefore results in an increase in e above 0.5.

11.6 Bat Performance Factor

Many attempts have been made over the years to set performance standards for baseball and softball bats. One of the original and more popular measures of performance is known as the Bat Performance Factor (BPF) which is a measure of the enhancement of the COR due to the trampoline effect. The USSA, NSA and Little League all use the BPF to regulate bat performance. The maximum COR for a collision between a bat and a ball, when measured at several different points along the barrel to find the maximum value, is typically in the range 0.45–0.6 depending on the type of bat, but it also depends on the COR of the ball. Suppose that when a ball is fired at a massive rigid surface at 60 mph, the measured COR of the ball is e_0. If the same ball is fired at 60 mph at a stationary, free bat, and if it impacts at the sweet spot to minimize bat vibrations, then the COR of the ball–bat combination will have a different value, e. To measure the COR for this test, the recoil speed of the bat must be measured as well as the recoil speed of the ball. If the bat exhibits a trampoline effect then e will be larger than e_0 by a factor given by

$$\mathrm{BPF} = \frac{e}{e_0} \quad \text{(Bat Performance Factor)} \tag{11.7}$$

There are many aluminum and composite bats on the market that are advertised for sale with BPF values around 1.15 or 1.20. Wood bats have a maximum BPF value of about 1.0, but some are only about 0.9. Bats with a BPF greater than 1.0 will therefore perform better than wood bats. However, if the BPF is say 1.20, that value does not mean that the ball will be struck 20% faster than a wood bat. It just means that the COR is 20% larger. For example, if the COR of the ball has a value $e_0 = 0.50$, then the COR for the ball–bat combination will be 0.60. Depending on the impact point on the bat, and on the pitch speed and the swing speed, the batted ball speed will then be about 10% larger, as given by (11.6).

The BPF gives a rough guide to the performance of a bat, but it can be quite misleading and is not accurate enough to compare bats in a consistent way. One problem is that wood bats don't always have a BPF of 1.0. An impact on a wood bat differs from an impact on a heavy flat surface since the bat surface is curved, which reduces the COR, and since a bat is much lighter, which increases the COR. Another problem with the BPF is that it does not necessarily give a reliable measure of performance under actual playing conditions, since the test speed (60 mph) is much lower than the relative bat and ball speeds encountered in normal play. Bats and balls can collide at relative speeds above 140 mph in practice.

The BPF also depends on the particular ball used to test the bat. Some balls are heavier than others, some are relatively stiff and some are relatively soft. Similarly, some balls have a high COR and some have a low COR. The COR for a given bat and ball combination depends on the elastic properties of the ball as well as the elastic properties of the bat. The value measured for the BPF will therefore depend on whether the bat is tested with a high COR ball or a low COR ball, and whether it is tested with a stiff ball or a soft ball. Given that elastic energy is shared between a bat and a ball in proportion to the relative stiffness of the bat and the ball, a larger share of the elastic energy is given to the bat as the ball stiffness increases.

There is yet another potential problem with the BPF as a measure of performance. The BPF is simply a measure of the increased performance resulting from the trampoline effect. However, the performance of a bat also depends on its swing weight. This is particularly the case for bats with a small swing weight, less than about 8,000 oz in.^{2}. Consequently, two bats with the same BPF do not necessarily perform the same. The one with the larger swing weight will perform better, as shown in Figs. 11.3 and 11.4. This particular problem with the BPF is more of a potential than an actual problem for NCAA regulated bats given that most NCAA non-wood bats of a given length have a similar swing weight, as explained at the end of Sect. 11.4 and as shown in Fig. 11.1.

11.7 ASA and NCAA Performance Tests

The current regulations regarding bat performance have evolved as a result of many experimental and theoretical studies of the subject. References [1–31] include most, but not all of those studies. The rules governing NCAA baseball and fast pitch softball are relatively strict. The rules governing slow pitch softball are less strict, but still focus on limiting the batted ball speed. The rules governing youth bats are even less strict, since the batted ball speeds are not as high as those in adult games and since the primary object in youth sports is to encourage young players to enjoy the sport by making it relatively easy to hit the ball.

Early laboratory tests of baseball bats were conducted by striking a 70-mph ball with a bat swung at 66 mph. The current method is similar in principle but uses a ball fired from an air cannon at 136 mph at a stationary bat. The advantages are that the method is simpler (since there is no need for additional apparatus to swing the bat), safer (since the ball rebound speed off a stationary bat is a lot smaller than that off a swung bat) and more accurate. The same method is used to test softball bats, although the incident ball speed is reduced to 110 mph since the relative speed of the bat and the ball is generally lower in softball. In both cases, the bat is supported

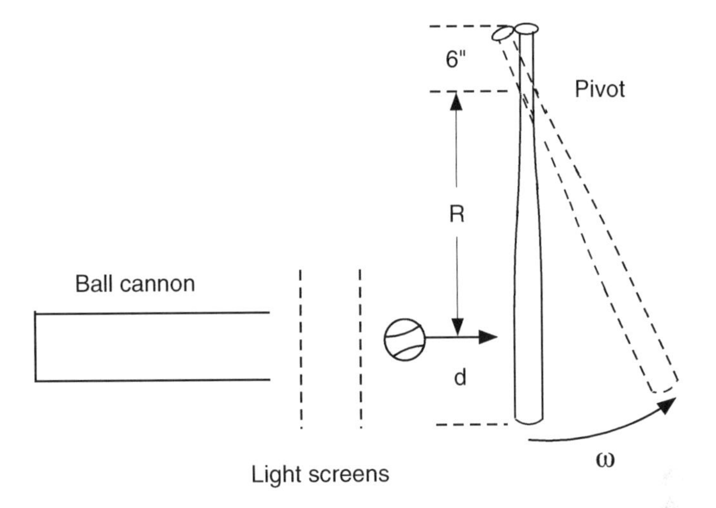

Fig. 11.6 Method used to measure the performance of baseball and softball bats

horizontally and allowed to swing freely about an axis in the handle 6 in. from the knob, as shown in Fig. 11.6. The incident and rebound speeds of the ball are measured using laser beams.

The rotational speed, ω, of the bat after the collision can also be measured to obtain additional information, such as a direct measurement of the COR or an indirect measurement of the bounce factor in cases where the ball rebound speed is too low for the ball to pass back through the laser beams. In practice, it is easier to measure the speed of the ball in the laboratory tests than it is to measure the speed of the bat. The main problem is that the bat vibrates as a result of the collision, so it is necessary to measure the bat speed carefully over several vibration cycles to subtract out the vibration part of the bat velocity, as described by Smith [27].

The bat is impacted at various points along the barrel to determine how the bounce factor varies along the barrel. The incident speed can be adjusted slightly for different impact locations, to simulate the variation in relative speed with impact point for a swung bat. Baseball tests for the NCAA are performed at the Baseball Research Center at the University of Massachusetts (see http://m-5.eng.uml.edu/umlbrc/index.htm) and softball tests for the ASA and NCAA are performed at Washington State University (see www.ssl.wsu.edu).

The results are interpreted in different ways by baseball and softball organizations. NCAA baseball regulations, since January 2011, require that the COR must not be larger than 0.5 at any point along the barrel. The NCAA regulation is one of three concerning bat performance, the other two being the Minus 3 rule and the Minimum MOI rule. All three rules must be satisfied for a bat to be certified as being compliant.

The ASA regulations require a calculation of the batted ball speed when the bat is swung rather than being initially at rest as in the actual test. For an approved bat, the calculated value of the batted ball speed must not exceed 98 mph, off any part of the bat, when the pitched ball speed is 25 mph and when the bat is swung at a speed given by (11.2). This regulation is easily interpreted by the average player, but what does it mean in terms of the properties of the bat? It can be interpreted either in terms of a maximum value of the bounce factor or a maximum value of the COR for any given bat. From (11.6), we see that q cannot exceed a value given by $98 = (1+q)V_{\text{bat}} + 25q$, so q must be less than $q_{\text{max}} = (98 - V_{\text{bat}})/(25 + V_{\text{bat}})$. For example, if $V_{\text{bat}} = 85$ mph then $q_{\text{max}} = 0.118$. Suppose also that $I_6 = 9,000$ oz in.^{2} and that the bat speed is 85 mph at a point located 22 in. from the standard pivot point, or 28 in. from the knob. Then the effective mass of the bat at that point is $M_e = I_6/22^2 = 18.6$ oz. The COR is given by

$$e = q + (1+q)m/M_e,$$

where m is the mass of the ball. If $m = 6.75$ oz then $e = 0.524$, but if $m = 7$ oz then $e = 0.539$. The ASA regulation therefore allows a player to use a bat with a COR that is greater than that of a typical wood bat, at least for a bat with a swing weight of 9,000 oz in.^{2}. A similar calculation for a 7,000 oz in.^{2} bat shows that the COR must be less than 0.56 at a point 28 in. from the knob. If the swing weight is 11,000 oz in.^{2} then the COR must be less than 0.507 at a point 28 in. from the knob. Despite the high COR allowed for a low swing weight bat, and despite the higher swing speed possible when using such a bat, the bat will perform no better than a high swing weight bat in terms of the resulting batted ball speed.

An important consideration in bat performance tests is the manner in which variations in ball properties affect the rebound speed of the ball. Balls used in the NCAA baseball Maximum COR test are first tested by firing them at 136 mph onto a standard wood bat before firing them at the bat to be tested. The procedures used in softball tests are similar [27,28], are refined from time to time in the light of continued research into the problem, and are based on calculations such as those given in Appendices 8.4 and 8.5.

It is not strictly necessary to measure the recoil speed of the bat to determine the ball–bat COR. A method to calculate the COR, using the ball speed data alone is described in Appendix 11.1. Results are shown in Fig. 11.7 for cases where the bat is pivoted about two different axes, and are compared with a case where the bat is freely suspended.

11.8 Hand-Held Bats

As shown in Fig. 11.7, the bounce factor, q, does not depend on how the handle is supported, but the COR does. Nevertheless, the variation in the COR is quite small, particularly in the sweet spot region where the COR is a maximum. In practice, the q value is measured for NCAA baseball bats using a 6 in. pivot point, and the COR is then calculated for a freely suspended bat to find the maximum value of the COR for that bat and ball.

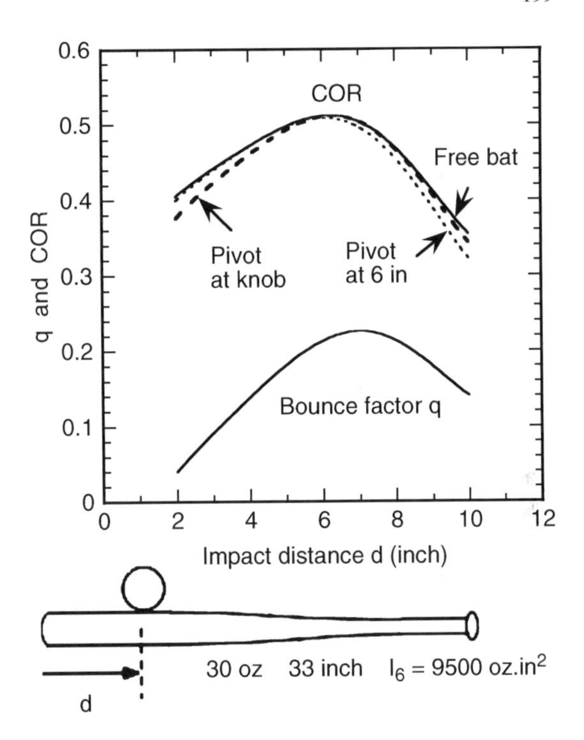

Fig. 11.7 COR vs. impact distance d for a given 30 oz, 33-in. bat supported in three different ways, either freely supported or pivoted about an axis at the knob end, or pivoted about an axis in the handle 6 in. from the knob. The COR for the pivoted bats was calculated from (11.13) using the assumed q profile shown in the lower part of the diagram

The variation of the COR with the support method is a subtle point, involving vibration of the bat. The bending wave generated by the impact takes a relatively long time to travel up and down the bat, with the result that the ball bounces before the reflected wave gets back from the handle end. The bounce factor, therefore, depends only on the properties of the bat in the vicinity of the point where the ball bounces, and is independent of the properties of the bat further along the handle. For that matter, the bounce factor is also independent of the size, shape, and weight of the person holding onto the handle.

When a ball strikes the barrel, the barrel bends locally in a section of the bat around the impact point. That section is initially about one foot long. By the time the ball bounces off the bat, the bend will have traveled a short distance along the barrel, but the handle will be totally unaffected by the time the ball bounces. By the time the bending wave reaches the handle, the ball will have traveled about 4 in. away from the bat. The fate of the energy in the bending wave depends on how the handle is supported, but it has no effect on the ball.

It takes several transits of the bending wave up and down the bat for the bending energy stored in the bat to get distributed in three possible ways. If the bat is freely supported then all three ways are available. Some of the energy in the bending wave will end up in translational energy of the bat, some will end up as rotational energy, and some will end up as vibrational energy. However, if the handle is held in a vice, then the bat can't rotate and it can't move away from the ball. The bat can only vibrate back and forth when it is struck. Consequently, all of the bending wave energy ends up as vibrational energy in the bat.

Given that the COR is a measure of the energy lost in the bat and the ball, in the form of heat in the ball and vibrations in the bat, the COR will be relative small if the handle is clamped. For that reason, most measurements of the COR in the laboratory are performed with a hinged or pivoted bat since the COR is then almost the same as it is with a free bat or for a hand-held bat. There is a particular impact point on the barrel, known as the center of percussion, where a pivoted bat behaves in an identical manner to a free bat since there is then no force on the pivot axis. In that case, the COR for a pivoted bat is identical to the COR for a free bat. Even at other impact points along the barrel, the COR for a pivoted bat is still very close to that for a free or a hand-held bat.

Appendix 11.1 COR for a Pivoted Bat

The COR for a bat and ball collision can be calculated just from v_1 and v_2, using some simple physics to calculate the bat recoil speed V_2 for a pivoted bat. Suppose that the ball exerts a force F on the bat at a point that is distance R from the pivot point. Then $FR = I_A \mathrm{d}\omega/\mathrm{d}t$, where ω is the angular velocity of the bat and I_A is the MOI about the pivot axis. Since F is also the force on the ball,

$$\int F\,\mathrm{d}t = m(v_1 + v_2) = mv_1(1+q) \tag{11.8}$$

and

$$R\int F\,\mathrm{d}t = I_A\omega_2, \tag{11.9}$$

where ω_2 is the angular velocity of the bat after the collision. The bat speed after the collision is, therefore, given by

$$V_2 = R\omega_2 = (1+q)\frac{mR^2v_1}{I_A} \tag{11.10}$$

The effective mass, M_e of the bat at the impact point is defined by the relation

$$F = M_e\frac{\mathrm{d}V}{\mathrm{d}t} = M_eR\frac{\mathrm{d}\omega}{\mathrm{d}t} = M_e\frac{FR^2}{I_A} \tag{11.11}$$

so

$$M_e = \frac{I_A}{R^2} \tag{11.12}$$

Since $e = (v_2 + V_2)/v_1$, we find from (11.10) that

$$e = q + (1+q)mR^2/I_A \tag{11.13}$$

and hence

$$q = \frac{e - m/M_e}{1 + m/M_e} \tag{11.14}$$

which is the same expression as that for a free bat. However, the effective mass for a pivoted bat is not exactly the same as that for a free bat. For a free bat, the actual rotation axis of the bat shifts toward the knob end of the bat as the impact point moves farther from the tip of the bat. Consequently, if q is the same, as it is, then e will be slightly different. The difference is shown in Fig. 11.7 for a typical 33 in. bat with $M = 30$ oz, a swing weight $I_6 = 9,500$ oz in.^{2} about an axis 6 in. from the knob, a ball mass $m = 5.25$ oz and for the *assumed* q profile shown in Fig. 11.7.

References

1. A. Nathan, Characterizing the performance of baseball bats. Am. J. Phys. **71**, 134–143 (2003)
2. L. Smith, Progress in measuring the performance of baseball and softball bats. Sports Technol. **1**(6), 291–299 (2008)
3. S.P. Hendee, R.M. Greenwald, J.J. Crisco, Static and dynamic properties of various baseballs. J. Appl. Biomechan. **14**, 390–400 (1998)
4. L. Smith, A. Nathan, J.G. Duris, A determination of the dynamic response of softballs. Sports Eng. (2010)
5. R.M. Greenwald, L.H. Penna, J.J. Crisco, Differences in batted-ball speed with wood and aluminium baseball bats: a batting cage study. J. Appl. Biomechan. **17**(3), 241–252 (2001); See also J.J. Crisco, R.M. Greenwald, J.D. Blume, L.H. Penna, Med. Sci. Sports Exerc. **34**, 1675–1684 (2002)
6. L.V. Smith, J. Broker, A. Nathan, in *A Study of Softball Player Swing Speed*, ed. by Subic, Trivailo, Alam. Sports Dynamics Discovery and Application (RMIT University, Melbourne, 2003), pp. 12–17
7. R. Cross, R. Bower, Effects of swing-weight on swing speed and racket power. J. Sports Sci. **24**, 23–30 (2006)
8. C.B. Daish, *The Physics of Ball Games* (English Universities Press, London, 1972)
9. G.S. Fleisig, N. Zheng, D.F. Stodden, J.R. Andrews, Relationship between bat mass properties and bat velocity. Sports Eng. **5**, 1–8 (2002)
10. K. Koenig, N.D. Mitchell, T.E. Hannigan, J.K. Clutter, The influence of moment of inertia on baseball/softball bat swing speed. Sports Eng. **7**, 105–117 (2004)
11. T. Naruo, F. Sato, An experimental study of baseball bat performance, in 2nd International Conference on Engineering of Sport, pp. 49–56 (1998)
12. J.J.Crisco, R.M. Greenwald, L.H. Penna, K.R. Saul, On measuring the performance of wood baseball bats, ed. by Subic, Haake, 3rd International Conference on the Engineering of Sport (Blackwell, London, 2000), pp. 193–200
13. M.M. Shenoy, L.V. Smith, J.T. Axtell, Performance assessment of wood, metal and composite bats. Compos. Struct. **52**, 397–404 (2001)
14. L.V. Smith, Evaluating baseball bat performance. Sports Eng. **4**, 205–214 (2001).
15. P.J. Drane, J.A. Sherwood, The effects of moisture content and workhardening on baseball bat performance, 4th International Conference on the Engineering of Sport, Kyoto Japan (2002)
16. J.J. Crisco, R.M. Greenwald, J.D. Blume, L.H. Penna, Batting performance of wood and metal baseball bats. Med. Sci. Sports Exerc. **34**, 1675–1684 (2002)
17. A. Nathan, Bat performance standards in NCAA baseball. Baseball Res. J. **32**, 11 (2003)
18. J. Ashton-Miller, M. Carroll, K. Johnson, A. Nathan, Ball exit speed ratio (BESR). Baseball Res. J. **32**, 12–14 (2003)

19. B. Thurston, It's a different game. Baseball Res. J. **32**, 3–8 (2003)
20. L.V. Smith, J.T. Axtell, Mechanical Testing of Baseball Bats. J. Testing Evaluat. **31**, 210–214 (2003)
21. S. Nabeshima, J.A. Sherwood, Comparison of the performance of US and Japanese aluminum baseball bats, in Proceedings of the 5th International Conference on the Engineering of Sport, vol. 2 (UC Davis, CA, 2004), pp. 73–79
22. K. Koenig, J.S. Dillard, D.K. Nance, D.B. Shafer, The effects of support conditions on baseball bat performance testing, 5th International Conference on the Engineering of Sport, vol. 2 (Davis, CA, 2004), pp. 87–93
23. R.L. Nicholls, K. Miller, B.C. Elliott, Numerical analysis of maximal bat performance in baseball. J. Biomechan. **39**, 1001–1009 (2006)
24. P. Gamache, G. Galante, G. Seben, A.J. Elbirt, Validating baseball bat compliance. Sports Eng. **10**, 157–164 (2007)
25. L.P. Fallon, R.D. Collier, J.A. Sherwood, T. Mustone, Determining baseball bat performance using a conservation equations model with field test validation, ed. by Subic, Haake, 3rd International Conference on the Engineering of Sport (Blackwell, London, 2000), pp. 201–212
26. J.A. Sherwood, T.J. Mustone, L.P. Fallon, Characterizing the performance of baseball bats using experimental and finite element methods, ed. by Subic, Haake, 3rd International Conference on the Engineering of Sport (Blackwell, London, 2000), pp. 377–387
27. J. Sherwood, P.J. Drane, An experimental investigation of the effect of use on the performance of composite baseball bats, in Proceedings of the 7th ISEA Conference, Biarritz, Paper 274 (2008)
28. J.W. Jones, J.A. Sherwood, P.J. Drane, Experimental investigation of youth baseball bat performance, in Proceedings of the 7th ISEA Conference, Biarritz, Paper 273 (2008)
29. L.V. Smith, C.M. Cruz, Identifying altered softball bats and their effect on performance. Sports Technol. **1**, 196–201 (2008)
30. F.O. Bryant, L.N. Burkett, S.S. Chen, G.S. Krahenbuhl, P. Liu, Dynamic and performance characteristics of baseball bats. Res. Q. Exerc. Sport **48**, 505–510 (1979)
31. B.C. Elliott, T.A. Ackland, Physical and impact characteristics of aluminium and wood cricket bats. J. Hum. Mov. Stud. **8**, 149–157 (1982)

Chapter 12
Bat Vibrations

12.1 Introduction

The bending of a bat at impact raises a number of interesting physics questions. Since the ball bounces after the bending wave travels about one foot along the bat, how much of the bat is actually involved in the collision? If the tip of the bat bends before the handle starts to bend, will the bend cause the bat to rotate or does it cause the bat to vibrate? And how does the bat "know" that it musn't vibrate when the ball is struck at the sweet spot, especially considering that the bat doesn't even know where it was struck until the bending wave gets to each end of the bat and reflects back to the impact point.

A more practical question is how the vibration of a bat affects its performance. The performance of a bat, as measured by the exit ball speed, depends mainly on its weight but it also depends on how that weight is distributed along the length of the bat and it depends on the stiffness of the bat. Weight and stiffness together determine how the bat vibrates. In this chapter, we will examine the nature and origin of vibrations in a bat and in other objects as well. We will then be in a better position to understand how vibrations affect the batter, the bat itself and the exit speed of the ball.

12.2 What is a Vibration?

A vibration is a back and forth motion that can be characterized by its frequency and its amplitude. Musicians refer to the vibration frequency of a musical instrument as its pitch, and they refer to its amplitude in terms of volume or loudness. The frequency or pitch refers to the number of back and forth cycles each second, measured in Hertz (Hz). If it takes $1/100$ s for a bat to complete one back and forth cycle, then it does so 100 times each second and the vibration frequency is 100 Hz.

It is just a fanciful figure of speech to describe the loudness control on a radio or TV set as the volume control. In physics, volume is measured in pints or gallons or liters or cubic meters. There is no such thing as a pint or a gallon of speech or music. A problem with loudness is that it depends on the distance between the listener and

R. Cross, *Physics of Baseball & Softball*, DOI 10.1007/978-1-4419-8113-4_12,

the source of the sound. At the source itself, a convenient measure of loudness is the amplitude of the vibration. The amplitude is typically less than 1 mm, meaning that the object that is vibrating moves back and forth over a total distance less than 2 mm. If an object vibrates first to the right by 1 mm and then to the left by 1 mm, we say that it vibrates with an amplitude of 1 mm. Some examples of vibrations include the following:

- A guitar string or a violin or piano string vibrates back and forth at a frequency that is typically around 100–500 Hz for the longer and thicker bass (low pitch) strings and typically up to about 4,000 Hz for the shorter and thinner treble (high pitch) strings.
- A vehicle vibrates up and down when it strikes a bump in the road, at about 1 Hz. It is the suspension system, usually constructed from heavy springs, that causes the vehicle to vibrate. The system is used to minimize the shock force transmitted to the passengers.
- The speaker in a TV set or stereo system or telephone vibrates back and forth at the same frequency as the sound being generated, to transmit those vibrations through the air to your ears as a sound wave.
- A wine glass or any other drinking glass vibrates at about 3,000 Hz when it is struck, giving out a musical note at that frequency.

12.3 How Do Vibrations Arise?

Vibrations require energy to get started, usually in the form of a short impact. A guitar string sits in a guitar minding its own business until it is pulled aside and released or plucked. Similarly, violin strings are bowed and piano strings are struck with a felt hammer to get them to vibrate. A bat vibrates back and forth rapidly when it is struck by a ball, sending out sound waves that are heard as a high frequency ping in the case of aluminum bats or sending out a crack or a thud off wood bats. The sudden compression and expansion of the ball is also responsible for some of the sound that is generated.

All vibrations arise in essentially the same way. Consider a guitar string, which is simpler and easier to understand than a bat. A guitar string is anchored at both ends and stretched or tensioned by an appropriate amount so that it vibrates at the correct frequency. It requires a force to pull the string sideways. The harder the pull the further the string stretches, just like the string in an archer's bow. If the string is pulled 1 mm to the right and then released, it accelerates rapidly back to the left, gaining speed as it does so. The string arrives back at its unstretched position with maximum speed and then overshoots that position because of its momentum, stretching the string to the left. The string slows down and comes to a stop when it reaches a point 1 mm to the left side of its final resting position. Exactly the same process occurs again, except that that the string accelerates back toward the starting point on the right side. When it gets back to the starting point it again comes to a stop, completing one whole cycle. Over time, the amplitude of the vibration decreases and the string will stop vibrating after perhaps 4 or 5 s, unless the person playing the guitar plucks the string again. The process is like a person on a swing, swinging backwards and forwards past the vertical position and coming to stop at the top of the swing, twice each cycle.

The time taken for one whole cycle is called the vibration period. The time taken depends on the force acting on the string and on the mass of the string. A large force acting on a string with a small mass will accelerate the string rapidly, so the string will complete one cycle quickly. If the string is thick and heavy, the string will accelerate more slowly and one complete cycle will take longer. The force on the string depends on how stiff it is, and that depends on high tightly it was tensioned. If it was stretched a long way lengthwise when it was tensioned, then it will take a large sideways force to pull it sideways. The string will therefore be quite stiff, and will return rapidly to its original position.

The time for one complete cycle does not depend on how far the string is pulled sideways. If it is pulled twice as far, the force on the string will double, so it accelerates twice as fast. But it has twice as far to go during one complete cycle, so it takes exactly the same time. The string will sound louder but its pitch or frequency will be unchanged.

The length of a guitar or piano string also determines how fast it vibrates. Consider two piano strings of the same diameter at the same tension, one being twice as long as the other. The longer string will be twice as easy to pull sideways so it is only half as stiff. The force on the longer string will be only half as large, assuming both strings are pulled sideways by the same amount. Furthermore, the longer string is twice as heavy. On both counts, the longer string will take longer to complete one cycle. In fact, it will take twice as long, not four times longer, for the following reason:

The force on the short string is twice as large, and it acts on only half the mass of the long string, so the acceleration of the short string is four times larger (given that $F = ma$ or $a = F/m$). Over any given time, the short string would travel four times farther than the long string. But it travels the same total distance as the long string during each cycle, and it does so in exactly half the time. Accelerating four times faster for half the time means that it reaches twice the speed of the longer string. Since it travels twice as fast it takes only half the time to travel the same distance. Mathematically, this result is expressed by the equation $s = \frac{1}{2}at^2$ where s is the distance traveled in time t when the acceleration is a.

12.4 Simple Vibration Formula

There is a simple mathematical relation that covers essentially all mechanical vibrations, regardless of whether it is is a string or a bell or a bat that is vibrating. If k is the stiffness of the object that is vibrating and m is its mass, then the time, T for one vibration cycle is given approximately by

$$T = 2\pi\sqrt{\frac{m}{k}} \tag{12.1}$$

The stiffness itself is defined by $k = F/x$ where F is the force required to push or bend or stretch the object by a distance x. The vibration frequency f is given by $f = 1/T$. For example, if $T = 1/2$ s, then $f = 2$ Hz. If $T = 1/100$ s then $f = 100$ Hz. So, if one object is half as stiff as another, and twice as heavy, then m/k is four times bigger, the time for one vibration cycle will be doubled and the vibration frequency will be halved. Doubling the length of a string, therefore, halves its vibration frequency (at the same string tension).

Equation (12.1) is strictly only valid for a localized mass m vibrating up and down or back and forth on the end of a spring with stiffness or "spring constant" k. For a bat, the barrel is stiffer than the handle, but the whole bat will have some average stiffness k that can be defined by (12.1) if we wish to do so. The value of k so defined will be similar in value to the stiffness defined in some other way. For example, if an 80 kg (176 lb) batter stands on a bat and the bat bends in the middle by 1 mm, then the stiffness of the bat $k = 80 \times 9.8/0.001 = 784{,}000\,\mathrm{N\,m^{-1}}$. For a bat of mass 1 kg, (12.1) gives $T = 0.007$ s so the bat will vibrate at about $1/0.007 = 143$ Hz.

The bat might actually vibrate at 150 Hz, partly because standing on a bat acts to bend the bat in a slightly different way. In any case, some parts of the bat vibrate a lot less than other parts so the effective mass of the bat involved in the vibration will be less than the actual mass. Equation (12.1) is, therefore, only approximately correct for extended objects like bats, but the manner in which T varies with overall mass and stiffness is correctly described by (12.1). For example, if one bat is twice as stiff and twice as heavy as another bat then m/k will be the same and both bats will vibrate at about the same frequency, depending on the exact shape of each bat.

Some objects have no stiffness and cannot vibrate. An example would be an unstretched length of string or a loose piece of cloth or a sock. All objects that do have some stiffness will vibrate when they are struck. A string becomes stiff in a sideways direction when it is stretched lengthwise. A bat has its own inbuilt stiffness and does not need to be stretched to develop any stiffness. Bending a bat acts to stretch it on the outside of the bending curve and to compress it on the inside of the curve. A considerable force is, therefore, required to bend it sideways due to the lengthwise stretch or compression of different parts of the bat.

The general rules regarding vibration are that (a) heavy objects vibrate slowly while light objects vibrate rapidly (b) stiff objects vibrate rapidly while flexible objects vibrate slowly, and (c) small or short objects vibrate rapidly while large or long objects vibrate slowly. An object that is very small, very light, and very stiff will vibrate extremely rapidly. Small quartz crystals can vibrate at 10 million Hz or more, although the quartz crystals used in digital watches normally vibrate at only 32,768 Hz $= 2^{15}$ Hz. An electronic circuit in the watch counts those 32,768 vibrations and then advances the time by 1 s.

12.5 Overtones

When a stretched string is pulled aside and released, it does not vibrate at just one frequency. It will usually vibrate at 10 or 20 different frequencies simultaneously. The lowest frequency is called the fundamental frequency. The other frequencies are described as overtones or harmonics. If the fundamental frequency is say 500 Hz, then the string will also vibrate at 1,000 Hz, 1,500 Hz, 2,000 Hz, 2,500 Hz, etc., but it will not vibrate at 600 Hz or 700 Hz. The higher frequencies arise because a stretched string can vibrate in many different ways, as shown in Fig. 12.1. In effect, it can vibrate over its full length, as in Fig. 12.1a, or over half its length as in Fig. 12.1b, or over 1/3 of its length as in Fig. 12.1c. In each case, the whole string vibrates but the vibration pattern can be repeated twice or three or more times along the whole length of the string.

It can be seen in Fig. 12.1 that some points along the string don't vibrate at all. These points are known as nodes. For the fundamental mode shown in Fig. 12.1a, there is a node at each end of the string where the string is tied down. The next mode, in Fig. 12.1b, has an extra node in the middle of the string, while the mode in Fig. 12.1c has a total of four nodes along the string. The string in Fig. 12.1c is tied down only at the two ends but it would behave in the same way if it was tied down at all four node points and if each of the three separate strings was plucked at the same time and in the correct direction to generate the vibration pattern shown in Fig. 12.1c.

There are several different ways of explaining the high frequency vibrations of a stretched string. In terms of string stiffness, the stiffness of a string depends on whether it is pulled aside and released in the middle of the string or towards one end. A relatively small force is needed to pull a string sideways by 1 mm in the middle. A much larger force is needed to pull the string 1 mm sideways if the force is applied near one end. It is the varying sideways stiffness along the string that allows a string to vibrate at many different frequencies.

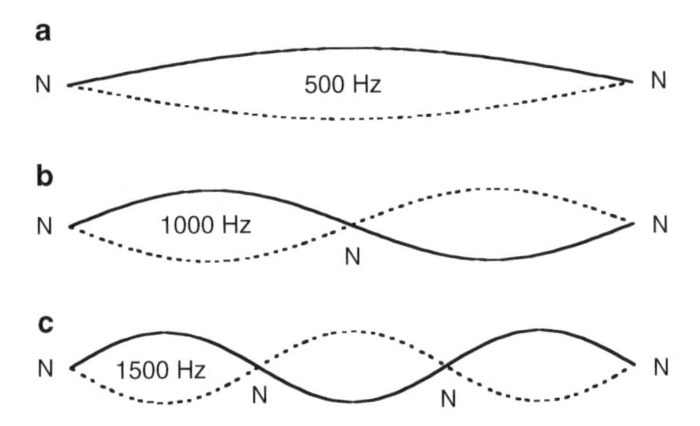

Fig. 12.1 A stretched string can vibrate at many different frequencies. If the lowest frequency is 500 Hz then it can also vibrate at 1,000 Hz, 1,500 Hz, etc. The *solid curves* show the string at one instant of time. The *dashed curves* show the position of the string one half period later. N denotes a vibration node where the string remains at rest

We have seen that if the length of a string is halved then its frequency will double. The situation shown in Fig. 12.1b is effectively two half length strings joined end to end so the frequency is twice that of the vibration in Fig. 12.1a. Similarly, the situation in Fig. 12.1c is effectively three short strings joined end to end and the frequency is three times that in Fig. 12.1a. The regular pattern of vibration of a string is created by a transverse bending wave that propagates along the string and reflects off each end. A regular pattern is created only at certain vibration frequencies and not at other frequencies. If the frequency has the correct value, then waves traveling left and right along the string at that frequency will interfere constructively, meaning that they add up to give the vibration patterns shown in Fig. 12.1. Waves traveling left and right at other frequencies all add up in such a way that they cancel each other over time. The vibration patterns shown in Fig. 12.1 are known as standing waves since they are created by left and right traveling waves but the resulting patterns themselves do not travel along the string.

The specific mixture of these different frequency components for a stretched string accounts for the distinctive differences in tone between a guitar and violin or piano even if they are all used to play the same note and therefore vibrate at the same fundamental frequency. Hitting or plucking a string does not correspond to any particular vibration frequency. However, what it *does* correspond to is a vibration over a whole range of different frequencies all at once. In other words, since the frequency of a short, sharp blow is not well-defined, it is actually equivalent to a broad and continuous spectrum of different frequencies. As a result, if the string is capable of vibrating at any of those frequencies, it will. The frequencies that are favored are those where the disturbances reflected from each end add together in sympathy.

12.6 Stiffness of a Uniform Beam

A bat vibrates in essentially the same manner as any other long, thin object. The simplest such objects are (a) a cylindrical rod that has the same diameter along its whole length or (b) a uniform rectangular cross-section bar or beam. The formulas for the stiffness of rods and beams can be found in text books on engineering mechanics. The stiffness depends on how the rod or beam is supported at each end and where the force is applied. It also depends on the stiffness of the material itself, as specified by a quantity called Young's modulus, E.

To calculate the properties of beams, it helps to define a quantity, I, called the area moment of inertia, related to the cross-sectional shape of the beam. For example, if a rectangular beam of width b, thickness a, and length L is bent in the thickness direction, then $I = ba^3/12$. For a circular rod of radius R, $I = \pi R^4/4$. For a hollow tube of outer radius b and inner radius a, we just subtract the value of I for an inner solid rod of radius a from the value of I for an outer solid rod of radius b to get $I = \pi(b^4 - a^4)/4$. For a tube with a very thin wall where $t = b - a \ll b$, $I = \pi t b^3$.

The overall stiffness and vibration frequency of a beam of any given cross section depends on the quantity EI known as the flexural rigidity of the beam. Obviously, a thick or large diameter beam will be relatively stiff, especially if it is made from a stiff material. The flexural rigidity combines both the inherent stiffness of the material and its cross-sectional shape into a single number to represent the overall stiffness.

The stiffness of various beams is shown in Fig. 12.2. A cantilever beam is one that is bolted down at one end and free at the other, like a diving board. A doubly clamped beam is bolted down at each end. A freely supported beam just rests on a support at each end.

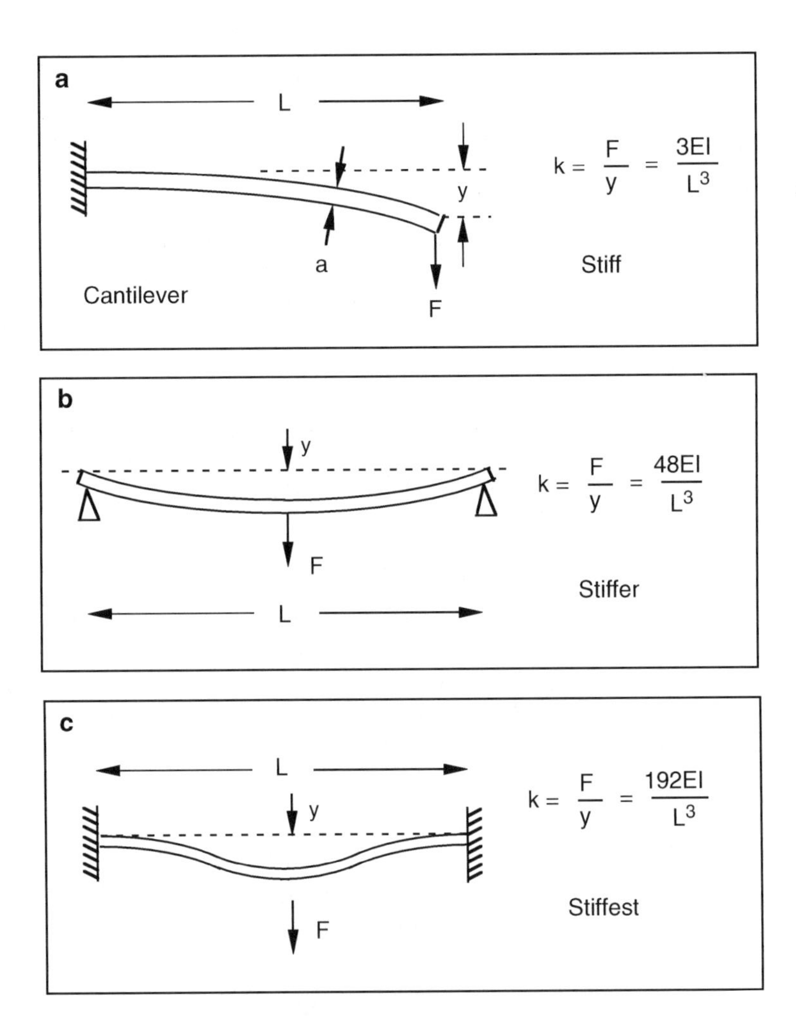

Fig. 12.2 The stiffness of a beam depends on how each end is supported. Regardless of the support method, halving the length increases the stiffness eight times. For a cylindrical rod, doubling its diameter increases its stiffness 16 times since $I = \pi R^4/4$ for a rod of radius R. When a bat vibrates, it bends in a manner similar to that shown in (**b**), at least for the fundamental (lowest frequency) mode. If the bat is clamped at the handle end, as in (**a**), the fundamental vibration frequency is much lower. If clamped at both ends as in (**c**), the fundamental vibration frequency is much higher

12.7 Bat Vibrations

In the case of a guitar string, you can see the string vibrate, you can hear it vibrate and you can feel it vibrate if you touch it. Some vibrations are good and some are bad, some are pleasant and some are annoying. It's the same with a bat. Every time someone hits a ball, the bat vibrates. Some bat vibrations are good, some are bad and some are annoying, especially when the batter mis-hits the ball and stings his hands. The "good" vibrations are the ones that give rise to the trampoline effect, sending the ball on its way with a little extra speed. "Bad" vibrations do exactly the opposite. Bad vibrations extract energy from the ball, remain trapped in the bat, don't give any of that energy back to the ball and sting the hands.

Bats can vibrate at several different frequencies at the same time. The low frequency vibrations are usually too low in frequency to be heard easily by the ear. If you put your ear near a bat and hit the bat gently with a ball, you will hear the low frequency vibrations as a humming sound. It is the low frequency vibrations that sting your hands and that extract energy from the ball. It is the high frequency ping vibrations coming from aluminum bats that are responsible for the trampoline effect, in much the same way that the ping from the strings of a tennis racquet are due to rapid vibrations of the strings. The strings of a racquet act as a trampoline when they eject the ball, and they vibrate for a second or so afterward with an audible ping sound.

Bats can vibrate in several different ways. The low frequency, barely audible vibrations correspond to a back and forth bending of the bat along its whole length. The bat bends into a banana or C shape, first in one direction and then in the opposite direction, at about 170 times per second or 170 Hz. The bat can also vibrate simultaneously at a higher frequency, around 530 Hz, by bending back and forth into an S shape. Both modes of vibration are shown in Fig. 12.3 for a wood bat. Different parts of the bat vibrate with different amplitude. For the 167 Hz mode, the ends and the middle section of the bat vibrate the most. There are two spots about 6.5 in. in from each end that don't vibrate at all. These spots are node points. For the 530 Hz mode there are three node points, one of them being about 5 in. from the barrel end, one being about 3 in. from the knob end and one near the middle of

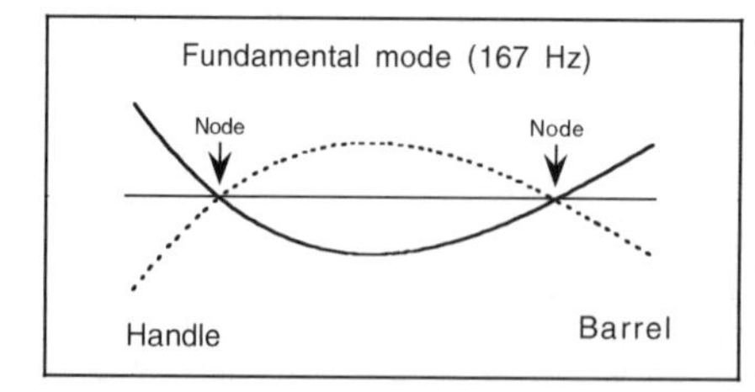

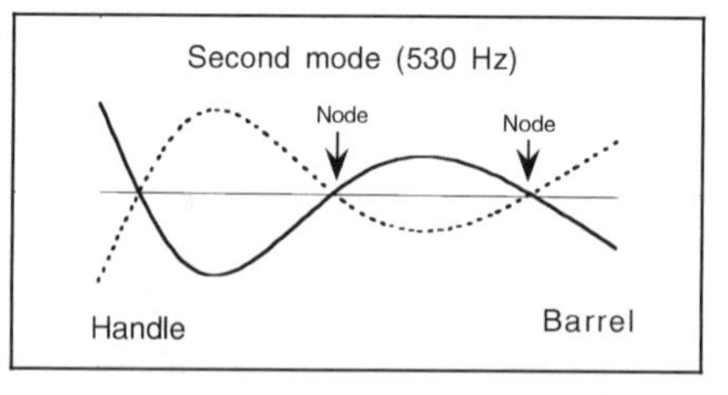

Fig. 12.3 The first two bending modes of a baseball bat. At one instant of time, the bat is bent in the shape of the *solid curve*. A short time later, the bat is completely straight but it shoots past the straight position and bends the other way, as indicated by the *dashed curve*. The handle is more flexible and bends farther than the barrel

the bat. The 167 Hz mode is called the fundamental mode, meaning that it is the mode with the lowest vibration frequency. The 530 Hz mode is called the second mode.

Other bending modes can be observed at even higher frequencies, but they usually play only a minor role in the behavior of a bat since their amplitudes are quite small compared with the fundamental and second modes. The number of vibration modes that is generated depends on the impact time between the bat and the ball, typically about 1 ms. That time is shorter than the time for one vibration cycle of the fundamental or second mode, but it is longer than the vibration period of all the higher frequency modes. The ball sits on the bat for too long to allow the high frequency bending modes to develop strongly, although hoop modes at frequencies above 1,000 Hz can still be heard with hollow bats. The ball damps out bending vibrations in much the same way that a guitar player can stop the vibrations of a guitar string simply by touching the string lightly with one finger. It is for that reason that piano hammers are designed to give a very short blow to a piano string. The hammer must not remain in contact with the string for too long, otherwise the high frequency modes will be strongly damped and the string will sound dead. High frequency bending modes in a bat can be observed and measured by bouncing a small steel ball bearing off a bat since the impact time is then only about 0.1 ms.

When a bat strikes a ball at a node point of the fundamental mode, then the fundamental mode is not generated at all, but the second mode is generated. Similarly, if the impact point is the node of the second mode, then the second mode is not generated, but the fundamental mode is generated. The farther the impact point is from a node, the larger is the resulting vibration. The vibration amplitude is also proportional to the relative speed of the bat and the ball. Vibrations of a bat are largest in amplitude when the ball is struck right at the tip of the bat. Such an impact usually stings the hands. Vibrations are weakest when the bat strikes the ball at one or other of the two node points about 6 in. from the tip of the bat. These are the sweet spots on the bat. In the sweet spot region between the two node points, vibrations are reduced to such a low level that the batter is almost unaware of the fact that he or she actually struck the ball.

The existence of a node point can be explained in a slightly oversimplified manner with reference to Fig. 12.4. We show in Fig. 12.4a a bat in a position at rest, with a ball approaching from the left. An impact near the tip of the bat will bend the tip to the right, as shown in Fig. 12.4b, and the bat will rotate away from the ball. An impact near the middle of the bat will bend the tip to the left, without any rotation of the bat if the impact is at the bat center of mass (CM). The whole bat will just recoil away from the ball without rotating. An impact at the node point about half way between the tip and the CM bends the bat locally at the impact point, but as the bending wave propagates toward each end, the bat quickly straightens out. The end result of an impact at the node point is that the bat translates and rotates away from the ball without vibrating.

In each case shown in Fig. 12.4 we have ignored the manner in which the bend develops and changes over time. The whole bat does not bend instantly into the positions shown. Rather, the bend occurs locally around the impact point while the rest

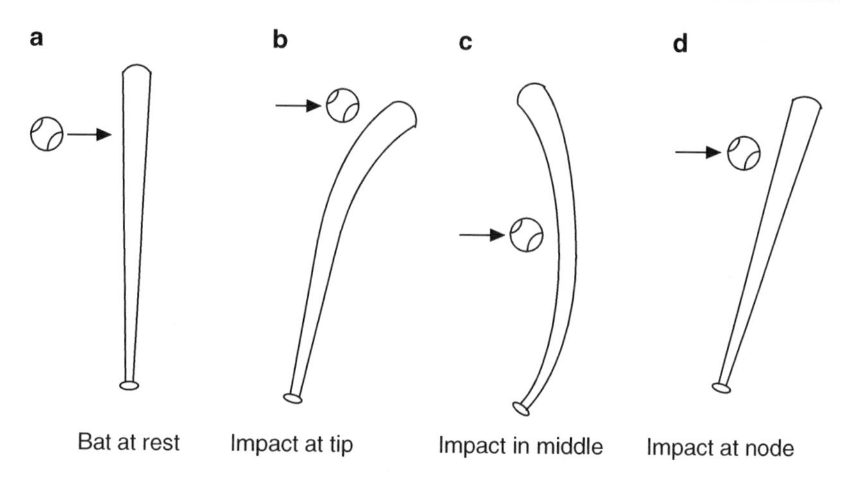

Fig. 12.4 A ball impacting near the tip of a stationary bat causes the tip to bend to the right. If the ball impacts near the middle of the bat then the tip bends to the left. An impact at the vibration node causes the bat to rotate without vibrating

of the bat remains straight. The local bend then propagates as a wave away from the impact point. As it does so, the bat assumes the shapes shown in Fig. 12.4. Some of the bending energy ends up as rotation of the bat, some as recoil motion of the whole bat and some as vibration of the bat. The energy that remains is retained by the ball as it bounces off the bat. If the bat is initially at rest then the ball bounces off the bat at only about 1/5th of its incident speed at most, depending on the impact point.

12.8 Hoop Modes

The high frequency ping of an aluminum bat is due to a change in bat shape around the circumference of the barrel. The barrel is circular in cross-section, but the shape of the barrel changes when it strikes a ball due to the large force acting on the barrel. The barrel squashes slightly into an elliptical or oval shape. Being springy, the barrel returns rapidly to its circular shape when the ball leaves the bat, but it overshoots and stretches out of shape in the opposite sense, as shown in Fig. 12.5. This mode of vibration is called a hoop mode, since it is how a circular hoop vibrates when it is dropped on the floor or struck on one edge. The vibration frequency is typically about 2,000 Hz for a single-walled aluminum bat, about 1,500 Hz for a double-walled bat and can be as low as 1,000 Hz for some composite bats. The vibration frequency depends on the stiffness of the bat across a diameter of the barrel. The thicker the wall, the stiffer it is and the higher is the vibration frequency. The softer the wall, the lower is the vibration frequency and the better is the trampoline effect. The bat is likely to break if the wall is too thin so the manufacturer has to compromise between the desire to make a high performance bat and one that lasts a reasonable length of time.

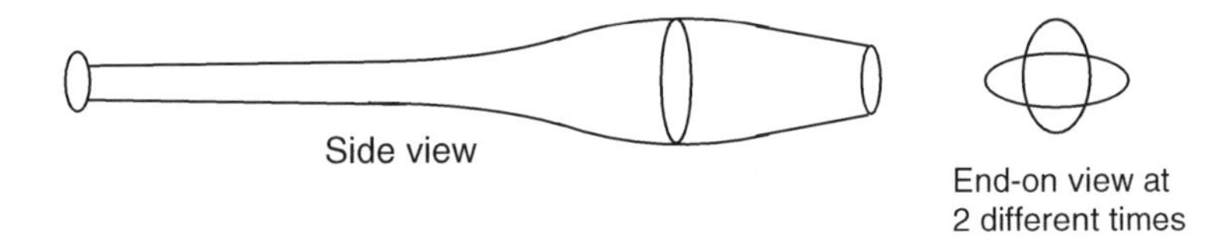

Fig. 12.5 The hoop mode of a bat. The barrel is squashed into an elliptical shape as shown by the end-on views at two different times during one vibration cycle. The side-on view shows the barrel at one of those times when the depth of the barrel (into the page) is smaller than its vertical height

Nice animations of the various vibration modes of a bat can be found at Dan Russel's web site at www.kettering.edu/~drussel/bats.html

12.9 Development of a Vibration

When a bat and ball collide, the bat bends locally at and near the impact point. The whole bat does not bend instantaneously. Rather, a small region around the ball bends, launching a bending wave that travels at about 1,000 ft s^{-1} along the bat, in both directions. Despite the high speed of this wave, the ball bounces off the bat before the wave reaches the handle. The ball remains in contact with the bat for only about 1 ms, or 0.001 s, by which time the bending wave will have traveled about one foot away from the impact point. The wave travels too fast for the eye to see, but the effect is the same as a wave on the surface of water.

Water waves travel slowly enough to see by eye. You can check that out for yourself. Partly fill a sink or a bath with water and wait for the water surface to stop moving around. Then let one drop of water fall into the middle of the artificial lake. A circular wave will propagate out from the impact point, bending the surface into a series of ripples. A similar thing happens to a bat but the bending wave travels a lot faster.

Suppose that a bat is 33 in. long and a bending wave travels along the bat at 1,000 ft s^{-1}. As a soon as the wave gets to the end of the handle, it reflects off the handle end and heads back to the impact point. At 1,000 ft s^{-1}, the wave takes 5.5 ms to make a complete 66 in. round trip from one end of the bat and back again. It does so 182 times in 1 s, which is the same as the vibration frequency of the bat. By the time the ball leaves the bat, after 1 ms, the handle end has not bent at all and doesn't even know that a wave is headed its way. There is no vibration of the whole bat until that wave reaches the end of the handle, well after the ball is on its way. Even if we attempted to measure what was happening, we would not see or recognize a vibration at 182 Hz until the wave had traveled a few times up and down the bat. It therefore takes about 10 or 15 ms after the initial impact before we can say that the bat is vibrating at 182 Hz (or at some other nearby frequency, depending on the bat mass, length and stiffness).

The initial impact does not send a 182-Hz vibration along the bat. It just sends a bending wave that cannot be recognized as being a wave of any particular frequency at all. In fact, the wave can be regarded as being something that consists of hundreds of different waves all at different frequencies, and they all add up to give the localized bend in the bat that was created by the impact with the ball. It turns out that high frequency bending waves always travel faster than low frequency waves. The speed of a transverse bending wave along a bat (or along a rod or beam) is given by $v = c\sqrt{f}$, where f is the frequency and c is a constant depending on the mass and stiffness of the bat (see Appendix 12.1).

The distance between wave peaks is shorter for a high frequency wave than a low frequency wave. As a result, the bat bends over a shorter distance for the high frequency waves. It is harder to bend a bat over a short distance than over its whole length, so the bat is stiffer for high frequency waves, which therefore travel faster. This phenomenon is known as dispersion, meaning that the wave speed depends on the frequency of the wave.

Dispersion is a nice descriptive word for the effect that it produces. Since the high frequency components of the original bend in the bat travel faster, they race ahead of the low frequency components. The bend therefore gets dispersed as it travels along the bat. If the bat happened to be 1 mile (5,280 ft) long, and if it was struck at one end by a ball, then the frequency components traveling at 1,000 ft s^{-1} would take 5.28 s to get to the far end. Higher frequency components traveling at say 2,000 ft s^{-1} would take only 2.64 s to get to the far end, while lower frequency components traveling at say 500 ft s^{-1} would take 10.56 s to get to the far end. A person with one ear on or near the end of the bat would hear a whistling sound decreasing in pitch, lasting about 8 s. The original impact might have lasted only 1/1000 s, yet the the bending wave that arrives at the other end is so dispersed or spread out that it arrives after a delay of about 2.6 s and continues to arrive for another 8 s. The effect is analogous to a marathon race. All the runners might start at the same time, within half a second. The first to finish might arrive 2.5 h later with legs moving rapidly and the rest arrive over the next hour or so with the last to arrive with their legs barely moving.

12.10 Experiment with a Brass Bar

An experiment illustrating wave dispersion is shown in Fig. 12.6. A 3.6 m long, 9 mm thick brass bar was suspended by means of a few lengths of string and a small piezo disk was taped to one end. A tennis ball dropped onto the piezo generated a voltage signal lasting about 5 ms, indicating that the ball bounced off the bar after 5 ms. The impact generated a bending wave that traveled along the bar, reflected off the end of the bar and arrived back at the start after a delay of about 30 ms. The impact contained strong frequency components from about 50 Hz to about 250 Hz, and weaker components outside that range. The low frequency components of the initial impact arrived much later than the high frequency components. A similar result was obtained by dropping a golf ball on the piezo. The impact was then only 1 ms in duration and it contained strong frequency components from about 500 Hz

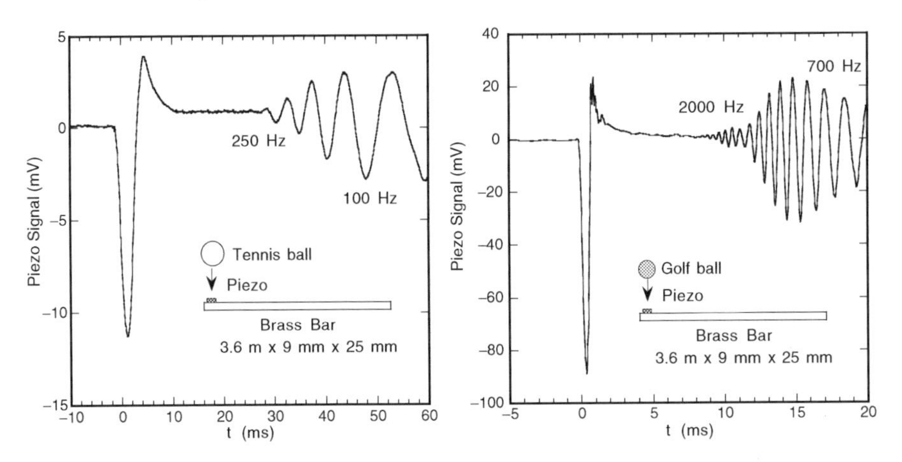

Fig. 12.6 Transverse wave propagation along a 3.6-m long brass bar observed by dropping a tennis or golf ball onto a piezo at one end. The contact time of the tennis ball was about 5 ms, as shown by the initial large signal. The contact time of the golf ball was about 1 ms. The wave reflected off the far end of the bar and was observed on its return, the high frequency components arriving first. The bar continued to vibrate for about 30 s afterwards, but only the low frequency standing wave modes survived

up to about 2,000 Hz. The 2,000 Hz components arrived after a delay of about 10 ms, while the 500 Hz components arrived after a delay of about 20 ms. Standing waves were generated after several trips up and down the bar, at many different frequencies, and they persisted for about half a minute after the initial impact. The equation describing transverse waves on a beam, or any other long object such as a bat, is given in Appendix 12.1.

12.11 Vibration Frequency of a Bat

The vibration frequencies of a stretched string are easy to calculate. If the fundamental mode is at frequency f then the next mode is at frequency $2f$, the next is at $3f$ and so on. If the length of the string is doubled then the frequency is halved. In the case of a baseball or a softball bat, the situation is more complicated. If the fundamental bending mode is at frequency f then the next mode is at a frequency about $2.8f$ depending on the exact shape of the bat. If the length of a bat is doubled then the frequency would drop by a factor of four. A doubling of bat length is of no practical interest but the point here is that a small change in bat length results in a large change in the frequency. The frequency is inversely proportional to the length squared. The reason for this is that the stiffness, k, of a bat is inversely proportional to its length cubed (as indicated in Fig. 12.2) and the mass, m, of the bat is proportional to its length. Since k/m is proportional to $1/L^4$, we see from (12.1) that f is proportional to $1/L^2$.

The speed of a transverse wave on a stretched string does not depend on the vibration frequency. It is for this reason that the various vibration frequencies are integer multiples and can be described as harmonics. A bat is more complicated, for two reasons. One is that the speed of a bending wave depends on the vibration frequency, even for a uniform cylindrical rod. The other problem is that a wave traveling along a bat speeds up when it encounters the stiff barrel and slows down when it encounters the flexible handle. The various vibration frequencies are not integer multiples and cannot be described as harmonics.

The hoop mode of a hollow bat has a vibration frequency around 1,500 Hz. Some bats are softer, with a lower hoop mode frequency, and some are stiffer, with a higher hoop mode frequency. The stiffness of the barrel can be estimated from (12.1). A typical baseball bat weighs 31 oz (0.88 kg), the barrel weighs about 0.6 kg, a 10 cm section of the barrel weighs about 0.15 kg and each half has a mass $m \sim 0.075$ kg. If that section vibrates at 1,500 Hz, then the vibration period is $T = 1/f = 0.67$ ms, so $k = 4\pi^2 m/T^2 = 6.6 \times 10^6\,\mathrm{N\,m^{-1}}$. In that case, the bat is about five times stiffer than the ball. If the ball squashes by say 15 mm when the bat strikes the ball, then the bat will squash by about 3 mm. This will help to propel the ball off the bat a fraction faster since some of the elastic energy of the collision is taken up by the bat and then given back to the ball. The details of this process are described in Chap. 13.

Measurements of the transverse stiffness and hoop vibration frequency for a range of aluminum tubes, obtained by the author, are shown in Fig. 12.7. The stiffness shown in Fig. 12.7 refers to short tubes cut to a length of 25 mm and measured in a materials testing machine. The values obtained are consistent with those found for real bats when compressed over a short length [1]. The transverse stiffness of a long tube is proportional to its length, provided the tube is compressed over its whole length. However, if a long tube is compressed over a short length near the middle of the tube, then that short section is typically about three or four times

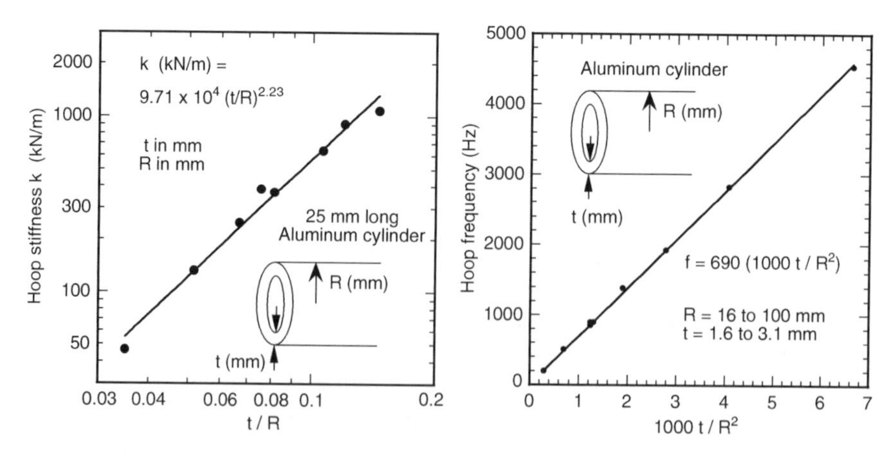

Fig. 12.7 Measured transverse stiffness and hoop frequency for a range of aluminum tubes of different wall thickness, t, and tube radius, R

stiffer than when the tube is cut to a length equal to the short section. Furthermore, the stiffness of a long tube compressed over a short length increases as the deformation increases, while the stiffness of a short tube remains constant if the deformation is relatively small.

Given that the mass of a tube is proportional and tR and that the hoop frequency was found to be accurately proportional to t/R^2, one would expect that the stiffness should be proportional to $(t/R)^3$. In fact, the stiffness was found to be proportional to $(t/R)^{2.23}$, suggesting that different tubes may have been made from different alloys or that some had been annealed or hardened.

Appendix 12.1 Transverse Waves on a Beam

The behavior of the long brass bar in Fig. 12.6, and the behavior of a bat when it impacts a ball can both be described mathematically by the transverse wave equation for a beam, which has the form

$$\rho A \frac{\partial^2 y}{\partial t^2} = F_o - \frac{\partial^2}{\partial x^2}\left(EI\frac{\partial^2 y}{\partial x^2}\right), \qquad (12.2)$$

where F_o is the external force per unit length acting on the beam, ρ is the density of the beam, A is its cross-sectional area, E is Young's modulus, I is the area moment of inertia, and y is the transverse displacement of the beam at coordinate x along the beam. For a uniform beam of mass M and length L, numerical solutions of (12.2) can be obtained by dividing the beam into N equal segments each of mass $m = M/N$ and separated in the x direction by a distance $s = L/N$. An impacting ball may exert a force acting over several adjacent segments, depending on the ball diameter and the assumed number of segments. If it is assumed that the ball impacts on only one of the segments, exerting a time-dependent force, F, then the equation of motion for that segment (the nth segment) is obtained by multiplying all terms in (12.2) by s, in which case

$$m\frac{\partial^2 y_n}{\partial t^2} = F - (EIs)\frac{\partial^4 y_n}{\partial x^4} \qquad (12.3)$$

assuming that the beam is uniform so that E and I are independent of x. The equation of motion for the other segments is given by (12.3) with $F = 0$. Equation (12.3) is easily interpreted. The left hand side is the mass times the acceleration of a segment of the beam in the y direction. F is the force of the ball acting on that segment. The last term is the force exerted on the nth segment by the two segments either side of the nth segment, arising from bending of the beam.

The boundary conditions at a freely supported end of the beam (or the barrel end of a bat) are given by $\partial^2 y/\partial x^2 = 0$ and $\partial^3 y/\partial x^3 = 0$.

Solutions of (12.2) yield not only the outgoing speed of the ball but also describe the behavior of the bat in terms of its rotation, translation and vibration. It is a very powerful equation. Solutions for aluminum beams are given by the author [2] and solutions for real bats are described by Alan Nathan [3]. Vibration modes of a cricket bat are described in [4].

A simple solution of (12.2) can be found for a sinusoidal wave propagating along a uniform beam at frequency f with a wavelength λ. The phase velocity of such a wave is given by $v = f\lambda$ or by $v = \omega/k$, where $\omega = 2\pi f$ and $k = 2\pi/\lambda$. The solution can be found if we describe y in terms of the standard wave equation $y = y_o \exp[i(kx - \omega t)]$ in which case $\partial^2 y/\partial t^2 = -\omega^2 y$ and $\partial^2 y/\partial x^2 = -k^2 y$. Substitution in (12.2) gives

$$\omega^2 = \frac{EIk^4}{\rho A} \tag{12.4}$$

and hence the phase velocity of the wave is

$$v = \frac{\omega}{k} = \left(\frac{EI}{\rho A}\right)^{1/4} \sqrt{\omega} \tag{12.5}$$

Consequently, high frequency waves propagate at a higher speed than low frequency waves. The velocity is high for stiff beams (with a large value of EI) and for light beams (with a low value of ρA). For a freely supported, uniform beam of length L, standing waves exist when $\lambda = 1.328L$ (for the fundamental mode) or when $\lambda = 0.800L$ (for the second mode) corresponding to $k = 4.730/L$ and $k = 7.853/L$, respectively. The fundamental mode for a uniform beam has a node at $x = 0.22L$ and another node at $x = 0.78L$, similar to the situation shown in Fig. 12.3 for a freely supported bat.

Solutions of the beam equation are shown in Fig. 12.8 for a uniform beam of mass 0.885 kg and length 0.84 m, similar in mass and length to a real bat, for impacts near the tip, sweet spot and middle of the beam. The beam was initially stationary and was impacted with a baseball of mass 0.145 kg and stiffness 1,000 kN m^{-1}, exciting both the fundamental and second bending modes. The behavior of the beam is very similar to that of a real bat [3] and it exhibits many of the features that we have previously described. In particular, the beam translates, rotates and vibrates in a manner that depends on the impact point on the beam, with the result that the COR varies strongly with impact position along the beam.

If the ball impacts near the tip then the beam rotates rapidly away from the ball, resulting in a relatively small impact duration, strong vibration of the beam after the impact is over and a low COR for the collision. A bending wave is seen to travel toward the handle end, but the ball bounces off the beam before the bending wave reflects back to the impact point. If the ball impacts at the middle of the beam, as shown in Fig. 12.8c, then a bending wave travels toward each end, but the ball bounces before the wave is reflected back to the middle of the beam. The second mode is not excited when the ball impacts at the node in the middle of the beam, and there is also no rotation of the beam. Nevertheless, the COR is low for impacts near the middle of the beam due to significant bending of the beam, resulting in a

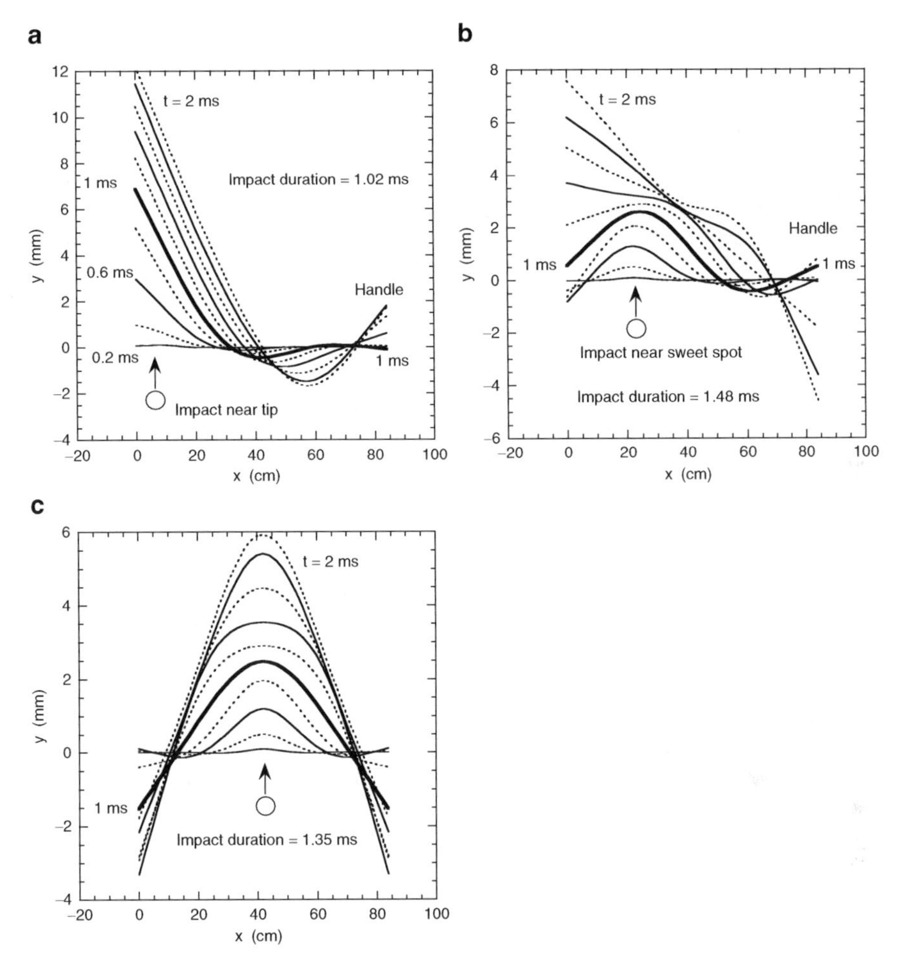

Fig. 12.8 Solutions of the beam equation for a baseball impacting on a stationary, uniform wood beam of mass 0.885 kg and length 0.84 m, with $EI = 1,000$ Nm, showing the position of the beam every 0.2 ms after the initial impact. The position of the beam after 1 ms is shown by the *thicker line*

strong fundamental mode vibration. That vibration is not evident in Fig. 12.8c since the position of the beam is shown only up to $t = 2$ ms. The fundamental vibration period for the beam is about 8 ms, so the position of the beam would need to be shown for 8 ms or more to see a complete vibration cycle.

The COR is largest for impacts near the sweet spot. The initial bend caused by an impact near the sweet spot does not result in any significant vibration of the beam. Instead, the beam straightens out as it translates and rotates, with the result that only a weak vibration persists after the collision is over. The vibration is due to excitation of the fundamental or second mode or both modes together, depending on the impact point and whether it coincides with the node of the fundamental mode or the node of the second mode.

References

1. L.V. Smith, C.M. Cruz, Identifying altered softball bats and their effect on performance. Sports Technol. **1**, 196–201 (2008)
2. R. Cross, Impact of a ball with a bat or racket. Am. J. Phys. **67**, 692–702 (1999)
3. A. Nathan, Dynamics of the baseball bat collision. Am. J. Phys. **68**, 979–990 (2000)
4. R. Brooks, J.S.B. Mather, S. Knowles, The influence of impact vibration modes and frequencies on cricket bat performance. Proc. IMechE, Part L J. Mater. Design Appl. **220**, 237–248 (2006)

Chapter 13
The Trampoline Effect

13.1 Introduction

The superior performance of non-wood bats is due to the fact that they are hollow rather than solid. The wall of a hollow bat vibrates like a drum when it strikes the ball, emitting a characteristic high pitched ping sound. The wall vibrates not while the bat and ball are in contact, but after the ball leaves the bat. The enhanced performance of a hollow bat is not due directly to the vibration of the wall of the bat. Rather, the enhanced performance is due to the fact that the wall bends slightly and then springs back as the ball leaves the bat. It is the springiness of the wall that gives rise to the improved performance of the bat. If the wall could somehow do its job more efficiently, it wouldn't ping at all. The wall would then give all its stored elastic energy back to the ball, shooting the ball out at high speed, and there would be no energy left in the wall for it to vibrate.

The wall does help to shoot the ball out, but the wall retains some of its stored elastic energy in the form of vibrational energy. Instead of giving all its energy back to the ball, by returning rapidly to its undisturbed position, the wall overshoots that position and then vibrates back and forth like a plucked guitar string, or like the membrane of a drum. The ping sound is a sign that the wall indeed retains some vibrational energy, although the sound level itself does not provide a good indication of the amount of that energy. Tennis strings also ping when they strike a tennis ball, yet they give back 95% of their stored elastic energy to the ball. Very little energy is retained by the strings since they are very light. The wall of a bat is much heavier than tennis strings and the vibrational energy content is proportionally larger.

13.2 Simple Trampoline Experiments

The spring action of the wall of a bat in enhancing the exit speed of the ball can be demonstrated by dropping a ball on various solid and springy surfaces. As indicated in Fig. 13.1, a baseball dropped from a height of 3 ft onto a hard wood floor will bounce to a height of about 13 in. Dropped onto some other surface, the ball will

R. Cross, *Physics of Baseball & Softball*, DOI 10.1007/978-1-4419-8113-4_13,

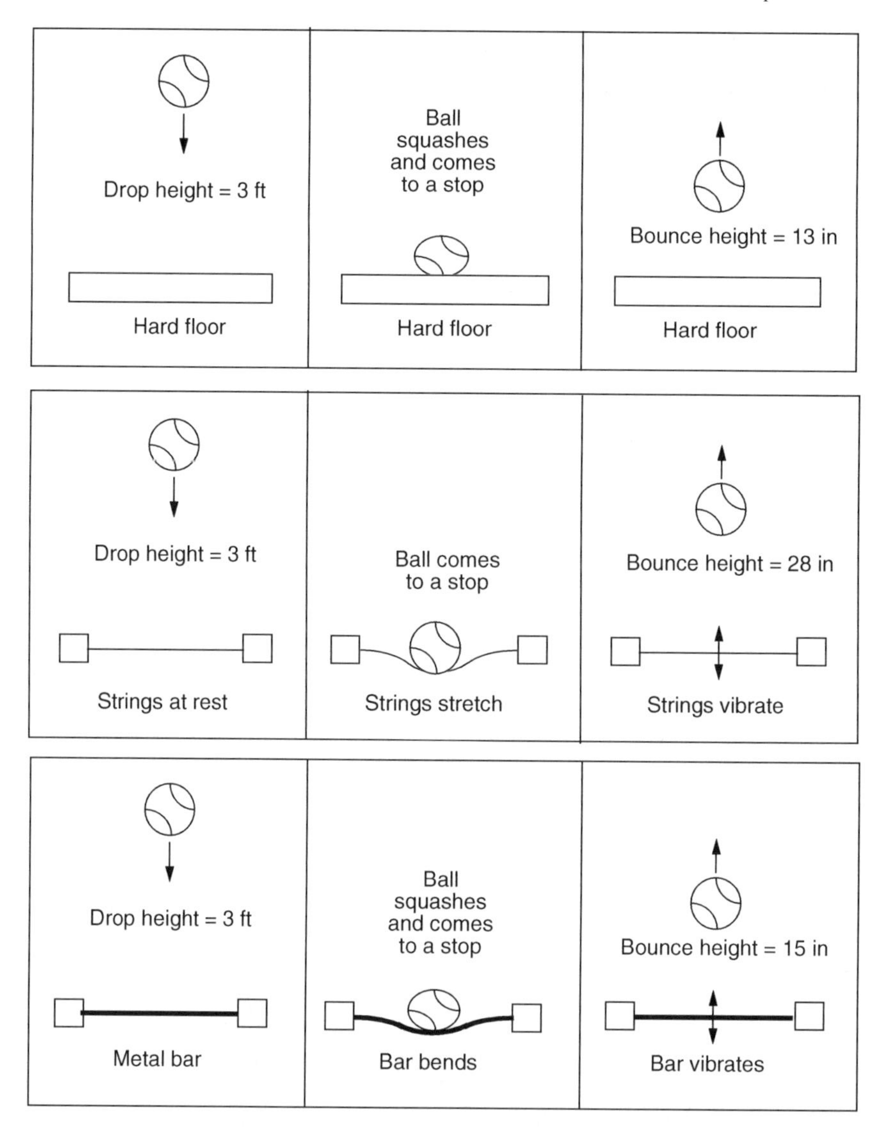

Fig. 13.1 When a ball is dropped onto a given surface, the bounce height depends on the mass and stiffness of that surface. A hard floor is heavier and stiffer than the ball. Tennis strings are lighter and softer. The bounce height off a flexible metal bar clamped at each end increases as the thickness of the bar decreases, and decreases as the length of the bar increases, despite the increase in flexibility of the bar in both cases (author's observations)

bounce to a different height. For example, if the ball is dropped onto soft earth or dry sand, the ball will bounce to a height of less than 1 in. Dropped onto a trampoline, the ball will bounce to a height of about 28 in. You can check this out yourself using either a real trampoline or the strings of a tennis racquet. If a racquet is placed on a solid floor and the frame or handle is held down under foot so that it can't move or vibrate, then a baseball dropped onto the strings bounces a lot higher than when it is dropped from the same height directly onto the floor. This effect is known as the trampoline effect. It is important in baseball and softball, and it is also important in golf and tennis. In golf, a springy club face will eject the ball off the club at a higher speed than off a very stiff club face. In tennis, the strings act as a very good trampoline, ejecting the ball at high speed and reducing the shock force transmitted to the arm.

People can jump much higher on a trampoline than on a wood floor because every jump on a trampoline adds to the height of the previous jump. The way this works is that the kinetic energy gained when a person falls back to the trampoline is used to stretch the trampoline, and the trampoline then returns that energy back to the jumper. The trampoline itself does not generate any extra energy. On a wood floor, the floor does not stretch so it has no elastic energy to give back to the jumper. A person jumping up and down on a wood floor will jump to the same height every time.

When a ball is dropped onto the strings of a racquet, the strings stretch and the ball squashes. The strings are not as stiff as the ball so they stretch further than the ball compresses. If the ball compresses by say 1 mm then the string plane will stretch by 4 mm if it is four times softer than the ball, or by 10 mm if it is ten times softer. The same effect happens when a ball is dropped on soft earth or dry sand. However, earth or sand does not spring back like the strings of a racquet. Strings are used in a racquet because they are highly elastic, meaning that whatever energy was used to stretch them is given back to the ball when the strings return to their original position. Dry sand has no elasticity and remains permanently deformed when a ball is dropped onto it or when someone steps onto the sand.

Elastic Energy Stored in Tennis Strings

When a ball is dropped onto a hard wood floor, all of the kinetic energy gained by the ball as it fell to the floor is used to compress the ball. The energy is stored as elastic energy in the ball, but 65% of that energy is lost immediately in a low speed bounce and only 35% remains to push the ball back off the floor. The ball therefore bounces to only about 35% of its drop height. If the ball is dropped onto the strings of a racquet, and if the strings are four times softer than the ball, then the strings will store 4 times more elastic energy than the ball. The ball loses 65% of the energy stored in the ball, but the strings lose only 5% of the energy stored in the strings.

Suppose that the kinetic energy of the ball is 100 units just before it hits the strings. Eighty units are used to stretch the strings, and 20 units are used to compress the ball. The strings then return $0.95 \times 80 = 76$ units back to the ball. The ball itself loses $0.65 \times 20 = 13$ units, and retains 7 units. The total energy of the ball when it bounces is $76 + 7 = 83$ units, so the ball bounces to 83% of its drop height. On a hard floor the ball bounces to 35% of its drop height. The ball therefore bounces off the strings more than twice as high as it does on a wood floor. This is a much more dramatic effect than that occurring when a ball bounces off a hollow bat, but it provides an interesting demonstration of the basic physics behind the trampoline effect.

Elastic Energy Stored in a Metal Bar

Before we examine the trampoline effect in a hollow bat, it is useful to consider briefly the situation shown in Fig. 13.1 where a baseball or a softball is dropped onto a flexible metal bar. This arrangement is more like an actual trampoline that is flexible in the middle and clamped around the edges. However, the results are surprisingly complicated. If the bar is thin and light then it behaves like the strings of a racquet or the fabric of an actual trampoline, and the ball bounces much higher than it does off a hard floor. If the bar is about as heavy as the ball, then the ball can slow down considerably when it impacts the bar, much like one billiard ball colliding with another. The ball and the bar can lose contact for a while, but then the bar vibrates back toward its original position and strikes the almost stationary ball in mid air. The result depends on whether the ball is moving toward or away from the bar when it is struck by the vibrating bar, so the bounce height varies strongly with the mass and stiffness of the bar. Given that a bat would recoil away from an incoming ball in this situation, since it is not clamped down at each end, we will not pursue the clamped bar case any further.

Elastic Energy Stored in the Wall of a Bat

The effect of bat stiffness on the trampoline effect can be calculated in the same way as in our calculation for tennis strings. The calculation is given in Appendix 13.1 and the result of that calculation is shown in Fig. 13.2. The result depends on the relative stiffness of the bat and the ball, which we can estimate crudely as follows. If you stand on a baseball or a softball, the ball will squash by about 1 mm. If you stand on a hollow bat, the bat will squash by less than 1 mm. If you stand on a solid wood bat, the bat will squash by only about 0.2 mm. If you stand on the strings of a racquet then the string plane will stretch about 15 mm.

The stiffness of a hollow bat depends on the wall thickness, the diameter of the bat and the material from which it is made. One indication of bat stiffness is the pitch of the ping sound it makes when a bat is struck by a ball or by any other solid object. A low frequency ping around 1,000 Hz indicates that the bat is relatively soft, while a high frequency ping, around 2,000 Hz or more, indicates that the bat is relatively stiff. As the wall thickness of a bat increases, the wall gets heavier but it also gets stiffer. The net result is that the ping frequency increases as the wall thickness increases. If the barrel diameter increases, the wall becomes softer (easier to bend) and the ping frequency decreases. Measurements [1] and calculations [2]

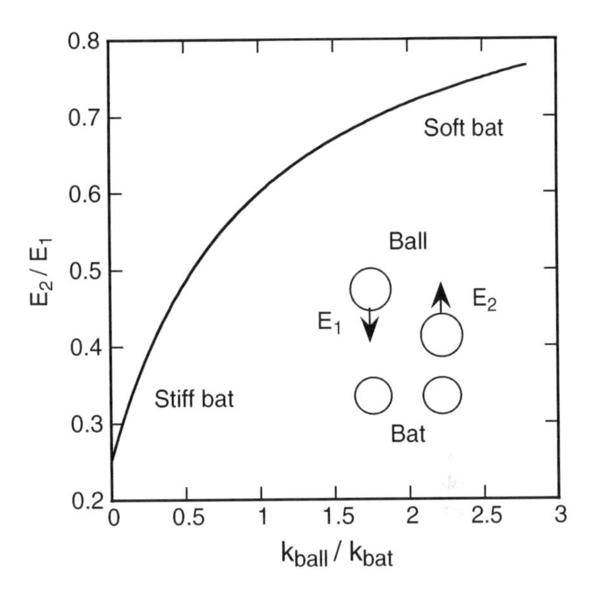

Fig. 13.2 Ratio of the rebound energy, E_2, to the incident energy, E_1, of a ball incident on a clamped bat. The rebound speed increases as the ball stiffness, k_{ball}, increases or as the bat stiffness, k_{bat}, decreases

both show that the COR for a collision between a bat and a ball increases as the wall of the bat becomes softer. The batted ball speed is therefore largest off bats that ping at low frequency.

Figure 13.2 shows a calculation for a ball of stiffness k_{ball} bouncing off a bat of stiffness k_{bat} when the bat is clamped to a floor so that it cannot recoil. E_1 is the kinetic energy of the ball before it bounces, and E_2 is the kinetic energy of the ball after it bounces. It was assumed that the ball loses 75% of its elastic energy during the bounce and the bat loses only 5% of its elastic energy. In other words, it was assumed that the wall of a bat is sufficiently light that it behaves like the strings of a racquet in returning 95% of its stored elastic energy back to the ball.

The ratio E_2/E_1 is shown in Fig. 13.2 as a function of the ball to bat stiffness ratio. It gives the ratio of bounce height to drop height for a ball dropped onto a hollow bat. If the bat is infinitely stiff, then $E_2/E_1 = 0.25$, which is the same result as that for a ball bouncing on a hard wood floor. If the bat and ball were equally stiff then $E_2/E_1 = 0.6$, meaning that the ball would bounce 2.4 times higher than it would off a hard wood floor.

In practice, the wall of a bat can be as heavy as the ball, and our assumption about small wall mass is not really a good one. A better, but more complicated calculation is given in the following section.

13.3 Trampoline Calculations for a Bat

The wall of a bat is a lot heavier and stiffer than the strings of a racquet, and it can potentially retain more vibration energy than it returns to the ball via the trampoline effect. In other words, a hollow bat designed to give extra energy to the ball might even extract energy out of the ball. We assumed in the previous section that the wall would return 95% of its stored energy, since that is how the strings of a racquet behave. But what if it is only 20% or perhaps only 10%? Then a thin wall bat would perform worse than a solid wood bat. To estimate the energy return from the wall of a bat, we need to determine or estimate the mass of that part of the wall that acts as a trampoline and we need to estimate its stiffness.

When a hollow bat strikes a ball, the wall bends inwards in a manner that depends on the wall stiffness. If the wall was very soft, made from say soft plastic or rubber, or if the wall was very thin, then the wall on the impact side would buckle and bend inwards as shown in Fig. 13.3. However, if the wall is relatively stiff, as it is with most modern bats, then the wall on the impact side will tend to flatten but it will not bend inwards unless the force on the wall is extremely large. The wall on the opposite side of the impact point will tend to remain circular during the impact, although it will start vibrating in and out after the ball bounces off the bat. The mass of the wall that is in direct contact with the ball is hard to estimate precisely, but we can at least make a reasonable estimate that it will be somewhere in the range from about 100–300 g for a typical aluminum or composite bat. The actual value makes a big difference to the batted ball speed, as we will now show.

The elastic behavior of the ball and the wall can be modeled as shown in Fig. 13.4. The ball is represented by a mass m_1 connected to a spring S_1 of spring

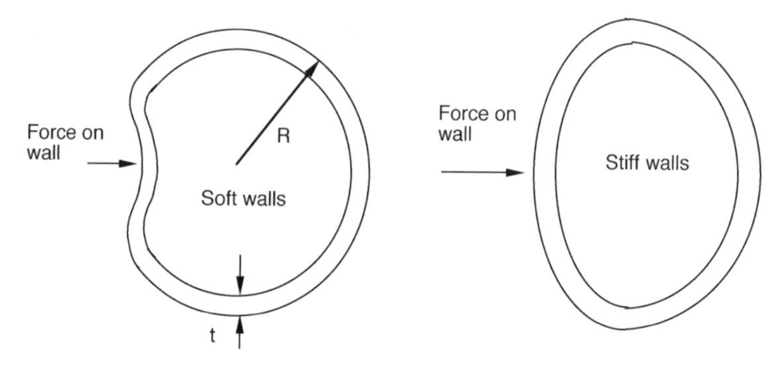

Fig. 13.3 When a hollow bat and ball collide, the wall will buckle if it is soft enough, or tend to flatten if it is relatively stiff

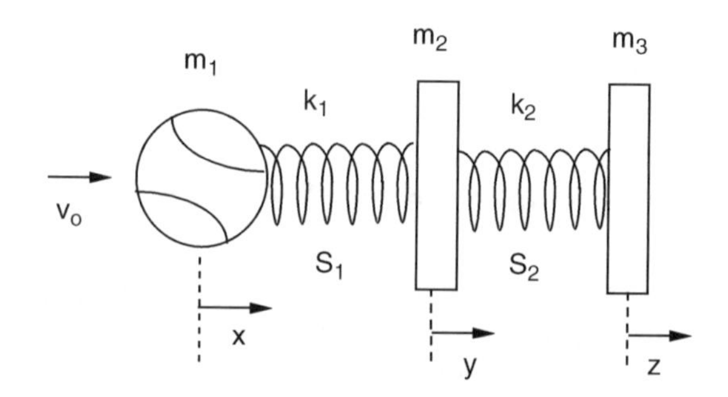

Fig. 13.4 Model used to describe the collision between a ball and the wall of a hollow bat

constant k_1. The wall in contact with the ball is represented by a mass m_2, and is connected elastically to a mass m_3 by a spring S_2 with spring constant k_2. The total mass of the bat at the impact point, $m_2 + m_3$, will be taken as the effective mass of the bat so that the speed of the bat and the ball after the collision will be consistent with results obtained previously in Chaps. 10 and 11. The advantage of splitting the effective mass into two separate parts is that energy loss due to bat vibrations can be described in a relatively simple and intuitive manner, without having to solve the more complicated beam equation outlined in Appendix 12.1.

We will assume that $m_1 = 0.145$ kg, which is the mass of a standard baseball. The stiffness of a baseball is typically about $1 \times 10^6\,\mathrm{N\,m^{-1}}$ (5,710 lb in.$^{-1}$) when it is compressed but the ball stiffness decreases when it expands. To account for the fact that the COR of a baseball is about 0.5 when it collides with a heavy, solid surface, there are several different ways that energy loss in the ball can be modeled. One way is to let $F = k_1 s$ while the ball is compressing, where s is the ball compression, and to let $F = k_3 s^7$ while the ball is expanding. In Fig. 9.4, the ball is shown to expand with an exponent s^6. The exponent s^7 results in a COR of 0.5 for a high speed impact with a rigid wall. The results of our calculation are not sensitive to the exact ball model. A more realistic ball model is described in Chap. 9. The model used to calculate the results in Fig. 13.5 is described in Appendix 13.2. The main points we wish to demonstrate here are that (a) energy loss in the ball can be reduced if the wall deforms elastically and (b) too much energy given to the wall results in a lower rather than a higher ball speed off the bat.

The calculations in Fig. 13.5 are shown for a case where the effective mass of the bat at the impact point is 0.6 kg. This is a typical value of the effective mass for an impact near the sweet spot of the bat. The impact side of the bat was allowed to vary in mass from 0.1 kg to 0.3 kg, the total bat mass being fixed at $m_2 + m_3 = 0.6$ kg. For each of the assumed values of m_2 and m_3, the COR was calculated as a function of the bat stiffness k_2 or as a function of ball stiffness k_1. When the bat is either very stiff or very soft, the results can be explained in the following manner.

If the bat is very stiff, then the two masses m_2 and m_3 behave as if they are connected by a rigid rod. That is, they behave like a rigid object of total mass $m_2 + m_3$. Since no elastic energy is stored in such an object, all of the elastic energy stored during the collision will reside in the ball. There will be no effect on the coefficient of restitution for the collision, and the ball will bounce no better or worse than it would off a wood bat. That is, the COR for the collision will be 0.5, as indicated by the results in Fig. 13.5a, b at large values of k_2.

If the spring S_2 is very soft, then the ball will collide with and bounce off m_2 as if m_2 were almost a free object, unconnected to anything else. The mass m_2 will then travel toward m_3, compressing the connecting spring S_2 and causing m_3 to accelerate in the same direction as m_2. While the spring remains compressed, m_2 continues to slow down and m_3 continues to speed up. Eventually, m_2 will either come to stop and reverse direction, or the spring will start to stretch in which case m_2 will speed up and m_3 will slow down. This tug of war between m_2 and m_3 will result in vibration of the walls of the bat, well after the ball is sent on its way. In this manner, elastic energy is transferred to the bat but can be retained as vibrational

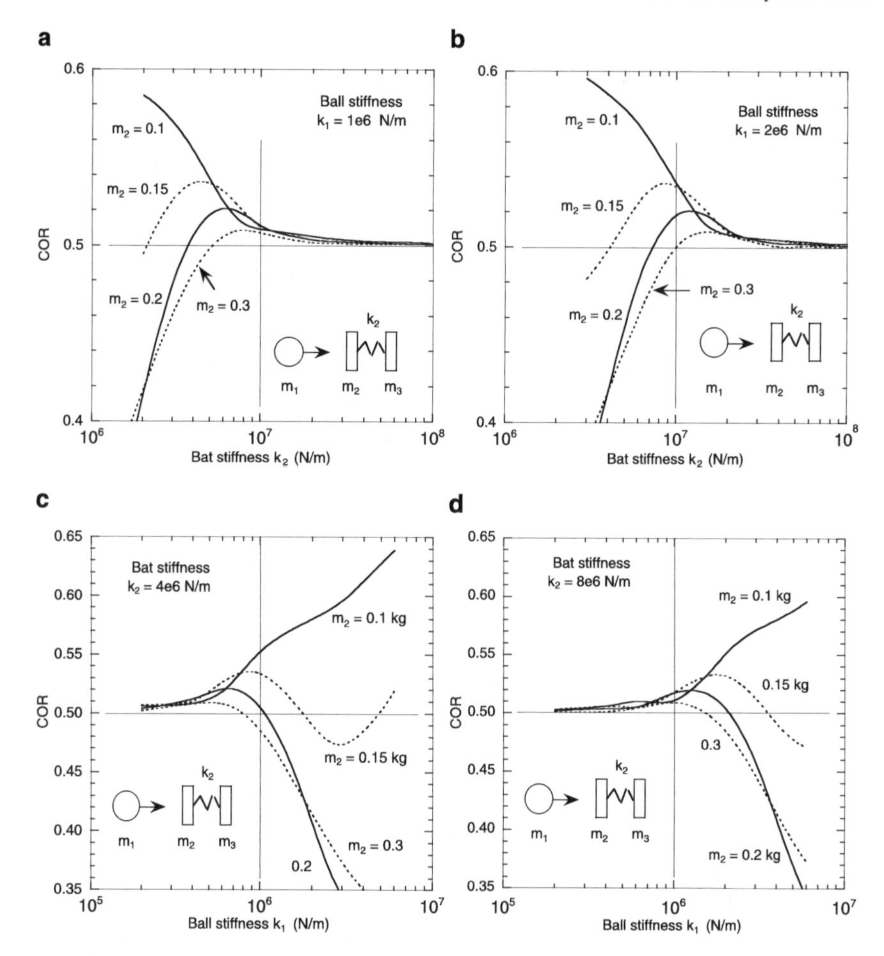

Fig. 13.5 Calculated values of the COR for the model shown in Fig. 13.4 when $m_2 + m_3 = 0.6$ kg and when (**a**) $k_1 = 1 \times 10^6$ N m^{-1}, (**b**) $k_1 = 2 \times 10^6$ N m^{-1}, (**c**) $k_2 = 4 \times 10^6$ N m^{-1} and (**d**) $k_2 = 8 \times 10^6$ N m^{-1}

energy in the bat. As a result, the bounce of the ball can be even worse than that off a wood bat and the COR for the collision will then be less than 0.5.

The COR when the bat is relatively soft depends on the mass m_2. If m_2 is small, then the wall behaves as a light trampoline, like the strings of a racquet. The wall vibrates and pings but the vibrational energy retained by the bat is small since m_2 is small. Furthermore, the mass m_3 is relatively large, so only a small fraction of the elastic energy stored during the collision is transferred to m_3. As a result, a large fraction of the stored elastic energy is given back to the ball and the COR is larger than 0.5. If m_2 is increased then the front wall behaves as a heavy trampoline and retains a large fraction of its elastic energy as vibrational energy. In addition, a larger fraction of the stored elastic energy is given to m_3, resulting in a stronger vibration of both sides of the bat. That is why, in Fig. 13.5a, b, the COR decreases well below 0.5 when m_2 is 0.2 kg or more and when the bat stiffness k_2 is less than about 5×10^6 N m^{-1}.

The most interesting results in Fig. 13.5a, b are obtained when the bat is about as stiff as the ball or up to ten times stiffer, with k_2 in the range from 10^6 to $10^7\,\mathrm{N\,m^{-1}}$. In that region, the COR can be significantly larger than 0.5 as a result of the trampoline effect. The magnitude of the effect depends on both the ball stiffness and the bat stiffness and also on the ratio of m_2 to m_3. The largest increase in the COR occurs if the ball makes two separate collisions with the bat. The first collision causes the ball to slow down until it almost comes to a stop. This sets the mass m_2 in motion in such a way that it bounces off m_3, reverses direction, and then collides with the ball. The result is a high speed bounce of the ball off the bat. However, the result is more of academic interest than practical interest, for two reasons. The first is that this large COR region occurs when k_2 is around 1×10^6 or $2 \times 10^6\,\mathrm{N\,m^{-1}}$, in which case the bat has about the same stiffness as the ball. Given that a baseball can squash in half during a high speed collision, the result would be disastrous for the bat. Aluminum bats are not designed to squash in half. If they did they would not recover. The result would be a mangled bat. Even if they did or could recover, the resulting exit speed of the ball would be much too high to pass the required performance limits. For these reasons, the double bounce results were not included in Fig. 13.5.

From a practical point of view, the region of interest in Fig. 13.5 is the region where k_2 is larger than about $4 \times 10^6\,\mathrm{N\,m^{-1}}$. A bat of this stiffness has a hoop frequency of about 900 Hz, which is about the lowest frequency that one finds with modern bats. The vibration frequency of the two masses connected by spring S_2 is given by

$$f = \frac{1}{2\pi}\sqrt{k_2\left(\frac{1}{m_2} + \frac{1}{m_3}\right)} \tag{13.1}$$

For example, if $k_2 = 5 \times 10^6\,\mathrm{N\,m^{-1}}$, and if $m_2 = m_3 = 0.3\,\mathrm{kg}$ then $f = 919\,\mathrm{Hz}$. For the same value of k_2, $m_2 = 0.1\,\mathrm{kg}$ and $m_3 = 0.5\,\mathrm{kg}$, $f = 1{,}233\,\mathrm{Hz}$. Higher values of k_2 correspond to higher vibration frequencies. The bat–ball collision model here is somewhat oversimplified, but it does account for the hoop frequencies observed in practice, and it also accounts for the enhanced values of the COR obtained with hollow bats.

13.4 COR vs. Ball Stiffness

The variation of the COR with ball stiffness is shown in Fig. 13.5c, d for two different values of the bat stiffness. When the ball is relatively soft, there is no significant trampoline effect since almost all of the elastic energy is stored in the ball and very little elastic energy is stored in the bat. Consequently, the COR is essentially the same as that for a wood bat. In terms of limiting the performance of aluminum bats, a simple solution would therefore be to use balls that are softer than those currently used. An additional advantage would be that a softer ball would exert a smaller force on a player struck by such a ball.

An increase in the COR is observed when the ball stiffness is increased so that the bat is about ten times stiffer than ball. The COR continues to increase as the ball becomes stiffer, provided the mass of the front wall is less than the mass of the ball, as indicated by the result when $m_2 = 0.1$ kg. However, if m_2 is equal to or larger than the mass of the ball then the COR will decrease as the ball stiffness is increased. Given that m_2 can be decreased by making the wall thinner and more flexible, the design problem in producing a high performance bat is to ensure that the wall is not so thin that the bat will break.

13.5 Trampoline Effects in a Wood Bat

The mass-spring model in Fig. 13.4 can also be used to describe the collision of a ball with a wood bat. In the case of a wood bat, $m_2 + m_3$ in Fig. 13.4 is equal to the effective mass of the bat at the impact point and k_2 is determined by the frequency of the fundamental vibration mode, about 170 Hz for most bats. The model provides an accurate description of the collision when the fundamental mode is the dominant mode, as it is when the impact speed is low or when the bat strikes the ball at the node of the second mode. A more complicated model would be needed, with additional springs and masses, if higher frequency modes contribute to energy losses. The simpler model, already described in Sect. 13.2, provides useful insights into the process by which energy is lost to bat vibrations, either as a result of bending or as a result of hoop vibrations.

When a bat impacts with a ball, the exit speed of the ball and the change in speed of the bat both depend on the effective mass of the bat at the impact point. The bat also vibrates. The amount of energy lost as a result of bending vibrations depends, in the mass–spring model, on the separate values of m_2 and m_3 and on the spring stiffness k_2. In effect, m_2 represents the mass of the bat in the immediate vicinity of the impact point, and m_3 is the mass of neighboring parts of the bat which are connected elastically to the impact region. The neighboring parts of a bat exert a force on the impact region due to bending of the bat, that force being zero at the node point in the barrel. Since there are no bending vibrations for an impact at the node point, m_3 is zero at the node point. Near the tip of a bat m_2 and m_3 are both relatively small and vibration losses are a maximum. Near the handle end of the barrel, m_2 and m_3 are both relatively large and vibration losses are also large.

Calculations based on solutions of the beam equation [3] show that energy losses in a wood bat, due to bending vibrations, are reduced to a minimum when the impact occurs near the node of the fundamental or the second bending mode. At low impact speeds where the impact duration is relatively long (about 2 ms), the minimum occurs at the node of the fundamental mode since that is the only mode of significance excited by the collision. At high impact speeds where the impact duration is less than 1 ms, the minimum occurs near the node of the second mode. Energy losses in the ball also vary with impact position along the barrel and are close to a maximum just where energy losses in the bat are a minimum.

The latter result requires an explanation. It would be more convenient for the batter if ball losses could be reduced to a minimum at the same impact point where bat losses are a minimum. If that happened then the outgoing ball speed could increase substantially. In fact, energy losses in the ball are not only a maximum close to the minimum vibration loss point but they are the main source of energy loss in the collision between a bat and a ball. The energy lost in the ball is even greater than the kinetic energy of the outgoing ball.

There are a number of competing effects that occur in the collision of a bat and a ball, all of which we have described previously, but that need to be invoked to explain energy losses in the bat and the ball. One of those effects is the trampoline effect. The main effects are as follows:

1. When a bat is swung at a ball, the impact point that results in maximum outgoing ball speed is close to the point where vibration losses are a minimum. Part of the reason is that bat speed is greatest at the tip of the bat and least at the handle end. The maximum outgoing ball speed is therefore a maximum near the far end of the bat. It is not a maximum at the very tip of the bat because the effective mass of the bat is a minimum at the tip and because energy losses due to bat vibrations are a maximum at the tip.
2. The outgoing ball speed is a maximum at the point where maximum force is exerted on the ball. The ball therefore squashes most at that point, stores more elastic energy, and loses more energy than at any other impact point.
3. The COR is a maximum close to the point where vibrational losses in the bat are a minimum, despite the fact that ball losses are a maximum near that point. The COR measures the fraction of the stored elastic energy that is lost. That fraction is about 75% for the ball, regardless of the amount of energy stored in the ball and regardless of the impact point on the bat. Consequently, the main factor that influences the COR is the loss of vibrational energy in the bat.
4. The COR can be increased above that for an impact on a massive, rigid wall if the ball impacts on a relatively soft part of a bat since then more elastic energy is stored in the bat and less in the ball. However, the COR is increased only if more elastic energy is returned to the ball than is retained in the bat. For that to happen, the bat must have a hollow wall so that only a relatively light section of the bat is involved in the process. The barrel of a wood bat is stiffer and heavier than the wall of a hollow bat, and it usually retains more elastic energy than it gives back to the ball. An exception is the sweet spot of the bat where almost no elastic energy is stored in the bat so the COR remains essentially unaffected.
5. The softest parts of the barrel of a wood bat are near the tip and near the handle end since the tip is not supported by anything beyond the tip and since the handle end of the barrel is supported only by a relatively thin handle. The barrel bends most in the softest parts, so the bat vibrates most for impacts near the tip and near the handle end. The trampoline effect in those soft regions acts to reduce both the force on the ball and energy loss in the ball, and it increases vibrational energy losses in the bat, but the losses in a wood bat outweigh the gains in the ball, with the result that the COR and the outgoing ball speed are both reduced in the softer parts of the barrel.

Appendix 13.1 Bounce Off a Clamped Bat

Suppose that a ball of stiffness k_1 is dropped onto a bat of stiffness k_2 and that the bat is clamped to the floor and cannot recoil away from the ball. If the ball compresses by an amount x_1 then the force F on the ball is given by $F = k_1 x_1$. For example, if $x_1 = 0.1$ in and $k_1 = 2{,}000\,\text{lb in.}^{-1}$ then $F = 200\,\text{lb}$. The amount of work needed to compress the ball is given by $\frac{1}{2}k_1 x_1^2$. This gives the amount of elastic energy stored in the ball.

If the bat compresses by x_2 then the force on the bat is $k_2 x_2$ Since the forces on the ball and the bat are equal (but act in opposite directions) $k_1 x_1 = k_2 x_2$.

The elastic energy stored in the bat is $E_{\text{bat}} = \frac{1}{2}k_2 x_2^2$. The ratio of the energy stored in the ball to that in the bat is

$$\frac{E_{\text{ball}}}{E_{\text{bat}}} = \frac{k_1 x_1^2}{k_2 x_2^2} = \frac{k_2}{k_1}$$

For example, if the bat is twice as stiff as the ball then the elastic energy stored in the bat will be half of that stored in the ball.

If the ball has kinetic energy E_1 when it collides with the bat, then the total elastic energy stored in the bat and the ball will be given by

$$E_1 = E_{\text{ball}} + E_{\text{bat}} = E_{\text{bat}}\left(1 + \frac{k_2}{k_1}\right)$$

so

$$E_{\text{bat}} = \frac{k_1 E_1}{k_1 + k_2}$$

Similarly,

$$E_{\text{ball}} = \frac{k_2 E_1}{k_1 + k_2}$$

If we assume that the ball loses 75% of its stored energy and the bat loses only 5% of its stored energy, then the total kinetic energy remaining after the collision is

$$E_2 = \frac{(0.25k_2 + 0.95k_1)E_1}{k_1 + k_2}$$

Since the bat is clamped, it has no kinetic energy after the collision, so E_2 is the kinetic energy of the ball after the collision. If $r = k_1/k_2$ then

$$\frac{E_2}{E_1} = \frac{0.25 + 0.95r}{1 + r}$$

The ratio E_2/E_1 is plotted as a function of r in Fig. 13.2.

Appendix 13.2 Trampoline Model

The collision between a ball and the elastic wall of a bat can be modeled as shown in Fig. 13.4, assuming that part of the front wall of the bat, of mass m_2, moves radially towards the rear wall, of mass m_3. The total wall mass, $m_2 + m_3$ is equal to the equivalent mass of the bat at the impact point. During the compression phase, the force F_1 on spring S_1 is $F_1 = k_1(x - y)$ and the force F_2 on spring 2 is given by $F_2 = k_2(y - z)$ where k_1 is the stiffness of the ball, k_2 is the stiffness of the bat, $x - y$ is the compression of the ball and $y - z$ is the compression of the bat. F_1 acts to the left on the ball and to the right on the front wall, while F_2 acts to the left on the front wall and to the right on the rear wall. Hence

$$m_1 \frac{d^2x}{dt^2} = -F_1 \qquad m_2 \frac{d^2y}{dt^2} = F_1 - F_2 \qquad m_3 \frac{d^2z}{dt^2} = F_2 \tag{13.2}$$

These relations can be solved numerically assuming that the bat is initially at rest, with $dy/dt = dz/dt = 0$ and that the ball is incident at speed $dx/dt = v_o$. The initial coordinates can be taken for convenience as $x = y = z = 0$.

If spring S_1 begins to expand during the collision, we can let $F_1 = k_3(x - y)^7$ so that the ball has a COR $= 0.5$ when it impacts on a rigid wall. At maximum compression, when $x - y = s_o$, the peak force $F_1 = k_1 s_o = k_3 s_o^7$, so $k_3 = k_1/s_o^6$. The area enclosed by the ball hysteresis curve in this case then indicates that 75% of the stored elastic energy in the ball is lost and that the COR is therefore 0.5.

Figure 13.6 shows an alternative model that can be used to account for energy loss in the ball. In this case, a series dashpot is included. The force on the dashpot is the same as the force on the spring S_1 so $k_1(x - w) = k_d\, d(w - y)/dt$ where $s = x - w$ is the compression of the spring and $w - y$ is the compression of the dashpot. We can solve to find s using the relations

$$\frac{d^2s}{dt^2} = \frac{d^2x}{dt^2} - \frac{d^2w}{dt^2}$$

$$\frac{d(w - y)}{dt} = \frac{k_1 s}{k_d} \qquad \frac{d^2w}{dt^2} - \frac{d^2y}{dt^2} = \frac{k_1}{k_d}\frac{ds}{dt}$$

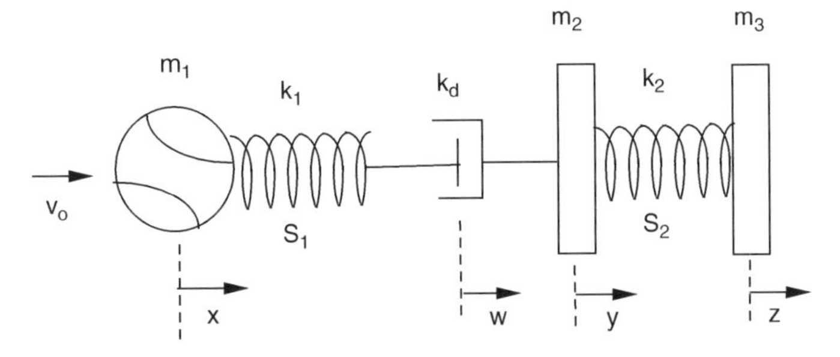

Fig. 13.6 Alternative model used to describe the collision between a baseball and the walls of a hollow bat. This model includes a series dashpot to account for energy losses in the ball

so

$$\frac{d^2s}{dt^2} = \frac{d^2x}{dt^2} - \frac{d^2y}{dt^2} - \frac{k_1}{k_d}\frac{ds}{dt}$$

Since

$$m_1\frac{d^2x}{dt^2} = -k_1 s \qquad m_2\frac{d^2y}{dt^2} = k_1 s - k_2(y - z)$$

we obtain the following equation for s:

$$\frac{d^2s}{dt^2} + \frac{k_1}{k_d}\frac{ds}{dt} + k_1\left(\frac{1}{m_1} + \frac{1}{m_2}\right)s = \frac{k_2}{m_2}(y - z)$$

which describes a damped oscillation with a forcing term. The corresponding equations for y and z are

$$m_2\frac{d^2y}{dt^2} = k_1 s - k_2(y - z) \qquad m_3\frac{d^2z}{dt^2} = k_2(y - z)$$

The damping term k_d can be chosen so that the ball bounces with COR $= 0.5$ off a very stiff bat, and the equations for s, y and z can be solved numerically to find the COR when the ball bounces off a flexible bat. The two different ball models give essentially the same results. The results in Fig. 13.5 were obtained using the slightly more complicated dashpot model depicted in Fig. 13.6.

References

1. D.A. Russell, Hoop frequency as a predictor of performance for softball bats, in Proceedings of the 5th International Conference on the Engineering of Sport, vol. 2 (UC Davis, CA, 2004), pp. 641–647
2. A.M. Nathan, D.A. Russell, L. Smith, The physics of the trampoline effect in baseball and softball bats, in Proceedings of the 5th International Conference on the Engineering of Sport, vol. 2 (UC Davis, CA, 2004), pp. 38–44
3. A.M. Nathan, Dynamics of the baseball bat collision. Am. J. Phys. **68**, 979–990 (2000)

Chapter 14
The Sweet Spot of a Bat

Some of the longest home runs I've hit, I didn't actually realize they were going that far. Everyone says, What does it feel like to hit the ball that far? Actually, there's no feeling at all. I know when the ball meets the bat whether or not it's left the park. It's a nice easy thing.

– Mark McGwire

14.1 Introduction

An interesting aspect of a bat and ball collision, from both a practical and a physics point of view, is that the batter can exert a huge force on the ball without feeling any particular discomfort at the handle end of the bat. The force on the ball is much larger than the force exerted by the batter on the bat. The batter swings the bat by exerting a force of around 50 lb on the handle. The bat then magnifies that force by a factor of about 100 to 5,000 lb or so when it collides with the ball, enough to squash the ball almost in half. Only a small fraction of that huge force gets back to the handle to trouble the batter. If the ball is struck at the sweet spot of the bat, then the batter feels almost no effect at all. It seems almost as if Newton's third law does not apply to baseball or softball bats. Newton's third law says that for every action there is an equal and opposite reaction. If there is a force of 5,000 lb on the ball then there will definitely be a force of 5,000 lb acting back on the bat. How come the batter is blissfully unaware of that fact?

Newton's third law is not violated when a batter swings a bat, nor when the bat strikes a ball. The subtlety here is that there are two separate pairs of forces acting on the bat. There is an equal and opposite force between the bat and the ball, acting at the barrel end, and a different equal and opposite force between the bat and the batter's hands, acting at the handle end. Fortunately for the batter, the force at the handle end is a lot smaller than the force at the barrel end. In this respect, a bat acts in the same manner as a hammer. You can hammer a nail using a small force on the handle to generate a much larger force on the nail. You only notice the large force at the heavy end of the hammer if you accidently hit your thumb instead of the nail.

The different forces acting on the bat and the batter can be explained by a simple analogy. Suppose you punch a brick wall with your bare fist as hard as you can. The brick wall will probably survive intact and you will probably injure your hand. In that case, the effect on your fist is not exactly the same as the effect on the brick wall. Newton's third law has nothing to do with the consequences of the actions and reactions. The law is concerned only with the equal and opposite forces. The force

R. Cross, *Physics of Baseball & Softball*, DOI 10.1007/978-1-4419-8113-4_14,

of the brick wall on your fist is equal to the force of your fist on the wall, but it acts in the opposite direction. If you hit the wall, the wall will hit you back with the same force. The wall doesn't stop to think about it. It doesn't give you a chance to run away before it hits you back. It reacts instantly. That force, acting on weak bones, can break those bones. The same force, acting on a strong brick, is not enough to break the brick. In the process, while injuring one hand, you will feel nothing in your other hand or your feet. The force and the reaction force are both exerted at the point of impact. If you kick the wall you won't injure your fist. You will injure your foot. It is the same with a bat and a ball.

When a bat collides with a ball, the force of the ball on the bat does not act on the handle, nor does it act on the batter's hands. It acts on the bat at the point of impact on the barrel. The barrel is a lot heavier than the ball, with the result that the barrel will slow down a fraction but it will continue moving in the same direction as before the collision. It is like a heavy truck slamming into a basketball. The truck will slow down a fraction, but it won't come to a stop or reverse direction. It's the same with the barrel of a bat. The question is, what happens to the handle of the bat when the barrel slows down a fraction? If we can answer that question then we can figure out the force that is exerted by the handle on the batter's hands.

The situation is shown in Fig. 14.1. Before the bat strikes the ball, there is essentially zero force on the ball, although the air does slow it down a fraction as it approaches the bat. The only force on the bat is the force exerted by the batter on the handle, around 50 lb. The handle exerts an equal and opposite force on the hands. That 50-lb force doesn't bother the batter, but it does slow down his swing speed a fraction, particularly if he is using a heavy bat. He could swing his arms a fraction faster if he didn't have a bat in his hands.

On impact, the force on the ball increases rapidly as the ball squashes, up to about 5,000 lb. That force brings the ball to a complete stop before it turns around

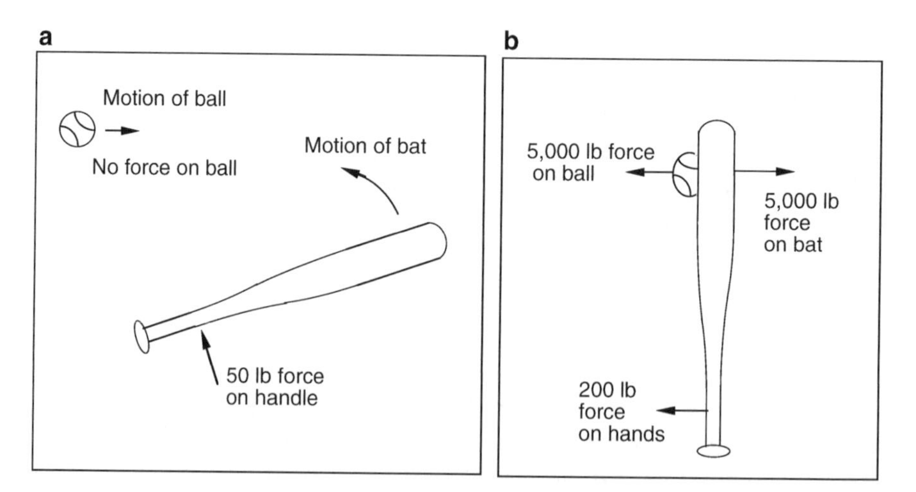

Fig. 14.1 (**a**) The batter exerts a force of about 50 lb on the bat in order to swing it. (**b**) The bat exerts a force of about 5,000 lb on the ball, and the ball exerts an equal and opposite force on the bat

and heads back in the opposite direction. There is an equal and opposite force of 5,000 lb acting on the bat at the impact point. That force does two things to the bat. It slows down the barrel a fraction and it changes the rotation speed of the bat. The knob end of the bat can either slow down or speed up, depending on exactly where the bat strikes the ball. If the bat and ball collide near the far end of the bat, then the knob will jerk forward, toward the pitcher. The jerking motion of the handle exerts a force on the batter's hands. If the batter has a good grip then all that happens is that the batter feels a very short, 200 or 300 lb tug on his hands and arms. That tug force does not last long enough to pull the handle out of the batter's hands. However, there is an equal and opposite force of around 200–300 lb exerted by the batter's hands on the handle. The latter force affects the motion of the bat, as we will see shortly. If it wasn't for the fact that the batter maintains a firm grip on the handle, then the whole bat would fly off toward the pitcher, rotating rapidly.

It turns out that there is an impact point on the barrel where the force on the batter's hands is reduced to such an extent that the batter feels almost nothing. That impact point is known as the sweet spot of the bat. However, the exact nature of the sweet spot has been hard to pin down in terms of the physics of the problem [1–9]. Some researchers claim that the sweet spot corresponds to the center of percussion (COP) of the bat. Others claim that it is either the spot where bat vibrations are reduced to a minimum, or it is the spot where the batted ball speed is a maximum. We will now consider each of these claims in turn.

14.2 The Center of Percussion

All striking implements have a COP, including bats, clubs and racquets. The COP is the impact point that results in minimum shock being transferred to the hands. The basic physics is illustrated in Fig. 14.2 where we show a ball incident from the left on a stationary bat. The COP for a bat swung toward the ball is located at the same point as that for a stationary bat, but the explanation is slightly more complicated. The bat can be held in a stationary position by suspending it by the handle with a length of string. The point labeled CM in Fig. 14.2 is the center of mass of the bat.

If a ball strikes the bat in line with its CM then the whole bat will move forward without rotating, as shown in Fig. 14.2a. The barrel, the CM and the handle all move forward at the same speed. However, if the ball strikes the bat near the end of the barrel, then the bat will move forward but it will also rotate, as shown in Fig. 14.2b. The barrel moves forward faster than the CM and the handle moves backward. You can try this experiment yourself in a slightly different way. Place a ruler on a horizontal table with a slippery surface and flick the ruler with one finger at a point in the middle of the ruler. If you strike it exactly in the middle (at its CM) then the ruler will slide across the table without rotating. But if you flick it to the left or right of the CM, then the ruler will rotate in a manner similar to that shown in Fig. 14.2b.

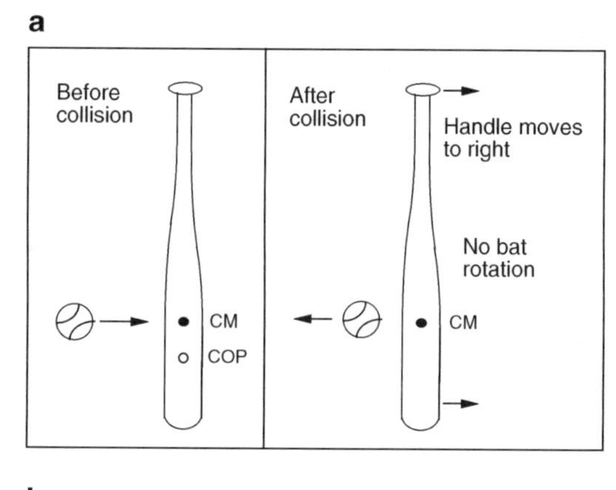

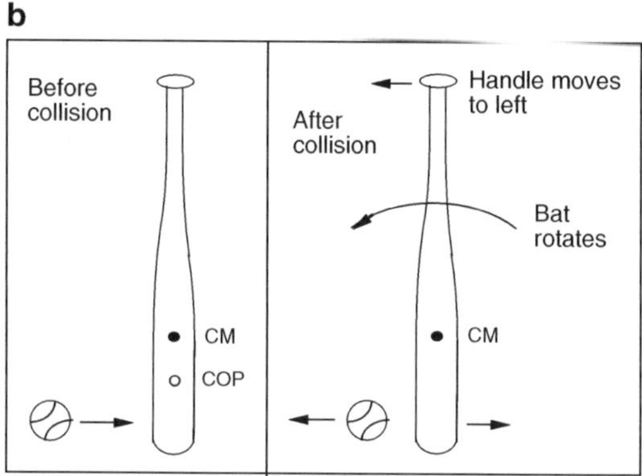

Fig. 14.2 (**a**) If a bat is suspended freely by a length of string and is struck by a ball at its center of mass (CM) then the whole bat moves to the right without rotating. (**b**) Struck at a point near the end of the barrel, the barrel moves to the right but the handle moves to the left since the whole bat rotates counter-clockwise. When struck at the COP, the bat rotates about an axis through the handle, so there is no left or right motion of points on that axis

Given that the handle moves forward in Fig. 14.2a, and backward in Fig. 14.2b, then there must be an impact point along the barrel, between the tip and the CM, where the handle does not move at all. This impact point is the COP. An impact at the COP causes the whole bat to rotate about an axis through the handle. That is, every point along the bat rotates, except for points on the axis. A point located 1 in. from the end of the handle has a slightly different COP to a point 2 in. from the end of the handle, so it is not possible to strike a ball in such a way that the whole handle remains at rest. But striking at a point say 6 or 7 in. from the end of the barrel will generate significantly less handle motion than an impact at a point only 1 or 2 in. from the end of the barrel.

It is possible to calculate the location of the COP for any given axis along the handle, and the answer is given by (10.29) in Chap. 10. Such a calculation ignores an important practical detail. In practice, a bat is never freely supported or held by a length of string. It is always held firmly by two hands. If a ball strikes a stationary bat at some point along the barrel, then the hands will prevent the bat handle moving freely in the manner shown in Fig. 14.2. In fact, the bat will rotate about about an axis through the hands, regardless of the impact point. Consequently, the simple situation shown in Fig. 14.2 is not really relevant in practice. The behavior of a hand-held bat is more complicated.

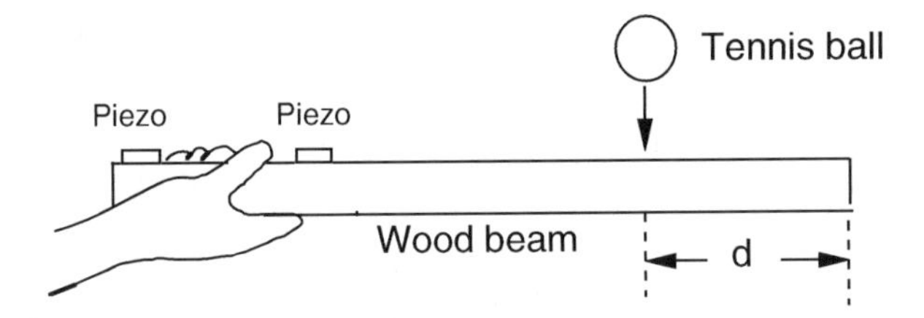

Fig. 14.3 Arrangement used to measure motion of the handle end of a wood beam when struck near the other end. Small piezo disks were used as accelerometers to measure the acceleration and velocity of the handle at points 2 and 16 cm from the end of the handle

To determine what does happen, the author conducted a simple experiment in 2003 by bouncing a tennis ball off a rectangular wood beam held at one end by one hand. A flat wood beam was used rather than a baseball bat so that two accelerometers could be conveniently attached to the beam, one either side of the hand, as shown in Fig. 14.3. The beam was 72 cm long, 4 cm wide, 1.9 cm thick, and weighed 328 g. The accelerometers were in the form of 25 mm diameter piezo-electric disks, only 0.3 mm thick, and they had a negligible effect on motion of the beam.

As its name implies, an accelerometer measures the acceleration of an object at the point where the accelerometer is attached. Different parts of an object can accelerate at different rates, especially if the whole object is rotating or vibrating or both. Indeed, that is exactly what happens to a bat when it is struck by a ball. The bat rotates and it vibrates, and it does so faster than than can be seen or measured conveniently with a video camera. The best way to measure the motion of a bat under these conditions is to attach an accelerometer at several different points on the bat. The acceleration $a = \mathrm{d}v/\mathrm{d}t$ where v is the velocity of the bat at the point where the accelerometer is attached. Using a computer, it is easy to integrate the acceleration signal to measure the velocity. Some of the results are shown in Fig. 14.4 (when the beam was freely suspended by a length of string) and Fig. 14.5 (for the case where the beam was hand held). In both cases, the velocity is shown at points 2 cm and 16 cm from the end of the beam, labeled as v_2 and v_{16} respectively. The ball impacted at a distance d from the other end. Results are shown for $d = 4$, 16, 22, 28 and 56 cm.

The velocity at each of the two measurement points is zero until the ball hits the beam, at time $t = 0$. The velocity traces were artificially shifted up or down the vertical axis so they do not overlap and can, therefore, be seen more clearly. When the beam is freely suspended, the impact causes the beam to vibrate strongly and it also causes the beam to rotate. When the beam is struck 4 cm from one end, the struck end moves away from the ball and the other end moves toward the ball. In that case, the rotation axis is near the middle of the beam, and the velocity v_2 is larger than v_{16}. If the ball impacts at $d = 16$ cm then there is no vibration of the beam but the beam still rotates, about an axis roughly 14 cm from the other end (since v_2 is large and positive while v_{16} is small and negative). An impact at $d = 22$ cm

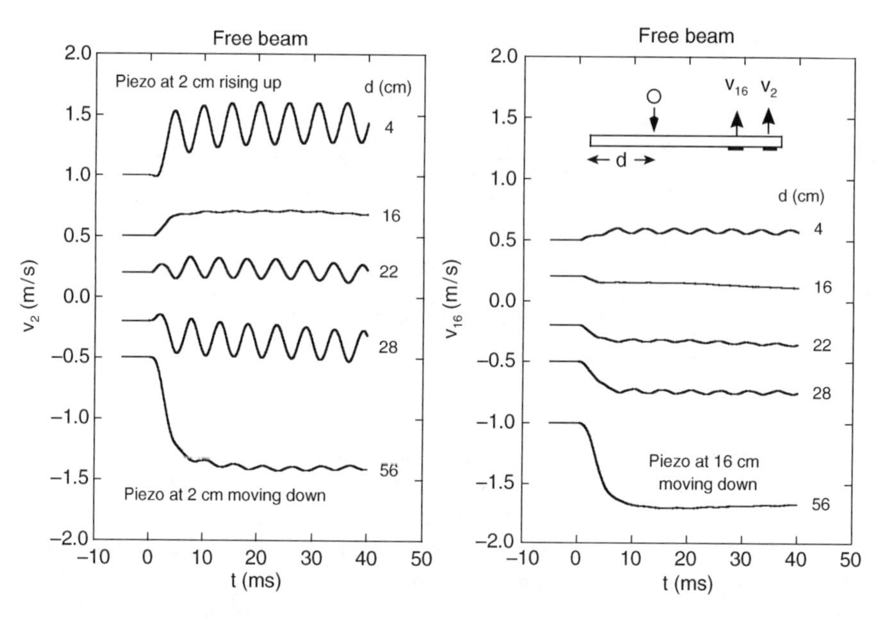

Fig. 14.4 Velocity of the handle at points 2 and 16 cm from the end of the handle when the beam was suspended freely (not hand-held)

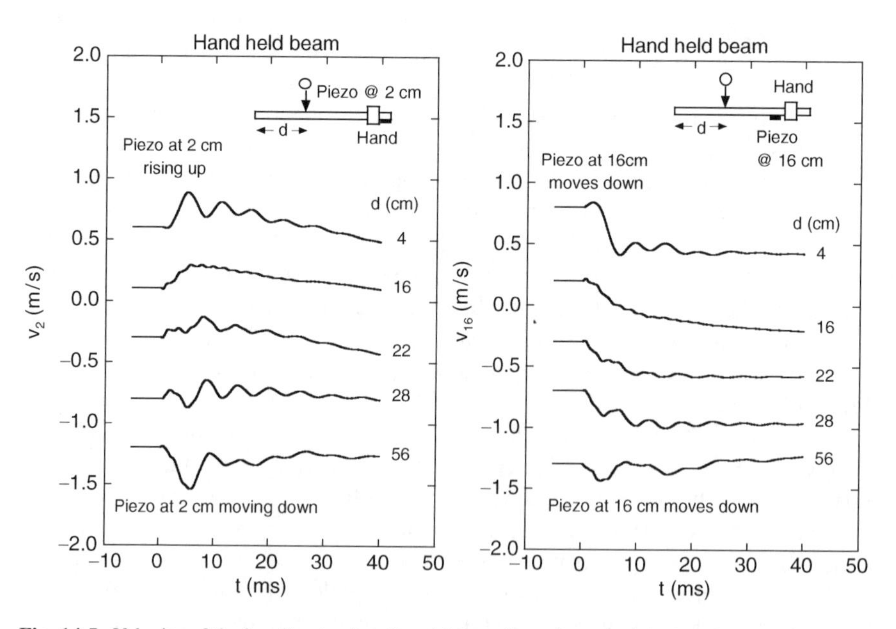

Fig. 14.5 Velocity of the handle at points 2 and 16 cm from the end of the handle when the beam was held in one hand

causes the beam to vibrate but the average value of v_2 is zero, meaning that the beam rotates about an axis 2 cm from the other end, where one of the two piezos is located. The COP for this beam, with respect to an axis 2 cm from one end, is therefore located 22 cm from the other end.

The corresponding results for the hand held beam are shown in Fig. 14.5. The effect of the hand is quite dramatic. The beam still vibrates but not nearly as much. One effect of the hand is therefore to dampen the vibrations. The same type of effect occurs when a musician touches a vibrating guitar string. If the beam was held by two hands then the vibrations would be even more strongly damped. The hand also has a dramatic effect in shifting the COP. For example, consider the results for an impact at $d = 4$ cm. For the free beam, v_2 and v_{16} are both positive. For the hand-held beam, v_2 is positive but v_{16} is negative, meaning that the rotation axis lies under the hand. Similarly, when $d = 16$ cm, or even when $d = 22$ or 28 cm, v_2 is positive and v_{16} is negative, meaning that the rotation axis still lies under the hand. Regardless of where the ball strikes the beam, the beam rotates about an axis through the hand.

The effect of the hand can be understood by considering the situation shown in Fig. 14.6. Here we show the beam (A) connected by a hinge to another beam (B) that is pivoted about its far end. Beam B simulates a human arm holding beam A, and the hinge represents the flexible wrist joint. A ball striking beam A near its free end will cause beam A to rotate but beam B will restrict the amount of rotation. As a result, the rotation axis is shifted to a point near the hinge joint, effectively under the hand. In the case of a free beam, the rotation axis would be closer to the middle of beam A. Consequently, the concept of the COP, for a hand-held beam or bat is strongly modified by the hand and the arm holding onto it. Nevertheless, sudden rotation of a bat will occur when it is struck by a ball, and the handle will exert a sudden, enhanced force on the hand and the arm. Such an effect will occur at all impact points along the barrel of the bat, so there is no particular impact point along the barrel that will lead to any significant reduction of the force on the hands or arms. The sweet spot of a baseball bat is not associated directly with any such spot. To understand why the sweet spot exists, we need to examine the effect of the impact point on bat vibrations.

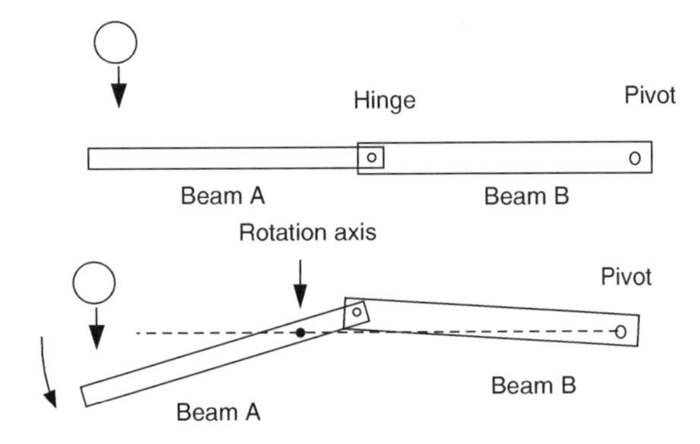

Fig. 14.6 The effect of the hand and forearm (Beam B) is to shift the rotation axis to a point under the hand, regardless of the impact point. The hinge joint represents the wrist

14.3 Beam Vibrations

As shown in Figs. 14.4 and 14.5, there is an impact point on the wood beam, 16 cm from one end, where the beam rotates without vibrating. At least, there are no vibrations recorded by the accelerometers located 2 and 16 cm from the other end. In fact, no vibrations are recorded anywhere along the beam. The impact point here is known as a node point. By holding the beam at one end, it is easy to establish that the node point is the sweet spot of the beam. The "feel" of the beam is just right when the ball strikes the node point, since there are no vibrations transmitted to the hand or the arm.

The whole beam vibrates when it is struck near the end of the beam, but it does not vibrate at the node point when the beam is struck near one end. In Fig. 14.4, the vibrations recorded 2 cm from the end of the beam are much larger than the vibrations recorded 16 cm from the end of the beam, regardless of the impact point. The actual node point is 15.8 cm from the end of the beam, which means that the vibration level drops to zero there, while the vibration level at 16 cm is very small but not quite zero. The vibration levels are largest at each end of the beam and in the middle of the beam. These points are described as anti-nodes.

The wood beam used to obtain the results in Figs. 14.4 and 14.5 was 72 cm long. When the beam was freely supported, its COP was located 24 cm from one end of the beam when the axis of rotation was at the other end of the beam. The node point was located 15.8 cm from the end of the beam. Consequently, it was easy to pick the correct sweet spot, about 16 cm from the impact end, just by holding onto one end and feeling the effects of the impact subjectively. The situation with a real bat is more complicated but the end result is the same. That is, the sweet spot of a real bat is located about 7 in. from the end of the barrel, and it coincides with the impact point that minimizes the level of vibrations in the whole bat, including the handle.

14.4 Bat Vibrations

There are two complications with a real bat. The first is that the COP of a freely suspended bat is only 1 or 2 cm away from the node point, so it is very difficult to tell from the subjective feel whether the sweet spot is the node point or the COP or perhaps represents the combined effects of both points. The other complication is that a real bat has two node points located a few cm apart. When a bat is struck by a ball, the bat can vibrate at two different frequencies simultaneously. It can vibrate at a frequency of about 170 Hz, and it can also vibrate at a frequency of about 530 Hz. In a high speed collision there may be a third vibration present at an even higher frequency, but the third mode is not observed in low speed collision experiments. The 170-Hz vibration has a node point about 17 cm from the barrel end, while the 530 Hz vibration has a node point about 13 cm from the barrel end. If the bat strikes the ball at a point 13 cm from the barrel end, then the bat vibrates only at 170 Hz. If the bat strikes the ball at a point 17 cm from the barrel end, then the bat vibrates only at 530 Hz. If the bat strikes the ball at some other point, then the bat vibrates at both frequencies. The sweet spot could in theory be located at either of the two node points or over a narrow region encompassing both node points. Subjective measurements indicate that it is located close to the node point of the 170-Hz vibration. It seems that the hand and arm are more sensitive to low frequency vibrations than to high frequency vibrations. Either that, or the 170-Hz vibration is larger in amplitude than the 530-Hz vibration. Both effects may be important.

Measurements with accelerometers attached to a real baseball bat are shown in Figs. 14.7–14.11. The bat was a Louisville Slugger model R161 wood bat of length 84 cm (33 in.) with a barrel diameter of 6.67 cm (2 5/8 in.) and mass 0.885 kg (31 oz). Two sets of measurements were made, one where the bat was suspended freely by a length of string tied to the knob, and one where the bat was held firmly by two hands in the normal manner. In the latter case, the ball was suspended as a pendulum bob and struck at low speed by the bat. The impact speed was much

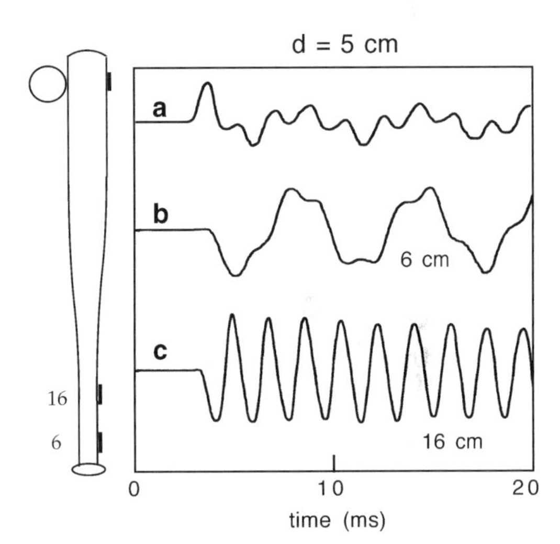

Fig. 14.7 Vibrations of a freely suspended baseball bat struck 5 cm from the end of the barrel

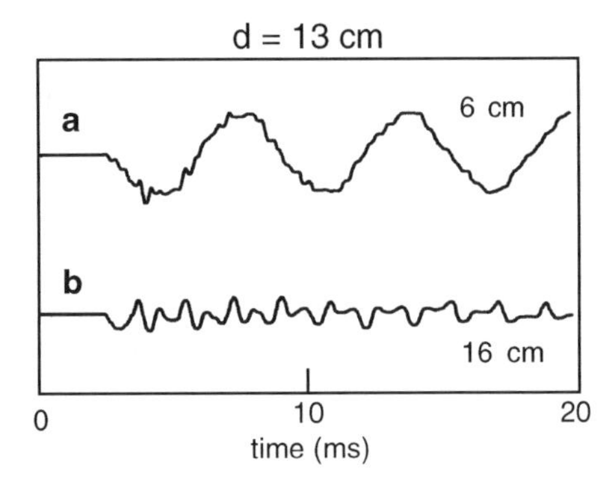

Fig. 14.8 Vibration of a freely suspended baseball bat struck 13 cm from the end of the barrel

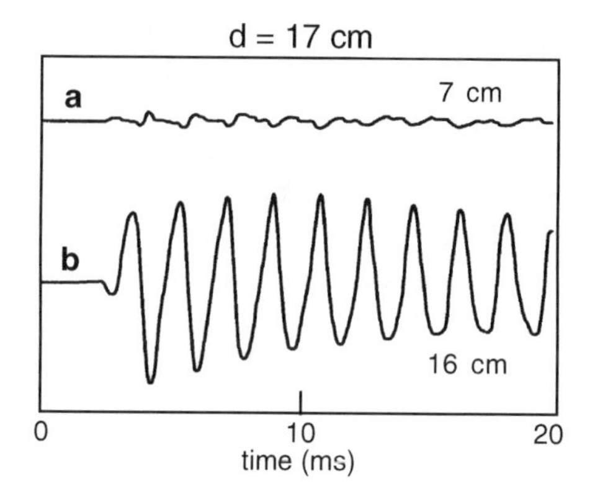

Fig. 14.9 Vibration of a freely suspended baseball bat struck 17 cm from the end of the barrel

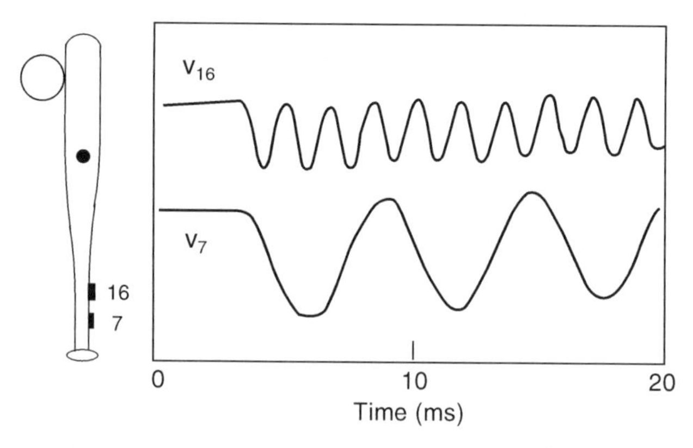

Fig. 14.10 Measured velocity of the handle at points 7 and 16 cm from the handle end when the bat was struck with a baseball at a point 5 cm from the end of the barrel

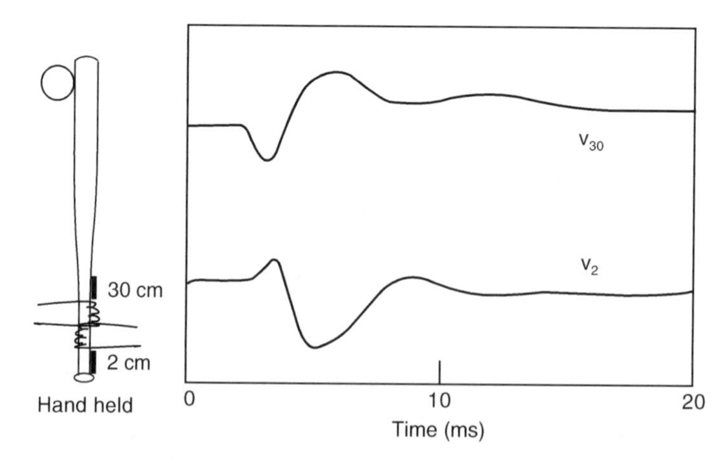

Fig. 14.11 Measured velocity of the handle at points 2 and 30 cm from the handle end when the bat was struck 5 cm from the end of the barrel. The handle was held with two hands

smaller than in a real game of baseball, but the low speed had no significant effect on bat vibrations apart from the vibrations being smaller than usual and the fact that the third mode was almost entirely absent. The free bat vibrations are easier to interpret since they are not damped by the hands.

Figure 14.7 shows the acceleration of the free bat at points 6 cm and 16 cm from the knob when struck by a baseball 5 cm from the barrel end. Also shown is the acceleration of the bat at the impact point, reduced by a factor of 5 compared with the handle results. Two vibration frequencies can be seen, one at 167 Hz (period 6.0 ms) and one at 530 Hz (period 1.89 ms). The 167-Hz vibration is the lowest frequency vibration possible for this bat and is called the fundamental mode. The 530 Hz vibration is next available frequency and is called the second mode. Even higher frequency vibrations can be observed when the bat is struck by a small golf ball or a ball bearing, but only the lowest two frequency vibrations are seen when a bat is struck by a baseball at low speed. In Fig. 14.7, the 167-Hz mode is not seen at the point 16 cm from the knob since this mode has a node point 16 cm from the knob. The 530-Hz mode is seen at the point 6 cm from the knob since this mode has a node point nearby, about 7 cm from the knob.

The 167-Hz mode has another node point located 17 cm and from the barrel end, while the 530-Hz mode has two extra node points, located 13 and 47 cm from the barrel end. Consequently, if the ball strikes the bat 17 cm from the end of the barrel, the 167 Hz vibration is not observed anywhere along the bat. Only the 530-Hz vibration is seen, but it is not seen at its node point 7 cm from the knob, as shown in Fig. 14.8. Similarly, if the ball strikes the bat 13 cm from the barrel end, then only the 167 Hz vibration is observed, except at its node point 16 cm from the knob, as shown in Fig. 14.9.

It is interesting to consider the handle velocity of the freely suspended bat when it is struck at a point on the barrel. A typical result is shown in Fig. 14.10 for a case where the bat was struck 5 cm from the end of the barrel. These results were obtained by integrating the accelerometer signals measured 7 and 16 cm from the knob. The whole bat rotates about an axis 30 cm from the knob end, giving a negative handle velocity which is largest near the knob and smallest at the 30 cm point. The velocity waveforms have a negative dc component due to rotation and translation of the bat and also have an ac component due to vibration of the bat. The 167 and 530-Hz vibrations were both generated by the impact, but they are not observed at their respective node points.

The situation changes dramatically when the bat is held in both hands. In that case, the vibrations are almost completely eliminated, as shown in Fig. 14.11. The handle velocity is minimized for an impact in the sweet spot region, about 15 cm from the end of the barrel. It is difficult to determine, just by inspection of the velocity data, whether handle motion is reduced because the impact occurs near the vibration node or whether it is reduced because the impact is close to the COP. However, the results in Fig. 14.5 indicate that the primary effect is the former, given that the COP shifts toward the middle of the bat when the handle is hand held.

14.5 Results for an Aluminum Bat

A similar set of results to those just described for the wood bat was obtained for an Easton BK7 aluminum baseball bat. The aluminum bat was impacted with a low speed golf ball to reduce the impact duration to 1.0 ms, which is close to but still slightly longer than the duration for a high speed impact with a baseball. The shorter impact duration resulted in a stronger third mode vibration than observed for the wood bat, but it was still smaller in amplitude than the fundamental and second modes. Furthermore, the third mode was strongly damped in the aluminum bat, whereas the fundamental and second modes were only lightly damped. The second mode frequency was 615 Hz for the aluminum bat, with a node point 11 cm from the tip of the barrel, and the fundamental mode frequency was 178 Hz with a node point 16 cm from the tip of the barrel. The whole bat vibrated when it was impacted by the ball, the handle vibrations being slightly larger than those in the barrel. The whole bat also rotated and translated, the handle velocity being much smaller than the barrel velocity when the ball impacted along the barrel.

A piezo accelerometer was attached to the handle of the aluminum bat, 67 mm from the knob, and the bat was impacted at the same low speed every 1 cm along the barrel. The bat was freely suspended and the accelerometer signal was integrated to obtain a measurement of the handle velocity. The results were similar to those shown in Fig. 14.10, in that the handle velocity had both a dc and an ac component, both being about equal in amplitude for all impact points along the barrel.

Results of the experiment are shown in Fig. 14.12. The velocity was not calibrated so the velocity scale is given as AU (arbitrary units). The velocities in Fig. 14.12 are correctly proportioned but the scale factor was not measured. The dc and ac components were typically less than 1 m s^{-1}, the golf ball being incident at about 3 m s^{-1}. Of greater interest was the fact that the dc velocity of the handle was negative for impacts with $d < 16$ cm and it was positive for impacts with $d > 16$ cm. The COP was, therefore, located at $d = 16$ cm, with respect to the axis 67 mm from the knob. The COP coincided with the fundamental node in this experiment.

The fundamental vibration mode was about four times larger in amplitude than the second mode. The amplitude of the ac component of the handle velocity was therefore a minimum at $d = 16$ cm rather than at $d = 11$ cm. An impact at $d = 11$ cm excited the fundamental mode relatively strongly, whereas an impact at $d = 16$ cm did not excite the fundamental mode at all and the second mode remained relatively weak.

Given that the ac component of the handle velocity is about equal to the dc component for both the wood and aluminum bats, energy loss due to vibration of a bat is quite significant for both types of bat, especially for impacts well removed from the node of the fundamental mode. This result is more clearly evident when examining handle velocity data rather than the handle acceleration data since bat kinetic energy and vibrational energy are both proportional to velocity squared rather than being proportional to acceleration or displacement. Energy loss is significant even at the node of the second mode (at $d = 11$ cm) due to strong excitation of the fundamental

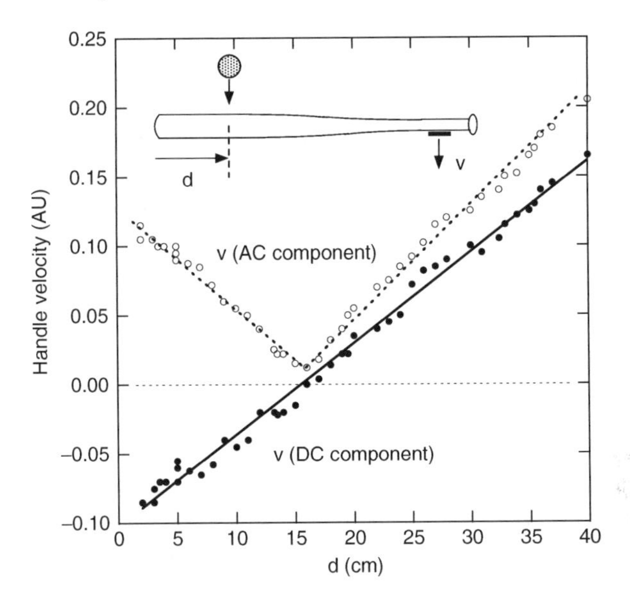

Fig. 14.12 Measured velocity of the handle, 67 mm from the knob, for a freely supported aluminum baseball bat impacted by a golf ball at a distance d from the tip of the barrel. The velocity has a dc component (shown by *solid dots*) due to rotation and translation of the whole bat, plus an ac component (amplitude shown by *open circles*) due to vibration of the whole bat

mode. Consequently, the node of the fundamental mode is more likely to represent the sweet spot perceived by batters than is the node of the second mode, at least for low speed impacts. Calculations at high batted ball speeds [10], where the impact duration can be as short as 0.6 ms, indicate that the second and even the third mode contributes significantly to the overall vibrational energy loss in a bat, in which case the spot that results in minimum handle vibration shifts away from the fundamental node toward the node of the second mode.

As a result of vibration losses in the bat and energy losses in the ball, the COR for the aluminum bat was about 0.4 near the tip of a bat and about 0.6 at the node of the fundamental mode when measured for low speed impacts (see Project 10). The COR is not necessarily a maximum at the node of the fundamental mode since energy losses in the ball are generally larger than vibrational losses in the bat and since losses in the ball vary with impact location along the barrel due to changes in the impact force on the ball [10].

14.6 Size of the Sweet Spot

The size of the sweet spot of a bat depends on how it is defined. When a batter talks about the sweet spot, he or she is talking about the spot that feels best or the spot where no apparent force or sting is transmitted to the hands. In that sense, there is only one best spot and it is the size of a spot. That is, it is about 1 mm wide. If a batter attempts to measure the sweet spot by dropping a ball onto the bat in that region, then he or she might be able to detect a small difference in the feel say 1/2 an inch away from the best spot, but not 1/4 in. away, in which case the measurable sweet spot is about 1/2 in. wide.

Alternatively, the sweet spot could be defined as a region where the measured level of vibration in the handle is less than a specified value when a ball is dropped from a certain height. In that case, the sweet spot might be 1 in. wide on one bat and 3/4 in. wide on another bat, even though the best spot of all is still only 1 mm wide on both bats.

Not all researchers are agreed on the definition of the sweet spot. Some prefer to describe the sweet spot as the spot on the barrel where the ball exit speed is a maximum [8]. That definition, by itself, is not a particularly good one since the location of that spot depends on the speed of the bat and the speed of the incoming ball. For example, if the bat is not swung at all then the spot shifts toward the center of mass of the bat and is located at the spot where q is a maximum. If the bat strikes a stationary ball then the spot shifts toward the tip of the bat and is located at the spot where $(1 + q)V_{\text{bat}}$ is a maximum. Even if the bat and ball speeds are specified when defining the sweet spot in this way, the definition is one concerning maximum outgoing ball speed rather than one concerning the feel of the bat. Since the definition concerns a different physical quantity it should ideally be given a different name, perhaps "the hot spot" or "the high v spot" or some other user-friendly term denoting maximum exit speed. Nevertheless, at bat and ball speeds typical of those in a normal game, the spot where the batted ball speed is a maximum is quite close to the spot where the level of vibrations in the handle is reduced to a minimum.

Another definition of the sweet spot is that it is the zone between the nodes of the fundamental and second modes. That definition is probably as good as any, especially if the minimum vibration spot shifts from the fundamental node toward the node of the second mode as the impact speed increases or as the impact duration decreases. However, the distance between the two nodes should not be interpreted as the width of the sweet spot at any given impact speed. Furthermore, measurements of batted ball speed at different points along the barrel show that wood and metal bats can have very similar ball speed profiles even though the distance between the two nodes might be different for wood and metal bats [10–12].

Manufacturers like to advertise their bats as having a large sweet spot, and usually do so without actually defining what they mean by their sweet spot. For example, they might prefer to describe their sweet spot as the width of the region along the barrel where the ball exit speed is say 90% or more of the maximum value. In that case, they would be interpreting the sweet spot as a "high exit speed zone" rather than a "small vibration zone." Another manufacturer might prefer it to be the region where the ball exit speed is 80% or more of the maximum value, in which case the second manufacturer's sweet spot will be wider than the first. There is no industry standard for measuring the width of the sweet spot, or even defining what is meant by the sweet spot. In the absence of such a standard, all claims about the sweet spot need to be taken with a grain of salt or treated as possible advertising hype.

A useful indication of the best hitting zone would be one that makes use of the measured ball exit speed when the bat is submitted for approval. The maximum exit speed does not necessarily coincide with the sweet spot but it is normally close to it since that is where vibration energy losses are minimized. The standard bat approval test includes measurements of the exit speed at various points along the barrel in order that the maximum exit speed can be found. It should be a simple task to certify each bat tested with a measure of the width of the power zone, defined as the width along the barrel where the ball exit speed is say 90% or more of the maximum value. Bats could even be marked to show where the two 90% points are located. Alternatively, bats could be marked (and marketed) to show the location of the two points (if any) where the batted ball speed exceeds say 90 mph during the test measurements. That way, the consumer would be in a better position to make an informed choice as to which bat he or she might prefer to buy.

References

1. H. Brody, The sweet spot of a baseball bat. Am. J. Phys. **54**, 640–643 (1986)
2. L. Noble, H. Walker, Baseball bat inertial and vibrational characteristics and discomfort following ball-bat impacts. J. Appl. Biomech. **10**, 132–144 (1994)
3. R. Cross, The sweet spots of a tennis racquet. Sports Eng. **1**, 63–78 (1998)
4. R. Cross, The sweet spot of a baseball bat. Am. J. Phys. **66**, 772–779 (1998)
5. R.K. Adair, Comment on "The sweet spot of a baseball bat," by Rod Cross [Am. J. Phys. **66**(9), 772–779 (1998)]. Am. J. Phys. **69**, 229–230 (2001)
6. R. Cross, Response to "Comment on 'The sweet spot of a baseball bat'" [Am. J. Phys. **69**, 229–230 (2001)]. Am. J. Phys. **69**, 231–232 (2001)
7. R. Cross, Center of percussion of hand-held implements. Am. J. Phys. **72**, 622–630 (2004)
8. G. Vedula, J.A. Sherwood, An experimental and finite element study of the relationship amongst the sweet spot and vibration nodes in baseball bats, in Proceedings of the 5th International Conference on the Engineering of Sport September, vol. 2 (UC Davis, CA, 2004), pp. 626–632
9. S. Fisher, J. Vogwell, M.P. Ansell, Measurement of hand loads and the centre of percussion of cricket bats. Proc. IMechE Part L J. Mater. Design Appl. **220**, 249–258 (2006)
10. D. Russell, web site at http://paws.kettering.edu/~drussell/bats.html
11. L.V. Smith, Evaluating baseball bat performance. Sports Eng. **4**, 205–214 (2001)
12. A.M. Nathan, Dynamics of the baseball bat collision. Am. J. Phys. **68**, 979–990 (2000)

Chapter 15
Flexible Bat Handles

15.1 Introduction

In recent years, bat manufacturers have been experimenting with flexible bat handles in an attempt to increase bat performance. Some manufacturers claim that better performance is obtained with flexible handles, while other manufacturers claim that stiff handles are better. For example, in 2007, Easton advertised a bat using its flexible handle ConneXion technology, while Louisville Slugger were simultaneously advertising their stiff handle technology. Easton claimed that: "Acting like a hinge, the ConneXion provides the most efficient energy transfer from handle to barrel, resulting in maximum bat head whip for a quicker bat and more power through the hitting zone." Louisville Slugger claimed that: "The last thing you want at the moment of contact is for your bat handle to flex. When the handle flexes, the barrel can't. That reduces your trampoline effect. A stiff handle produces more barrel flex, resulting in a maximum trampoline effect and, ultimately, greater performance."

Since both manufacturers were claiming more power or greater performance, it seemed that both claims could not be correct. But it is possible that they are both correct in the sense that they might both result in bats with better performance than a standard wood bat. For example, one might increase performance by 0.1% and the other might increase performance by 0.2%. In the advertising world, factual details tend to be either especially selected or omitted entirely, leaving the customer full of hope but actually in the dark. If the increase in performance is indeed only 0.1% or 0.2% then it is not worth worrying about. But if it is 5% or 6% then that would be more interesting.

The Easton design is interesting because a bat with a hinge in the middle would indeed result in a greater barrel speed when the bat is swung. Professor Howard Brody at the University of Pennsylvania once made a tennis racquet with a hinged handle to prove that players don't need to grip the handle firmly. The racquet could be swung just like a normal racquet and the head swung around to strike the ball as required. It didn't work for volleys, since the hinged head swung backward when the handle was pushed forward rapidly, but it worked fine for controlled groundstrokes. Humans are designed with hinged wrists and elbows for a good reason. When a person swings a bat or a club, the upper arm rotates fastest at first and then transfers

R. Cross, *Physics of Baseball & Softball*, DOI 10.1007/978-1-4419-8113-4_15,

its rotational energy to the forearm. Near the end of the swing, the forearm slows down to transfer its rotational energy to the bat or the club. Each hinge allows the final bat or club speed to be greater than it would if the joint was rigid. For example, if you throw a ball without bending your elbow, the throw speed will be low. That is how a cricket ball is "bowled" rather than thrown to the batter. To make up for the lower ball speed, the pitcher in cricket (called the bowler) is allowed to run as fast as he can toward the batter before bowling the ball.

There are two potential problems with the Easton design. One is that a slightly flexible joint does not work in the same way as a completely flexible hinge. A bat with a flexible handle might bend significantly when pushed firmly onto the ground, but that doesn't mean it will bend by the same amount when it is swung through the air. In fact, high speed video of the swing of bats with flexible handles suggest that the bats bend by such a small amount that the increase in barrel speed must be very small. The second problem is that when the barrel strikes the ball, the barrel will tend to bend backward due to the hinge effect. Again, the backward bend will be only slight, but it might cancel any gain arising from the slightly higher swing speed of the barrel. If the two effects cancelled exactly then it would make no difference at all to bat performance whether the handle was stiff or flexible.

All bats have thin handles and thick barrels, so all bat handles are more flexible than the barrel end of the bat. That design locates most of the mass of the bat in the barrel where it is needed, and it provides a flexible handle for increased comfort. A solid rubber handle would be even more comfortable in terms of reducing the shock of the impact, but such a handle would be way too flexible. The barrel would get left behind when the batter swung the bat.

Cricket players have been aware of the advantages of flexible handles in cricket bats for more than 150 years. The blade is made from willow and the handle is made from cane. Not just ordinary cane but top quality, furniture grade, flexible Manau cane from the jungles of Sumatra. The first bat handles were made from solid cane, but they vibrated too much and stung the hands. Since 1856, cricket bat handles have been made as laminated strips of cane and rubber. In addition, one or more thin rubber grips are rolled onto the handle to provide a relatively soft grip. It was not until the late 1900s that the cane was banned as a physical punishment in most schools. The cane was used in preference to a wooden dowel so as not to sting the hands of the person weilding the punishment and since cane was less likely to break.

15.2 Stiff vs. Flexible Handles

In terms of bat performance, there is reason to suspect that flexible and stiff handles will perform about the same. The collision of a bat with a ball occurs at the fat end of the bat. The bat bends on impact, but it does so in an unexpected manner. If you were to support a bat at each end and stand on the barrel of the bat, then the bat would bend in the middle or slightly closer to the knob end where the handle is relatively thin. That is not where the bat bends when it strikes a ball. The bat bends right at the impact point, despite the fact that this point is in the thickest and stiffest part of the bat. It bends there because that is where the ball pushes on the bat, with a force of 5,000 lb or more. As soon as the impact commences, the bat starts to bend, the ball starts to squash, and a bending wave starts traveling in both directions along the bat at a speed of about 1,000 ft s^{-1}. By the time the ball bounces off the bat, about 1 ms later, the wave has traveled about 1 ft along the bat. The handle doesn't start to bend until the bending wave arrives at the handle, about 2 ms after the initial impact. Consequently, it is irrelevant whether the handle is stiff or flexible or fat or thin, at least in terms of the actual bounce of the ball.

There are only two ways that a flexible handle could affect the bounce of a ball off the bat. One is the action of the "hinge" in increasing the swing speed of the barrel, as described previously. The other is if the ball is much softer than a normal ball, thereby allowing the bending wave to reflect off the handle and arrive back at the impact point before the ball bounces off the bat.

To investigate the physics involved, a simple experiment was conducted by the author to measure the effect of changing the handle stiffness. Four different home-made bats were used. Not real bats, but experimental bats designed to examine the effect of handle stiffness. Two of the bats had a handle of zero stiffness. The handle was simply cut off and replaced with a length of string. One of the bats had a handle of essentially infinite stiffness. Not actually infinite, but it was so stiff that it behaved as if it were infinitely stiff. The experiment is described in the following section.

15.3 Measurement Technique

The performance of any particular bat depends to a large extent on how well the ball bounces off the bat. It is not necessary to swing the bat to determine how well the ball bounces. In fact, it is a lot easier to measure the bounce when the bat itself is not swung. A suitable arrangement is shown in Fig. 15.1. However, the effect of the whip action of a flexible handle in increasing the speed of the barrel can only be tested reliably by swinging the bat. That particular test was not attempted, although an estimate of the effect is given in Sect. 15.5. The arrangement in Fig. 15.1 was used to test whether the bounce speed was affected when the handle was made more flexible.

The arrangement in Fig. 15.1 was used to measure the bounce off a rectangular length of wood rather than an actual bat, but the method could be used just as well with a real bat. A length of wood was chosen with cross section 66 mm × 32 mm to compare the performance of that piece of wood when it was cut short and when the ball bounced off (a) the 66 mm wide surface and (b) the 32 mm wide surface. That way, one particular sample of wood could be used to compare the performance of long and short bats, as well as to compare the performance of stiff vs flexible bats. The measured performance was then compared with the calculated performance to ensure that the performance of each "bat" was properly understood.

The bat was suspended by two lengths of string, for two reasons. One is that a piece of string simulates an extremely flexible handle. To determine whether the

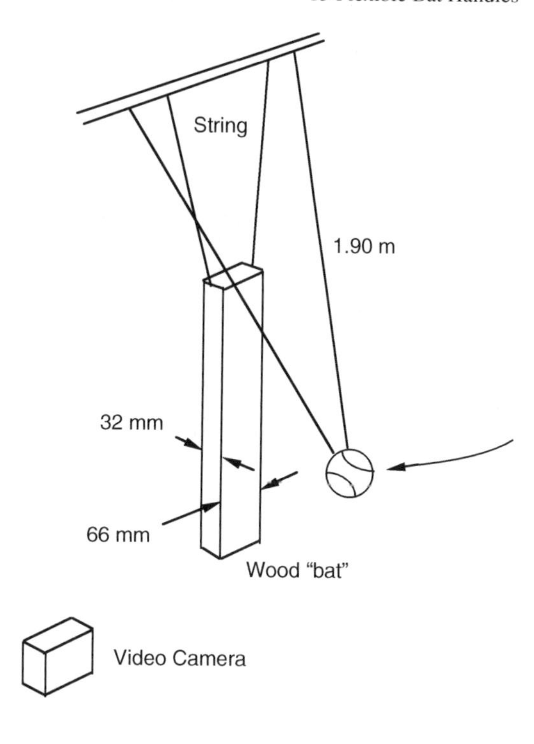

Fig. 15.1 Arrangement used to measure the bounce of a baseball off a wood "bat" made from a rectangular beam of wood. The ball was mounted as a 1.9-m long pendulum and a video camera was used to measure the bounce speed and distance

flexibility of the handle makes any difference to bat performance, a good test is to first see if a handle of zero stiffness makes any difference. The other reason for using string was that, in theory, bat performance should not depend on whether the handle is hand held, or held in a vice or completely free, and it should not depend on whether the handle is 6 in. long or 12 in. long.

A separate issue is how bats perform at low vs high incident ball speeds. Obviously, if a bat works well at an incident ball speed of 50 mph then it is likely to work well at a ball speed of 80 mph. Since ball speed wasn't the issue under investigation, the experiment was simplified to improve the convenience and accuracy of the measurements by using an incident ball speed of only 5 mph. The contact time of the ball was then 2 ms instead of the usual 1 ms or less that is more typical of a high speed impact. The length of wood was suspended with two lengths of string, using small nails hammered into the top end of the wood. Two lengths of string helped to prevent the wood spinning around if the ball collided near one edge. The ball was suspended in the same way, using two small nails hammered into the ball. That way the ball could be swung to impact the wood on a smooth part of the ball, without the ball twisting around and colliding on the stitching. If a ball bounces at low speed off the stitching then it can bounce off at an angle and rotate. The system was set up so that if the top end of the "bat" swung forward then it would not collide with the string supporting the ball. To avoid that problem, the support points at the top were more closely spaced for the bat support than they were for the ball support.

A video camera mounted on a tripod was located about 10 ft away to record the swing of the ball onto the bat and the bounce off the bat. The ball was withdrawn by a horizontal distance of 1.0 m, released so that it would collide with the middle of the 66 mm or the 32 mm wide face of the bat, and allowed to bounce off at right angles. If the ball collided a few mm off-center then the bat spun around and the ball did not bounce well. Those bounces were recorded on film together with the "good" bounces but were not analyzed. Only the good bounces were analyzed, where the bat recoiled smoothly without spinning around.

The analysis was relatively easy. The only measurement required was the rebound distance of the ball. It turns out (see Appendix 15.1) that the incident speed of the ball onto the bat is proportional to the initial horizontal displacement of the ball. Doubling the displacement from say 50 to 100 cm doubles the incident speed of the ball. Similarly, the rebound distance of the ball is proportional to the rebound speed. Since the bounce factor $q =$ rebound speed/incident speed, it is also equal to rebound distance/initial distance. For example, if the ball is released when it is 1.0 m from the bat, and if the ball bounces back 0.173 m, then $q = 0.173$. To measure q at different impact points along the bat, the bat was raised or lowered, keeping the pendulum length (and hence the impact speed) the same.

15.4 Experimental Results

Results for the four different "bats" are shown in Fig. 15.2, all using the same piece of wood, 66 mm × 32 mm in cross section and 84.8 cm long. The stiffness of each bat was varied by impacting on either the 66-mm wide face or the 32-mm wide face and then by shortening the bat to a length of 50 cm and repeating the experiment. The bats had the following additional properties, where $L =$ length of wood, $M =$ mass of wood and $f =$ fundamental vibration frequency of the bat.

(a) $L = 84.8$ cm, $M = 1.20$ kg, $f = 185$ Hz, impact on 66 mm wide face
(b) $L = 84.8$ cm, $M = 1.20$ kg, $f = 378$ Hz, impact on 32 mm wide face
(c) $L = 50.0$ cm, $M = 0.70$ kg, $f = 558$ Hz, impact on 66 mm wide face
(d) $L = 50.0$ cm, $M = 0.70$ kg, $f = 1,120$ Hz, impact on 32 mm wide face

Most wood bats vibrate at a fundamental frequency of about 180 Hz. The stiffer and lighter the bat the higher is the vibration frequency. A low vibration frequency implies that the bat is relatively flexible and/or relatively heavy. Consequently, one might expect that a bat with a very flexible handle would vibrate at a relatively low frequency. However, when the handle was replaced by flexible string, the vibration frequency increased by a large amount since the vibration frequency increases rapidly as the bat is made shorter. A short length of wood is much harder to bend than a long length of wood (of the same cross-section) so short bats are stiffer than long bats and they vibrate at a higher frequency.

Also shown in Fig. 15.2 are theoretical calculations based on the above bat properties, for the ball used in this study. The ball had a mass of 145 g and a COR of

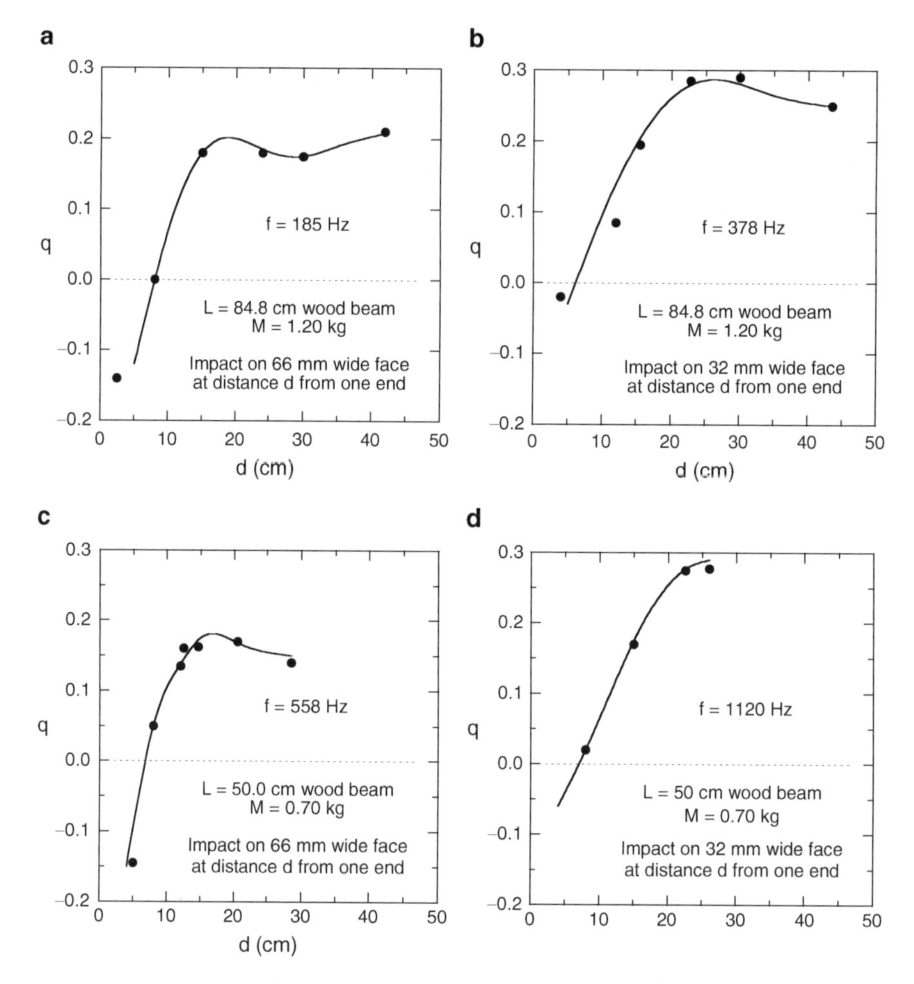

Fig. 15.2 Measurements of the bounce factor, q, vs impact distance, d, along a bat for four wood "bats" of different length and stiffness. The *black dots* are the experimental points and the *curves* are theoretical predictions

0.55 when impacting at low speed on a granite block. The agreement between theory and experiment is excellent. The theoretical curves were obtained by solving the bending wave equation for a flexible beam, as described in Appendix 12.1.

The most interesting result in Fig. 15.2 is that the bounce of the ball is essentially the same off all four bats when the impact distance d is between 10 and 15 cm. That is a typical range of impact points on a bat. The sweet spot is usually about 16 cm from the tip of the bat. A batter will occasionally strike the ball 5 cm or 20 cm from the end of the bat. To show how closely the four results agree, the four theoretical curves in this region are plotted in Fig. 15.3. In the region 10–15 cm, there is only a small loss of energy associated with bat vibrations. Outside this region the loss of energy due to bat vibrations is especially noticeable for the impact on the 66-mm wide surface where the value of q is significantly less than it is on the stiffer 32-mm

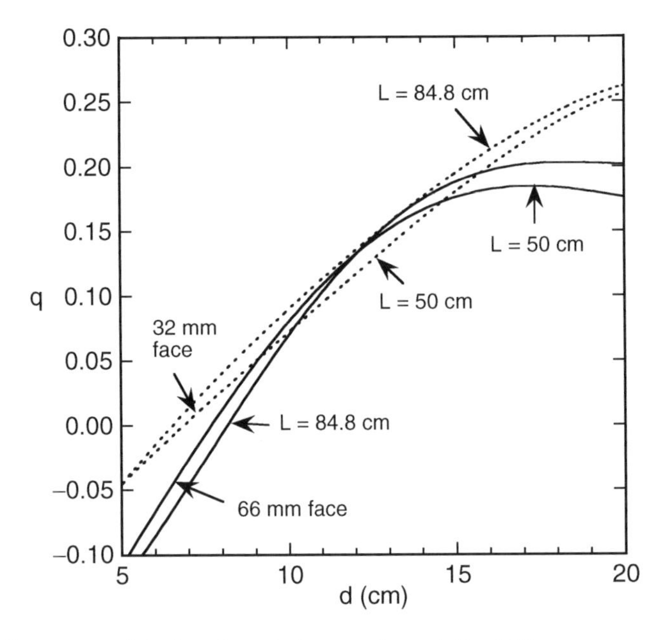

Fig. 15.3 Comparison of the four theoretical curves shown in Fig. 15.2

wide face. The overall length of the bat has a small effect on q but it is not as large as one might expect intuitively. That is, one can cut the handle off the 84.8-cm bat and the ball bounces almost as well, provided the impact is within the last 20 cm of the end of the bat. Cutting the handle off completely will make it impossible to swing the bat in the usual way, but the main point here is that it makes very little difference to the performance of a bat whether the handle is stiff or flexible or completely missing. The main factors affecting the performance of a bat are the mass and stiffness of the barrel.

The results in Figs. 15.2 and 15.3 can be explained qualitatively in two quite different and apparently unrelated ways. The first is to consider each bat as a semi-rigid beam that rotates when struck by a ball and that vibrates by an amount that depends on the stiffness of the bat. This was the approach used in Chap. 10 where we calculated the effects of a collision between a ball and a bat. Alternatively, we can regard each bat as being a flexible beam and describe the results in terms of the bending wave that travels along each bat.

If we regard each bat as a semi-rigid beam, then the effective mass at the tip of the bat is smaller than the effective mass further along the barrel. As a result, the bounce factor, q, increases as the impact point moves away from the tip. Vibrational losses are relatively weak near the sweet spot, which is located 18.6 cm from the tip of the longer bats and 11 cm from the tip of the shorter bats. The bounce factor near the tip is lower for impacts on the wider (66 mm) bats than for impacts on the narrow (32 mm) bats since the wider bats are more flexible and therefore vibrate more, with greater vibrational energy losses. Similarly, for impact points near $d = 20$ cm, the bounce factor is lower for the 66 mm wide bats than for the narrow bats due to greater vibrational energy losses.

A slightly surprising result is that vibrational energy losses are relatively large near $d = 20$ cm for the long, wide bat, despite the fact that the impact is close to or coincides with the sweet spot at $d = 18.6$ cm. The explanation in this case is that the fundamental vibration mode is excited only weakly for impacts near its node point, but the second vibration mode, at a frequency of 515 Hz, is excited strongly by the impact. The second vibration mode for the long, narrow bat or for the short bats is not exited by an impact anywhere along these bats since the vibration frequency of the second mode was higher than 1,000 Hz for these bats. When the impact duration is about 2 ms, as it was in these experiments, the only vibrations that can be excited are those with a frequency less than about 750 Hz.

The results in Figs. 15.2 and 15.3 can also be explained in terms of the bending wave generated by the impact. When the ball strikes the bat, the resulting bending wave propagates away from the impact point in both directions. When the bending wave reaches the tip of the bat it reflects back toward the impact point and causes the bat to move even further away from the ball. The bounce is therefore relatively weak for impacts near the tip of the bat, especially for impacts on the wide face. The bat is much stiffer for an impact on the narrow face, and the bounce factor is larger since the bat does not bend and move away from the ball as much. The significance of this result can be demonstrated more clearly if the tip of a bat is clamped in a vice. In that case, the ball bounces much better for impacts near the tip of the bat since the bending wave reflected off the vice causes the bat to move toward the ball rather than away from the ball. A bending wave reflecting off the free end of a bat does not change sign when it reflects off the end, whereas a bending wave reflecting off the clamped end of a bat changes sign. That is, a bend in the negative direction (away from the ball) gets reflected from a clamped end as a positive bend (toward the ball).

For impacts further from the tip, the bending wave reflected off the tip arrives back at the impact point later in time and has a smaller effect on the bounce. Consequently, the bounce factor increases as the impact point moves further away from the tip. If the bending wave arrives after the ball bounces off the bat then it has no effect on the bounce of the ball and the bounce factor will then be independent of the location of the impact point. That is why, in Fig. 15.2, the bounce factor is relatively constant in the impact region from 15 to 40 cm, at least for the 84.8 cm bats. For the 50 cm bats, the bounce factor decreases at distances greater than 25 cm from the tip since the bending wave from the handle end then arrives back at the impact point before the ball has left the bat.

The two explanations here seem to be unrelated, but they are equivalent. When a ball impacts a bat, the bat bends locally in the region of the impact point. The rest of the bat remains undisturbed until the bending wave arrives. If the ball bounces off the bat before the bending wave reaches the handle end, then there is no motion of the handle at all during the collision, meaning that the bat as a whole is not rotating or vibrating during this time. It is only after the bending wave travels up and down the bat a few times that we can regard the bat as a semi-rigid body that rotates and translates as a whole and that also vibrates at a well-defined frequency. The energy transferred from the ball to the bat starts out as a localized bending of the bat and is then redistributed throughout the bat, after the ball bounces, in the form of rotational, translational and vibrational energy, in proportions that are determined by conservation of energy and momentum. The explanation of the results in Figs. 15.2 and 15.3, in terms of the energy losses due to bat vibrations, are valid explanations but they are based on events occurring well after the ball has bounced off the bat.

15.5 Bat Bending Calculation

An approximate estimate of the amount that a bat will bend when it is swung can be obtained from the results shown in Fig. 12.2 and the estimated stiffness of a bat. We noted in Sect. 12.4 that a bat will bend by about 1 mm in the middle if it is supported at each end and a person stands in the middle of the bat. If the handle is very flexible then the bat might even bend by about 10 mm in the middle. In that case, the stiffness of the bat would be about $8 \times 10^4\,\mathrm{N\,m^{-1}}$ rather than our previous estimate of about $8 \times 10^5\,\mathrm{N\,m^{-1}}$. The relevant formula given in Fig. 12.2 indicates that $48EI/L^3 = 8 \times 10^4\,\mathrm{N\,m^{-1}}$ for the flexible bat. The cantilever stiffness of the bat is then $3EI/L^3 = 5 \times 10^3\,\mathrm{N\,m^{-1}}$. If the bat handle was held in a vice, and a person stood on the tip of the barrel, the tip would bend by 160 mm or just over 6 in. It is unlikely that anyone would want to try that experiment since the bat might break. However, we can use the result to estimate the bend when the bat is swung.

Suppose that the tip of the bat accelerates from 0 to $30\,\mathrm{m\,s^{-1}}$ (67 mph) in 0.15 s. Then the average acceleration of the tip is $30/0.15 = 200\,\mathrm{m\,s^{-2}}$. A bat can be regarded as consisting of a long, thin handle with an additional mass of about 0.6 kg attached to the far end. The force on the additional mass is then $0.6 \times 200 = 120\,\mathrm{N}$ or about 27 lb. If that force was applied to the barrel when the handle was held in a vice then the tip of the barrel would bend by about 1 in. The actual bend of the bat when it is swung will be less than that since a batter cannot exert a vice grip and since the inertial force on the barrel is not applied at the very tip but along the whole length of the barrel. Consequently, the bat might bend by 1/4 to 1/2 in. over its whole length. The bat will remain bent during the whole swing unless the handle decelerates rapidly over the last 0.05 s or so. Either way, the result would be a negligible additional speed of the tip. If the tip whips forward by 1/2 in. in 0.05 s, then the tip speed increases by only $10\,\mathrm{in.\,s^{-1}}$ or 0.57 mph.

15.6 Conclusion

The experimental results obtained with the four "bats" show that the rebound speed of a ball does not depend to any significant on the flexibility of the handle. Even if the handle is replaced with a length of string, the ball still bounces just as well as a ball bouncing off a bat with a stiff handle. The question remains as to whether a bat with a flexible handle will cause the barrel to whip around at higher speed when the

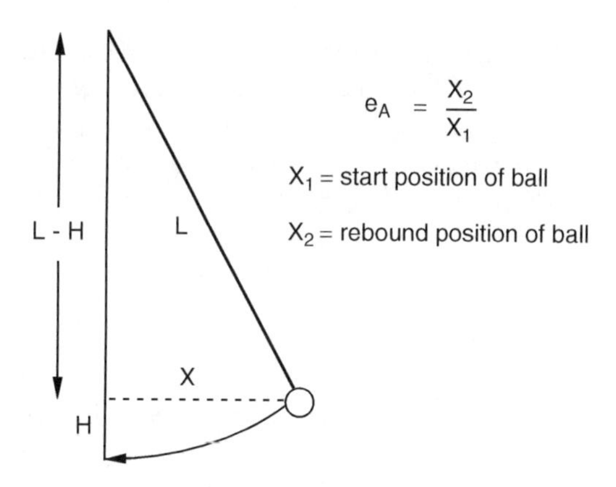

Fig. 15.4 Geometry of the pendulum method of measuring the bounce factor q

bat is swung. Video film indicates that the effect, if it exists, is too small to observe clearly. An estimate of the effect, based on the stiffness of a bat and the inertial force acting on the barrel, also indicates that whip action will not result in a significant increase in bat speed.

Appendix 15.1 Formula for q

If a pendulum of length L is withdrawn a horizontal distance X, as shown in Fig. 15.4, then it will rise a distance H given by $L^2 = X^2 + (L - H)^2$ so $X^2 = 2LH - H^2$. Provided that H is a lot less than L, then $H = X^2/(2L)$. When the ball is released, its potential energy (PE) is mgH. At the bottom of its swing, its PE is zero and its kinetic energy is $\frac{1}{2}mv^2 = mgH$. Hence, $v = X\sqrt{g/L}$. The ball speed is therefore proportional to X. Sir Isaac Newton used this method to measure collisions between different balls, since he didn't have a video camera to measure the ball speed directly.

Further Reading

http://paws.kettering.edu/~drussell/bats.html, by Dan Russell, has an extensive discussion of the properties of bats with flexible handles, as well as many other interesting articles on softball and baseball.

Chapter 16
Ball Bounce and Spin

16.1 Introduction

Almost all major sporting activities are played with a ball. The rules of the game always include rules about the type of ball that is allowed, starting with the size and weight of the ball. The ball must also have a certain stiffness. A ball might have the correct size and weight but if it is made as a hollow ball of steel it will be too stiff and if it is made from light foam rubber with a heavy center it will be too soft. Similarly, a ball needs to bounce properly. A solid rubber ball would be too bouncy for most sports, and a solid ball of plasticene would not bounce at all. The correct size, weight, stiffness and bounce for each ball used in each sport has evolved over many years of trial and error. That evolution settled down 30 or 40 years ago to something that everyone was more or less happy with.

We say "more or less" because there are concerns these days, in all sports, that modern technology is changing the nature of the game. Bats, racquets and clubs have evolved faster than the ball, with the result that balls are being struck harder, faster and further every year. A ball that is struck faster will also spin faster, changing the flight of the ball through the air and the distance it travels. In golf and tennis, clubs and racquets have evolved with larger heads using lighter and stronger materials such as graphite and titanium, making it easier for the average player to strike the ball and to swing the implement faster. In baseball and softball, hollow aluminum or composite bats have completely replaced wood bats in most leagues since they are stronger and last longer. In addition, a hollow bat is softer than a solid bat and distorts the ball less on impact. The result is a decrease in the energy lost in the ball so the ball speed off the bat is increased. However, modern technology is viewed by some as a two-edged sword, the main problem being that the potential for serious injury is increased when a player is struck by a faster ball.

The spin of a ball plays a major role in all ball sports. There are no specific rules in baseball or softball about spin generation. Suppose that a manufacturer comes up with a clever bat or ball design that doubles the amount of spin generated by the batter. The result would be a dramatic increase in the number of home runs each game since a ball launched with extra backspin stays in the air longer and travels farther than one launched with little or no backspin. Something like that happened

R. Cross, *Physics of Baseball & Softball*, DOI 10.1007/978-1-4419-8113-4_16,

in tennis in the late 1970s. An inventor found a way to use a special arrangement of strings in a racquet that doubled the amount of spin on the ball. It was called spaghetti stringing. It caused the ball to fly through the air in a strange way and it resulted in even stranger bounces when the ball hit the court. The International Tennis Federation had no choice but to ban it. But they still haven't come up with a rule that specifies the allowed amount of ball spin.

In golf, manufacturers have been able to manufacture golf balls that spin faster than usual by increasing the weight of the core and reducing the weight of the outside layers. That way, the ball has a smaller moment of inertia, meaning that it spins faster when struck than a ball of the same weight but with a light core and with heavy outer layers. Recent rule changes regarding the design of the grooves in a club head have been introduced to limit the amount of spin that can be achieved.

The detailed construction of a baseball has not changed for many years and is regulated by the rules. Nevertheless, it might be possible to alter the surface of the ball or the bat in some manner so that the bat gets a better grip on the ball and spins it faster. We examine in this chapter some of the physics issues involved in the way a ball bounces and the amount of spin generated when the ball bounces.

16.2 Bounce Off a Heavy Surface

The interaction between a bat and a ball is a surprisingly complicated process when one starts looking at the details of the collision. The situation is complicated by three factors. One is that the bat surface is curved and the ball tends to wrap itself around the bat. Another is that the bat is moving toward the ball when the ball bounces. The third is that the bat itself is relatively light and gets knocked around by the ball. The bat bends, squashes, vibrates, twists in the hands, and slows down during the collision.

To simplify matters we will concentrate in this chapter on a much simpler situation where a ball bounces off a fixed, flat, heavy surface such as a solid timber or concrete floor. That way we can focus on the behavior of the ball itself. We will then be in a much better position to ask questions such as, what difference does it make if the surface is curved, or if it is moving toward the ball or if the surface is relatively light? In other words, how does the bounce of a ball off a bat differ from the bounce off a heavy floor? As a preview of where we are headed in the next chapter, the short answers are

(a) An impact with a curved surface is essentially the same as an impact on a flat, inclined surface. An impact above or below the central axis of the bat causes the ball to deflect toward the sky or the ground, and it changes the amount of spin of the outgoing ball compared with a head-on impact.
(b) The outgoing ball speed is equal to the speed of the bat plus the bounce speed of the ball.
(c) Since the bat gets knocked around by the ball, the bounce speed off a stationary bat is reduced compared with the bounce off a very heavy surface. As an

extreme and obvious consequence, if you hit the ball with a long, thin pencil or broomstick instead of a bat, the ball would slow down slightly and pass straight through to the catcher. In order for the ball to bounce forward off a stationary bat, the bat needs to be heavier than the ball, as it normally is.

16.3 Vertical Drop of a Spinning Ball

If a smooth spherical ball is dropped vertically onto a heavy, smooth horizontal surface, and if the ball is not spinning when it is dropped, then the ball will bounce vertically without spin. In practice, the ball will bounce with a small amount of spin since (a) all balls and all surfaces are at least slightly rough, (b) balls used in sport are rarely perfect spheres, and (c) most surfaces are not perfectly smooth and horizontal. Nevertheless, if you drop a baseball or softball on a wood floor then the ball will spin quite slowly or not at all when it bounces.

It is interesting to compare that result with the vertical bounce of a non-spherical object such a football or a pencil eraser [1]. An eraser bounces well since it is made from an elastic rubber compound, but it usually spins rapidly when it bounces since it usually lands on one edge. If the upward force acting on the edge does not pass through the center of the eraser, then there will be a torque on the eraser causing it to spin and a sideways force causing the ball to bounce erratically. By contrast, the upward force at the contact point of a spherical ball acts through the center of the ball when the ball is dropped vertically, so there is no torque on the ball and no sideways force. Spherical balls therefore bounce in a predictable manner, which is why most balls used in sport are spheres rather than cubes.

When a batter strikes an incoming ball that is spinning, the outgoing spin will usually be quite different from the incoming spin. To understand why the spin changes, it helps to first consider a simpler situation where a spinning ball is dropped vertically onto a heavy, horizontal surface. Dropped from a height of say 3 ft, a spinning baseball will bounce to a height of about 13 in., the same height as a ball dropped without spin. Since a ball dropped vertically without spin bounces vertically without spin, there is no change in spin. However, there is a large change in spin when a spinning ball is dropped vertically. Furthermore, if the ball is spinning about a horizontal axis then the ball bounces off to one side, as shown in Fig. 16.1. If the ball is spinning about a vertical axis when it is dropped, then the ball bounces vertically with only a slight reduction in its spin.

The sideways deflection of a spinning ball is due to friction between the ball and the surface. Just before the ball contacts the surface, the bottom of the ball is moving down onto the surface and it is also moving in a horizontal direction because it is spinning in that direction. As soon as the ball contacts the surface, the bottom of the ball starts sliding across the surface, say from right to left. Friction between the ball and the surface acts to generate a horizontal force on the ball, acting a direction from left to right. The rate of spin is therefore reduced. The ball can stop spinning completely if there is enough friction, otherwise it is likely to bounce off

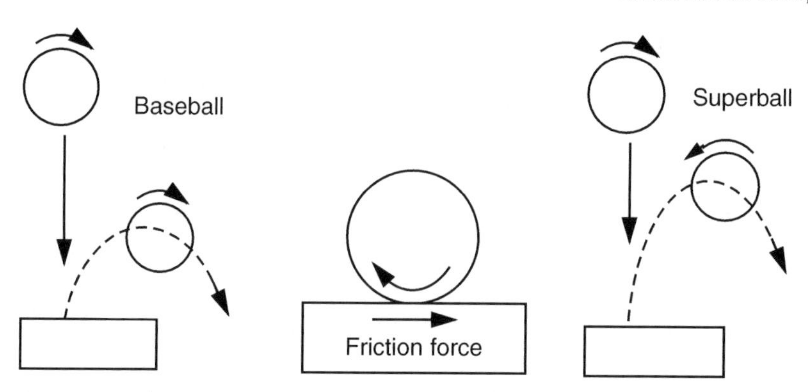

Fig. 16.1 A spinning ball bounces sideways off a horizontal surface, with reduced spin. Both effects are due to the horizontal friction force generated when the bottom of the ball slides on the surface. A superball reverses its spin direction when it bounces, and it bounces higher than a baseball

the surface before it has a chance to stop spinning. The ball, therefore, bounces with reduced spin and it bounces to the right since it receives a push to the right due to the friction force.

A spinning superball reverses its direction of spin when it bounces [2]. It does so for several reasons. The friction force on a dry ball is large enough to stop the bottom of the ball spinning. After the bottom of the ball comes to a stop, the top of the ball keeps spinning for a short time since the ball is sufficiently elastic to twist itself up slightly around its perimeter. As the ball untwists, the bottom maintains a firm grip on the surface and the top spins in the opposite direction. As the ball bounces up off the surface the bottom of the ball loses its grip and then the whole ball rotates in the reverse direction. It is a marvellous thing to watch. Most other balls are not sufficiently elastic for this to be obvious, but it can happen to a small extent with baseballs, golf balls and tennis balls. It can be made to happen by dropping a spinning ball on an elastic rubber surface. In that case it is the surface that gets twisted up slightly, in a direction parallel to the surface, and then the ball spins in the opposite direction as the surface untwists.

If the ball is spinning about a vertical axis when it contacts the surface, friction acts to reduce the rate of spin, so the ball bounces with reduced spin. If the friction force acts from left to right on one side of the contact area, then it acts from right to left on the opposite side since the ball spins in opposite directions on each side. Friction always acts in a direction to oppose sliding motion. There is no net horizontal force in any direction so the ball bounces vertically. Regardless of the direction of the spin axis, the change in spin as a result of the bounce will depend on how hard the ball impacts the surface. Dropped from a height of a few inches, the friction force will be relatively weak and the change in spin will be small. Dropped from a greater height, the friction force will be larger and the change in spin will also be larger. If you push a superball firmly onto a flat surface and try to drag the ball across the surface then the bottom of the ball tends to grip the surface firmly.

But if you try to rotate the ball around a vertical axis then the ball doesn't grip as well. The reason is that you can exert a torque by hand, on the edge of the ball, that is much larger than the torque exerted by the friction force near the axis. It is like loosening a nut with a long wrench, which is much easier than loosening the nut by hand.

16.4 Bounce Off an Inclined Surface

An important bounce event in baseball and softball, as well as in tennis and golf, is the bounce of a spinning ball off an inclined surface, as shown in Fig. 16.2. In tennis, a player can tilt the racquet head to vary both the rebound angle and spin of the ball. In fact, the modern game of tennis is dominated by the amount of spin that players impart to the ball. Players launch themselves off the court by belting the ball as hard as they can to spin the ball as fast as they can. Their opponent does the same, so the ball returns spinning furiously. If the player just taps or pushes the ball back at low speed, the ball will bounce off the strings at a strange angle.

In baseball and softball, the pitcher usually spins the ball rapidly so that it follows a strongly curved path through the air. The batter's main task is simply to connect with the ball, but if he is very good or just lucky, he can strike the ball above or below the axis to put even more spin on the ball. The ball impacts on the curved surface of the bat but the effect is the same as an impact on an inclined surface. The result is that the ball deflects skyward if the ball strikes above the long axis of the bat or it deflects down toward the ground if the ball strikes below the axis.

In Fig. 16.2, the ball on the left is incident with backspin, while the ball on the right is incident with topspin. The direction of the spin is the same in both cases, but the ball on the left is incident from left to right relative to the surface, and the ball on the right is incident from right to left. If the ball was incident without any spin then it would bounce to the right off the left hand surface and to the left off

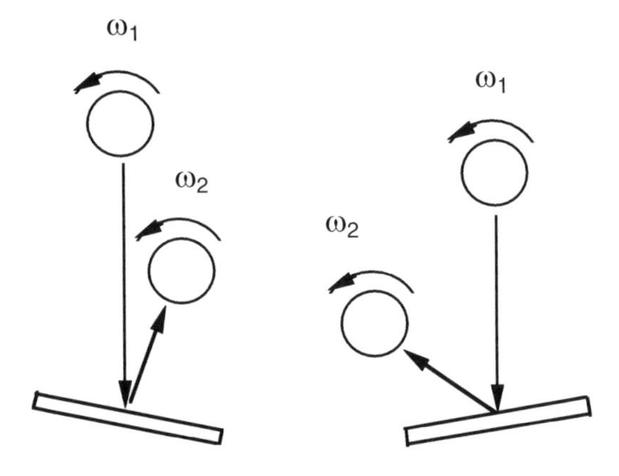

Fig. 16.2 Bounce of a spinning ball dropped onto an inclined surface. If the surface is tilted to the right, the ball bounces almost vertically. If the surface is tilted to the left, the ball bounces a long way to the left

the right hand surface. The effect of the counter-clockwise spin, on its own, is to deflect the ball to the left. The net result is an almost vertical bounce off the left hand surface and an exaggerated bounce to the left off the right hand surface. A tennis player therefore needs to be careful, when tilting the racquet, to tilt it in the right direction. A baseball player tends to get whatever comes, unless he is skillful enough to strike the ball exactly where he wants to. We will return to this later when we examine whether a curveball (incident with topspin) can be struck farther than a fastball (incident with backspin).

Oblique Bounce

When a ball without spin or with topspin is incident at an oblique angle on a flat, heavy, horizontal surface, the ball will bounce with topspin, as indicated in Fig. 16.3. Provided the surface is much heavier and stiffer than the ball then motion of the surface itself can be ignored and the bounce is determined mainly by the properties of the ball. Nevertheless, there is one property of the surface that does influence the bounce, and that is the smoothness or roughness of the surface. If the surface is slippery then the friction force on the ball will be relatively small. If the surface is rough then there will be a large friction force on the ball. The friction force on the ball acts in a direction parallel to the surface and has two effects. One is that it causes the ball to slow down in a direction parallel to the surface. The other effect is that the friction force exerts a torque on the ball and alters its spin. As soon as the ball first contacts the surface, it starts sliding along the surface as it continues to move in the same horizontal direction as it was moving before it hit the surface. The friction force acts backward in a horizontal direction, causing the ball to slow down in the horizontal direction, and causing it to rotate. The ball bounces up off the surface since a vertical force acting up on the ball is generated when the ball squashes against the surface. That force acts through the center of the ball, or close to it, so it does not generate any significant spin.

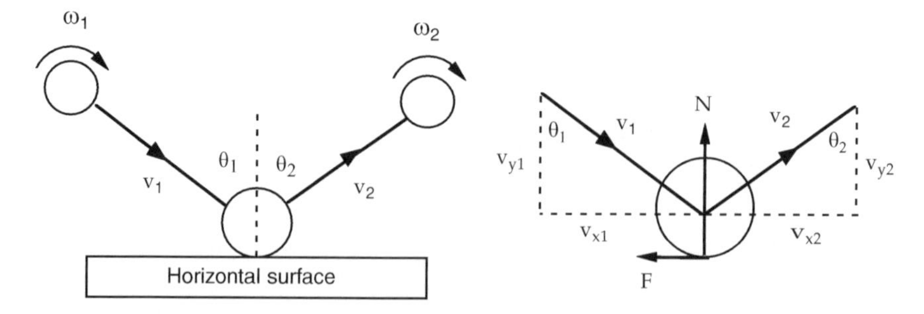

Fig. 16.3 Bounce of a ball incident obliquely on a horizontal surface. θ_1 is the angle of incidence and θ_2 is the angle of reflection. N is the vertical force on the ball and F is the horizontal friction force. v_1 is the incident speed and v_2 is the bounce speed

The behavior of a ball when it impacts obliquely on a surface depends on the angle of incidence and it also depends on whether the incident ball is spinning when it first contacts the surface. The behavior can be described in the following simplified terms, at least when the incident spin is zero or relatively small:

1. The angle of reflection is approximately equal to the angle of incidence. For example, if the ball is incident at 45° then it will bounce off the surface at an angle of about 45°.
2. The COR of the ball does not depend to any great extent on the angle of incidence. The perpendicular components of the incident and rebound speeds, v_{y1} and v_{y2}, are related by COR $= v_{y2}/v_{y1}$. The COR does vary slightly with the angle of incidence, but not very much. In other words, a bouncy ball will bounce well at any angle of incidence.
3. The ball bounces with topsin. The amount of spin depends on the angle of incidence, being a maximum when the ball is incident at about 60° to the perpendicular. Depending on the coefficient of friction (COF) between the ball and the surface, the spin might actually be a maximum when the angle of incidence is 50° or 70°, but 60° is typical. The spin is proportional to the incident speed, and also depends on the mass and diameter of the ball. Large and/or heavy balls spin slowly and small and/or light balls spin more rapidly. If you throw a basketball onto the floor at an oblique angle, it will bounce off the floor spinning slowly. If you throw a golf ball onto the floor at an oblique angle, it will bounce off the floor spinning rapidly.

Ball rotation can be explained in terms of the torque on the ball. If the friction force is F and the ball radius is R then the torque on the ball is $\tau = FR$. The longer this torque acts, the faster the ball will rotate. That is a mathematical explanation. A simple physical explanation is that when the ball first contacts the surface, there is no change in speed at the top of the ball since there is no force acting at the top of the ball. However, the bottom of the ball slows down rapidly, in both the vertical and horizontal directions. Compare this situation with a person who trips on a step while walking or running. The head keeps moving forward but the feet come to a sudden stop. As a result, the person rotates forward. That is essentially what happens to the ball in this situation. The bottom of the ball does not come to a complete stop instantly in the horizontal direction, but it slides and slows down rapidly in the horizontal direction.

In fact, if the ball is incident at any angle up to about 50° away from a perpendicular line to the surface, then the bottom of the ball *does* come to a complete stop. Not instantly, but shortly after the ball first contacts the surface. In other words, the friction force will be large enough to bring the bottom of the ball to a screeching stop, even though the top of the ball still rotates forward. If the ball is incident at an angle greater than about 50° to the perpendicular, then the friction force will not be large enough to stop the forward slide at the bottom of the ball, and the ball will bounce up off the surface before the bottom of the ball comes to a stop.

16.5 Slide or Roll or Grip?

The physics and mathematics of a bouncing ball has been described in many technical articles on the subject [3–9] and is outlined in the Appendix. The behavior of a ball when it bounces depends to a large extent on the angle of incidence and on the nature of the friction force between the ball and the surface on which it bounces. And that depends on whether the ball slides across the surface or whether it rolls along the surface or whether it grips the surface.

Sliding

A fast tennis serve hits the court at a glancing angle and slides along the court during the whole bounce period, leaving an elliptical mark on the court about 4 in long. In baseball, the equivalent type of bounce is one where the ball strikes the playing field at a glancing angle. The ball always slows down when it slides across the ground in this manner. A glancing angle collision also occurs when the ball strikes the bat near the top or bottom edge of the bat, in which case the ball can slide across the edge of the bat before bouncing off. It is easy to calculate the effect on the bounce under these conditions, using some simple physics. The calculation is shown in the Appendix.

Rolling

A ball rolling on a horizontal surface rolls a long distance before it comes to a stop. It does so because the bottom of the ball does not slide across the surface, and because there is almost no friction between the ball and the surface. When a ball rolls, any given point on the circumference rotates in a circular path around the axis at exactly the same speed as the axis itself travels along the surface. As a result, when a point on the circumference gets to the bottom of the ball, that point travels backwards around the axis at exactly the same speed as the axis moves forward. In other words, the point comes to a complete stop in the horizontal direction when it arrives on the surface. That is why there is almost no friction on a rolling ball, or on a rolling wheel.

It is easy to calculate the rotation speed of a rolling ball. A baseball is about 9 in. in circumference or 2.9 in. in diameter. If a baseball rolls on a horizontal surface then it completes one revolution when it rolls 9 in. along the surface. If the ball is rotating at 10 rev s^{-1} then it rolls forward at 90 in s^{-1} or 7.5 ft s^{-1}. At any other rolling speed the rotation rate can be simply proportioned. For example, if the ball was rolling at 75 ft s^{-1} (51 mph) then it would be spinning at 100 rev s^{-1} or 6,000 rpm. A ball struck by a bat usually spins at around 2,000 rpm but it can sometimes spin as fast as 6,000 rpm if the ball strikes the edge of the bat. A ball coming off a bat at 2,000 rpm tends to roll around the bat at about 17 mph during the impact. It doesn't roll right around the whole bat but during the 1/1,000 s impact time it can roll through an angle of about 10°. Furthermore, the ball doesn't actually roll but it does something similar, as we will describe shortly. What actually happens is that the contact area of the ball grips the bat while the rest of the ball keeps rotating.

When the brakes of a vehicle are applied, the wheels start rotating at a lower speed and the vehicle rolls to a stop. If the brakes are applied rapidly, the rotation speed of the wheels decreases faster than the vehicle can stop, so the tires start sliding on the road and the vehicle then slides or skids to a stop. A similar thing happens to a ball bouncing obliquely on a surface, but in reverse. If the ball is not spinning when it first contacts the surface, then the bottom of the ball starts sliding or skidding along the surface. Friction slows the ball in the horizontal direction and causes the ball to start rotating. While the friction force is acting in this way, the ball continues to decelerate in the horizontal direction but it spins faster. In fact, it is possible for the bottom of the ball to come to a complete stop before the ball bounces up off the surface. That is, the ball can arrive at a rolling condition where the bottom of the ball is stationary for an instant because the whole ball is rolling forward. The ball then grips the surface.

Grip

If a ball started to roll during its bounce on a horizontal surface then the friction force on the ball would drop to zero and the ball would roll along the surface with no further change in its spin or horizontal speed. In fact, measurements show that the friction force can indeed drop to zero but as soon as it does so it immediately reverses direction. Consequently, a ball does *not* roll when it bounces. It does something else. When the bottom of a ball slides to a stop, the ball grips the surface. That is, the bottom of the ball gets temporarily stuck on the surface while the top of the ball continues to rotate. As a result, the ball twists out of shape. The ball also squashes in the vertical direction since a large force is exerted on the ball in a direction perpendicular to the surface. It is that large perpendicular force pressing on the bottom of the ball that causes the ball to grip the surface. Since the ball twists out of shape, additional elastic energy is stored in the ball that would not arise in the case of a rigid, rolling ball. The end result is that when the ball bounces, it spins faster than one would normally expect for a ball that entered a rolling mode.

The grip mode of a ball is due to static friction rather than sliding friction. Suppose that a heavy brick is at rest on a horizontal platform. There is then no horizontal friction force on the brick. If one end of the platform is lifted so that the platform is inclined at say 10° to the horizontal, then the brick will remain at rest unless the platform is very slippery. The brick remains at rest because static friction prevents it sliding down the platform. The brick, therefore, grips the platform as if it was stuck down with sticky tape. As the platform is raised, the friction force increases by just enough to prevent the brick sliding. At a sufficiently large angle, the brick loses its grip and then slides down the platform. In order for the brick to slide, the platform might need to be raised to an angle of 30° or more. The angle at which sliding starts is a simple measure of the friction force and of the COF. A smooth, slippery surface is one with a COF less than about 0.3. A non-slippery surface is generally rougher and has a COF greater than about 0.7. A baseball sliding on wood or aluminum has a COF of about 0.4 or 0.5.

In the case of a bouncing ball, the bottom of the ball retains its grip even though the rest of the ball is rotating. The ball maintains its grip until it is no longer able to do so and then it suddenly loses its grip and starts to slide backward on the surface. The friction force on the ball reverses direction during the latter part of the grip phase and it continues to act in the reverse direction when the ball starts sliding backward.

Figure 16.4 shows a measurement of the horizontal friction force acting on a tennis ball when it bounced off a thin sheet of emery paper taped to a ceramic piezo mounted on a block of wood. The block rested on cylindrical rollers so that the block could accelerate freely in the direction of the friction force. An accelerometer attached to the front end of the block recorded its acceleration, which was proportional to the friction force, F. The piezo on top of the block recorded the vertical reaction force, N. The ball was incident at low speed, at an angle of about 30° away from the vertical. The total impact duration was 6 ms. The ball slid along the block for the first 2 ms but then it gripped the block, with the result that the friction force reversed direction.

The process that allows the ball to spin faster than it can roll is complicated. Even though the middle of the contact region gets stuck when it grips the surface, the outer parts of the contact region continue to slide forward since they are not pressed as firmly onto the surface. The net result is that sliding friction acts backward on the ball for an extended period of time, causing the ball to spin faster and to slow down

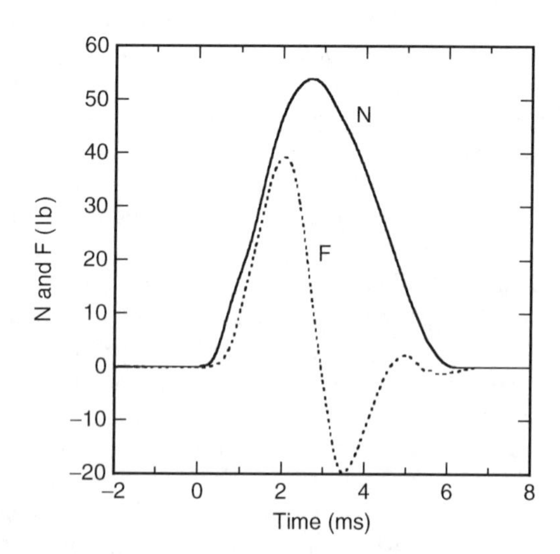

Fig. 16.4 The vertical force, N, and the horizontal friction force, F, acting on a tennis ball when it bounced off emery paper attached via a piezo to a block of wood. The ball started sliding along the block and then gripped the block, causing the friction force to reverse direction

in the horizontal direction more than one would expect for a rolling ball. When the friction force does eventually reverse direction then it acts to reduce the rate of spin but the ball still bounces with some of the enhanced spin acquired during the initial grip phase.

For many years, the physics of this process has been shrouded in mystery and has been subject to all sorts of weird and wonderful advertising claims. Anything that can be done to make a ball bounce faster or spin faster is at the heart of modern sports technology and the hype that surrounds it. This is especially true in golf and tennis, although similar claims are often made in relation to aluminum bats. The only way to counter the hype is to take careful measurements of ball speed and spin to determine whether there is any substance to the manufacturer's claims. Some progress has been made in this direction but a lot more still needs to be done.

16.6 Tangential COR

Suppose that a ball is incident obliquely on a horizontal surface, at speed v_1, and bounces at speed v_2, as shown in Fig. 16.5. The horizontal components of the ball speed before and after the bounce are v_{x1} and v_{x2}, respectively. The latter speeds refer to the speed of the ball center of mass (CM). Suppose also that the ball is incident with topspin and bounces with topspin, as shown in Fig. 16.3, with angular speeds ω_1 and ω_2, respectively. A point at the bottom of the ball will have a lower horizontal speed than the CM since the bottom of the ball is rotating backward. The horizontal speeds at the bottom of the ball are $s_1 = v_{x1} - R\omega_1$ before the bounce and $s_2 = v_{x2} - R\omega_2$ after the bounce, where R is the radius of the ball.

If the contact point has a vertical speed v_{y1} before the bounce, and v_{y2} after the bounce then we define $e_y = v_{y2}/v_{y1}$ as the COR in a direction perpendicular to

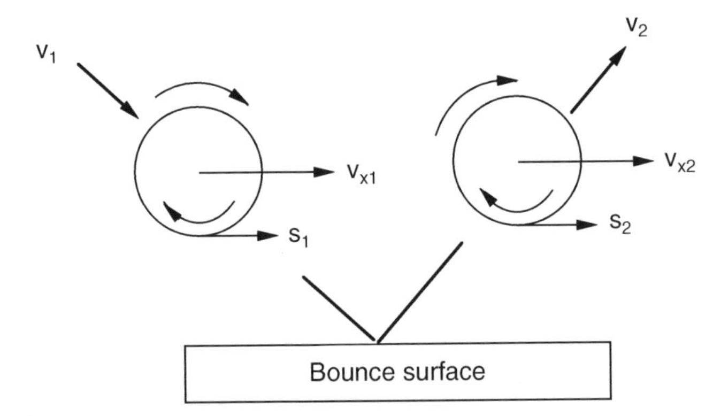

Fig. 16.5 If a ball is incident with topspin and bounces with topspin then a point at the *bottom* of the ball has a lower horizontal speed than the *middle* of the ball. The ratio $e_x = -s_2/s_1$ is called the tangential coefficient of restitution

the surface. In the same way, we can define the ratio $e_x = -s_2/s_1$ as the tangential COR. The minus sign is included so that e_x will be positive if the contact point reverses its direction of motion in the horizontal direction.

The most interesting feature of the bounce of a superball is that the COR is nearly 1.0 in both the vertical and horizontal directions. That is, the contact point reverses its direction of motion, without a significant change in speed, both vertically and horizontally. The spin can change by a large amount, but it does so in such a way that s_2 is approximately equal and opposite to s_1, regardless of the speed and spin of the incident ball. e_y is typically about 0.9 rather than 1.0, and e_x is typically about 0.6, but a superball is quite special in this respect. Most other balls have an e_x value that is only about 0.2 or 0.3, with the result that they don't spin as well as a superball when they bounce. Calculations showing the effects on bounce for several values of e_x are shown in Fig. 16.6.

A simple experiment that you can do yourself is to spin a baseball or a softball by hand so that it lands on a hard floor or concrete pavement. If the ball is given backspin then it can bounce with backspin or with topspin, depending on the angle of incidence. If you film the bounce with a video camera and then examine the film one frame at a time you can measure the speed and spin before and after the bounce. That information can then be used to calculate both e_y and e_x for the bounce. My own experience with a baseball is that e_x is relatively high for the first few bounces, around 0.3, and then drops to about 0.2 after about 10 bounces. I have not investigated this effect in enough detail to figure out why this happens. It seems that the ball surface is more elastic when the ball is relatively new and that repeated bounces act to harden the surface and to reduce its elasticity in a direction parallel to the

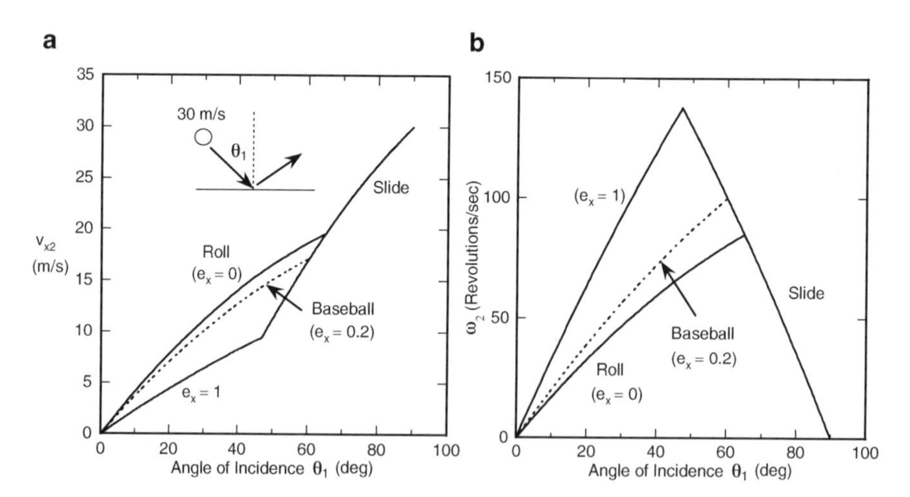

Fig. 16.6 Calculations for a ball incident at $30\,\text{m}\,\text{s}^{-1}$ on a horizontal surface, showing (**a**) the horizontal bounce speed, v_{x2}, and (**b**) the spin of the outgoing ball, vs. the angle of incidence, θ_1. The incident ball is assumed to have zero spin. At glancing angles, the ball slides throughout the bounce. For the sliding part of the bounce we assumed that $\mu = 0.4$ and $e_y = 0.55$. At small angles of incidence, closer to perpendicular incidence, the ball grips the surface during the bounce, in which case the bounce speed and spin depends on the value of e_x

surface. The end result is that a baseball can bounce with topspin when it is incident with backspin, but only for the first few bounces. After ten or more bounces the ball tends to bounce without any spin at all when it is incident with backspin.

Appendix 16.1 Ball Bounce Calculations

(a) Sliding

In Fig. 16.3, we show a ball incident at speed v_1 and bouncing at speed v_2 off a heavy surface. The horizontal component of v_1 is $v_{x1} = v_1 \sin\theta_1$ and the vertical component is $v_{y1} = v_1 \cos\theta_1$. Likewise, the components of v_2 are $v_{x2} = v_2 \sin\theta_2$ and $v_{y2} = v_2 \cos\theta_2$.

When the ball is sliding along the surface, a friction force $F = \mu N$ acts on the bottom of the ball in a direction parallel to the surface. N is the normal reaction force acting on the ball in a direction perpendicular to the surface, and μ is the coefficient of sliding friction (COF). For a baseball on wood or aluminum, μ is typically about 0.4 or 0.5.

If F and N are taken as average forces during the bounce, and if the ball remains in contact with the surface for a time T then the change in the horizontal and vertical components of the ball speed during the bounce are given by

$$F = -m\frac{dv_x}{dt} = \frac{m(v_{x1} - v_{x2})}{T} \tag{16.1}$$

and

$$N = m\frac{dv_y}{dt} = \frac{m(v_{y1} + v_{y2})}{T}, \tag{16.2}$$

where m is the ball mass. To avoid complicating the issue with negative numbers and signs, we have assumed here that all quantities are positive in the direction shown in Fig. 16.3. In particular, the ball reverses direction in the y direction, but we can still take v_{y1} to be a positive number if we want to. If the ball is incident in the vertical direction at speed $v_{y1} = 5\,\text{m s}^{-1}$ and bounces at speed $v_{y2} = 2\,\text{m s}^{-1}$, then the change in speed is $7\,\text{m s}^{-1}$, not $3\,\text{m s}^{-1}$, since the ball reverses direction. If the ball bounced with $v_{y2} = 5\,\text{m s}^{-1}$, then the change in vertical speed would be $10\,\text{m s}^{-1}$, not zero.

The coefficient of restitution, e_y, for a bounce on a heavy surface, is defined by

$$e_y = \frac{v_{y2}}{v_{y1}} \tag{16.3}$$

For a baseball bouncing on a hard wood or concrete surface, e_y is about 0.6 at low ball speeds and drops to about 0.5 at high ball speeds, even when the ball is incident at an oblique angle on the surface.

The three equations here can be combined to work out how the change in horizontal speed depends on the angle of incidence. Since $F = \mu N$, we find from (16.1) and (16.2) that

$$\mu = \frac{v_{x1} - v_{x2}}{v_{y1} + v_{y2}} \tag{16.4}$$

Using (16.3) and the relation $\tan\theta_1 = v_{x1}/v_{y1}$ we find from (4) that

$$\frac{v_{x2}}{v_{x1}} = 1 - \frac{(1 + e_y)\mu}{\tan\theta_1} \tag{16.5}$$

Equation (16.4) provides a useful way of measuring the COF, μ, for a baseball or a softball impacting on a particular surface. By filming an oblique bounce with a video camera, we can measure the change in bounce speed and angle and also measure the horizontal and vertical components of the ball speed. Substitution in (16.4) then gives the value of μ for the bounce. If we already know the values of μ and e_y, then (16.5) tells us the change in speed in the horizontal direction vs. the angle of incidence, as shown in Fig. 16.6a.

To work out the change in spin of the ball we can use the equation $\tau = I\,d\omega/dt$ where ω is the spin, I is the moment of inertia of the ball, and τ is the torque acting on the ball. For a solid ball of mass m and radius R, $I = 0.4mR^2$. The torque on the ball is given by $\tau = FR$, provided that the normal reaction force N acts up through the center of the ball. If not, then it will only be a small effect at low ball speeds and we needn't worry about it here. If the ball is incident with spin ω_1 and bounces at time T later with spin ω_2, then

$$\tau = FR = \frac{I(\omega_2 - \omega_1)}{T} = \frac{0.4mR^2(\omega_2 - \omega_1)}{T} \tag{16.6}$$

Using the expression for F in (16.1) we find that

$$\omega_2 = \omega_1 + \frac{v_{x1} - v_{x2}}{0.4R} \tag{16.7}$$

To see how the change in spin depends on the COF we can use (16.4) to show that

$$\omega_2 = \omega_1 + \mu\frac{(v_{y1} + v_{y2})}{0.4R} = \omega_1 + \frac{\mu\,(1 + e_y)\,v_{y1}}{0.4R} \tag{16.8}$$

The increase in spin is, therefore, proportional to μ, proportional to v_{y1} and inversely proportional to R. For a given ball and surface, the spin increases with the incoming speed of the ball and with a decrease in the angle of incidence, both of which act to increase v_{y1}. However, (16.8) remains valid only if the ball slides throughout the bounce. Sliding ball solutions are shown in Fig. 16.6b for a case where $\mu = 0.4$ and $e_y = 0.55$.

(b) Effects of Grip

The friction force acting at the bottom of a bouncing ball results in a decrease of the horizontal ball speed given by $F = -m\text{d}v_x/\text{d}t$. If the bounce duration is T then $FT = m(v_{x1} - v_{x2})$. The torque $\tau = FR$ on the ball results in a change of the angular speed, ω, given by $FR = I\text{d}\omega/\text{d}t$, so

$$FRT = I(\omega_2 - \omega_1) = mR(v_{x1} - v_{x2})$$

which can be rewritten as

$$I\omega_1 + mRv_{x1} = I\omega_2 + mRv_{x2} \tag{16.9}$$

For a spherical ball of radius R, I is given by $I = \alpha m R^2$ where $\alpha = 0.4$ for a uniform solid sphere and $\alpha = 2/3$ for a thin spherical shell. We define the tangential COR by

$$e_x = -\frac{(v_{x2} - R\omega_2)}{(v_{x1} - R\omega_1)} \tag{16.10}$$

Combining (16.9) and (16.10) we find that

$$\frac{v_{x2}}{v_{x1}} = \frac{(1 - \alpha e_x)}{(1 + \alpha)} + \frac{\alpha(1 + e_x)}{(1 + \alpha)}\left(\frac{R\omega_1}{v_{x1}}\right) \tag{16.11}$$

and

$$\frac{\omega_2}{\omega_1} = \frac{(\alpha - e_x)}{(1 + \alpha)} + \frac{(1 + e_x)}{(1 + \alpha)}\left(\frac{v_{x1}}{R\omega_1}\right) \tag{16.12}$$

While these equations look quite complicated, they become a lot less formidable when we substitute typical numbers. For example, we show in Fig. 16.6 some solutions for a solid ball with $\alpha = 0.4$ and with $e_x = 0$, 0.2 and 1.0. In the first case ($e_x = 0$) we find that

$$\frac{v_{x2}}{v_{x1}} = 0.714 + 0.286\left(\frac{R\omega_1}{v_{x1}}\right) \tag{16.13}$$

and

$$\frac{\omega_2}{\omega_1} = 0.286 + 0.714\left(\frac{v_{x1}}{R\omega_1}\right) \tag{16.14}$$

These equations simplify even further if the ball is incident without spin (i.e., $\omega_1 = 0$) since then $v_{x2}/v_{x1} = 0.714$ and $\omega_2 = 0.714 v_{x1}/R$. Since $R\omega_2 = v_{x2}$, we see that the case $e_x = 0$ corresponds to the special case of a rolling ball.

In the second case ($e_x = 0.2$) we find that

$$\frac{v_{x2}}{v_{x1}} = 0.657 + 0.343\left(\frac{R\omega_1}{v_{x1}}\right) \tag{16.15}$$

and

$$\frac{\omega_2}{\omega_1} = 0.143 + 0.857 \left(\frac{v_{x1}}{R\omega_1} \right) \tag{16.16}$$

Measurements of ball spin show that baseballs and tennis balls have a value of e_x about 0.2, while golf balls have a value about 0.1, at least when they grip the surface. At large angles of incidence all balls slide rather than grip, in which case e_x is not only negative but it also depends on the angle of incidence. The description of a bouncing ball in terms of e_x is useful only at sufficiently low angles of incidence (typically $\theta_1 < 50°$) that the ball grips the surface. At large angles of incidence, the bounce of a ball is best described in terms of the coefficient of sliding friction.

In the third case ($e_x = 1$) we find that

$$\frac{v_{x2}}{v_{x1}} = 0.429 + 0.571 \left(\frac{R\omega_1}{v_{x1}} \right) \tag{16.17}$$

and

$$\frac{\omega_2}{\omega_1} = -0.429 + 1.429 \left(\frac{v_{x1}}{R\omega_1} \right) \tag{16.18}$$

This case corresponds to an ideal superball. Most real superballs have e_x about 0.5 or 0.6.

One obvious conclusion from these results is that spin of the outgoing ball can be increased by changing the angle of incidence or by increasing the value of e_x. For a bat and ball collision, the angle of incidence is increased when the batter strikes the ball near the bottom of the ball rather than striking the middle of the ball. One way to increase e_x would be to coat the bat with a soft, flexible material like rubber. In that case, extra elastic energy would be stored in the rubber and then given back to the ball, increasing the values of both e_y and e_x. The ball would not only bounce at higher speed but it would also spin faster. That is why table tennis bats have a rubber surface.

(c) Oblique Bounce Off a Light Surface

Suppose that the surface in Fig. 16.3 is only slightly heavier than the incident ball. Then the ball will bounce off the surface at reduced speed since the ball transfers some of its kinetic energy to the surface. That is essentially the situation encountered when a ball strikes a stationary bat or a tennis racquet. If the ball makes a head-on collision with the bat then conservation of momentum for the collision, plus an estimate of the COR, tells us the recoil speed of the bat. We can then proceed as in Chap. 9 to calculate the bounce speed of the ball and the recoil speed of the bat.

However, if the ball does not strike the bat head-on then the ball will be deflected sideways by the curved surface of the bat, in which case we can treat the collision as an oblique bounce, as shown in Fig. 16.3. A similar situation occurs in tennis when the ball strikes the strings at an oblique angle and the racquet head recoils. Recoil motion of the bat or the racquet adds to the complexity of the problem, but there is a simple way around the problem. That is, we ignore the recoil motion and just measure what happens to the ball. We saw in Chap. 9 that it is very useful to define an apparent coefficient of restitution, e_A, or bounce factor, q describing the ratio of the ball speed after the collision to that before the collision. In the same way, we can measure e_A for an oblique bounce off a light surface, in which case (16.3) has the form

$$e_A = \frac{v_{y2}}{v_{y1}} \tag{16.19}$$

The value of e_A will depend on the mass of the bat, but once it is measured for a given ball speed we can then predict the outcome for other ball speeds. In a similar way, we can define an apparent tangential coefficient of restitution, e_T, using (16.10). That is, we simply replace e_x by e_T in (16.10), ignoring recoil motion of the bat. We can then measure e_T in terms of the tangential speed and spin of the ball before and after the collision, to predict the outcome for other values of the incident speed and spin of the ball. For that purpose, we simply use (16.11) and (16.12), replacing e_x by the measured value of e_T. The validity of this approach is demonstrated in the following chapter where we present measurements of the oblique bounce of a baseball off a hand-held bat.

References

1. R. Cross, The fall and bounce of pencils and other elongated objects. Am. J. Phys. **74**, 26–30 (2006)
2. R. Cross, Bounce of a spinning ball near normal incidence. Am. J. Phys. **73**, 914–920 (2005)
3. N. Maw, J.R. Barber, J.N. Fawcettt, The oblique impact of elastic spheres. Wear **38**, 101–114 (1976)
4. N. Maw, J.R. Barber, J.N. Fawcett, The role of elastic tangential compliance in oblique impact. J. Lubric. Technol. **103**, 74–80 (1981)
5. J. Bilbao, J. Campos, C. Bastero, On the planar impact of an elastic body with a rough surface. Int. J. Mech. Eng. Educ. **17**, 205–210 (1989)
6. R. Cross, Grip-slip behaviour of a bouncing ball. Am. J. Phys. **70**, 1093–1102 (2002)
7. R. Cross, Measurements of the horizontal and vertical speeds of tennis courts. Sports Eng. **6**, 95–111 (2003)
8. R. Cross, Oblique impact of a tennis ball on the strings of a tennis racquet. Sports Eng. **6**, 235–254 (2003)
9. R. Cross, Impact of a ball on a surface with tangential compliance. Am. J. Phys. **78**, 716–720 (2010)

Chapter 17
Ball Spin Generated by a Bat

It's unbelievable how much you don't know about the game you've been playing all your life.

– Mickey Mantle

17.1 Introduction

When a batter swings at a ball, the result is often disappointing for the batter. Striking the ball is better than missing it completely, but a miss-hit is not much better. Both are very common. Only rarely does a batter strike the ball cleanly, or head-on. By a head-on strike we mean one where the bat strikes the ball squarely in the middle of the ball, in the meaty part of the bat, and at right angles to the path of the ball. Such a strike will be satisfying in the sense that the ball will then come off the bat at the maximum possible speed. However, it might head straight back to the pitcher. Usually, this is not what the batter wants. Perhaps he wants to hit the ball slightly late or slightly early so that the ball heads off in the general direction of first or third base. Alternatively, he might prefer to strike the ball below its center to project the ball up into the air or to impart additional spin to the ball.

Whenever a ball is struck above or below its center, the ball comes off the bat at reduced speed, at a different angle and with a different amount of spin than a strike in the middle of the ball. We can describe the process as "scattering" since this is the term commonly used to describe an event where an incoming object strikes some other object and heads off in a different direction. Physicists have been conducting scattering experiments for more than 100 years to probe the structure of the atom and the nucleus. By firing a high speed electron or proton or neutron at an atom, it is possible to split the atom or its nucleus into many pieces to determine its structure. The pieces fly off in all directions. Essentially, the same experiment can be conducted with a baseball and a bat, not to smash the bat or the ball to pieces, but simply to find out what happens to the ball.

Depending on where the bat strikes the ball, the ball can retrace its path or head down into the ground or up over the pitcher's head or straight up into the air or straight through to the catcher. The ball can be scattered through any angle varying from zero to $\pm 180^\circ$, as shown in Fig. 17.1. If the ball heads straight back to the pitcher we say that the scattering angle is zero. Straight up is 90°, straight down is -90° and straight through to the catcher would be 180°.

R. Cross, *Physics of Baseball & Softball*, DOI 10.1007/978-1-4419-8113-4_17,

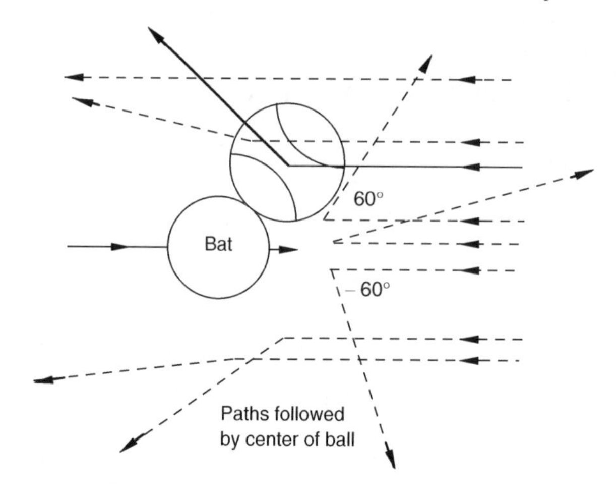

Fig. 17.1 Depending on where the bat strikes the ball, the ball can head straight back to the pitcher, up over the pitcher's head, or straight back to the catcher. Scattering angles of 60° and −60° are shown here to explain what we mean by the scattering angle

The sort of question that we can ask about the scattering process is the following. Suppose that the ball is pitched horizontally at 80 mph, the bat is swung horizontally at 70 mph and the ball is struck 1 in. below the middle of the ball. At what angle will the ball come off the bat, at what speed and how fast will it be spinning? To answer those questions we need more information. We need to know the properties of the bat and the ball, we need to know the impact distance from the tip of the barrel and we also need to know how fast the ball was spinning before it hit the bat and in what direction it was spinning. Given all that information, we should have a reasonable chance of figuring out the answer. However, we need even more information. In particular, we need to know whether the ball slides across the bat during part or all of the collision, or whether it rolls or whether it grips the bat. There is so much information we need to know to answer the original question that there must be a better way of figuring out the answer. Indeed, there is. We can measure what happens and then we will know for sure.

17.2 Scattering Experiment

The author conducted an experiment like this in 2005 in collaboration with Professor Alan Nathan and the results were published in the American Journal of Physics in 2006 [1]. To simplify the procedure, the experiment was conducted at low ball speeds ($4\,\mathrm{m\,s^{-1}}$ or 9 mph) partly so the experiment could be done safely indoors and partly because a high speed video camera was not available at the time. Ordinary video cameras work fine for measuring the speed and trajectory of baseballs, but they are not fast enough to capture the spin rate of a high speed ball. If a ball rotates at say 1,800 rpm or 30 complete revolutions in 1 s and is filmed at a rate of 30 frames s^{-1}, then the ball will be captured at the same rotation angle in every frame and it will seem like the ball is not rotating at all. To measure the rotation speed, the ball would need to captured at a rate of at least 100 frames s^{-1}. Such cameras are available, but tend to be very expensive. Casio introduced two very cheap versions in 2009 (the EX-FS10 and EX-FC100) both under $400, and both can record at 210, 420 or 1,000 fps.

The experiment consisted of dropping a baseball onto a bat by hand from a height of about 3 ft, and filming the event with an ordinary video camera. The experiment could easily be repeated by anyone with a computer and with suitable motion analysis software to examine the video film one frame at a time. By measuring the horizontal and vertical distances traveled by the ball from one frame to the next, and by measuring the angle through which it rotates, it is easy to calculate the speed of the ball, the spin of the ball, and the scattering angle.

The geometry of the experiment is shown in Fig. 17.2. To simplify the experiment even further, the bat was not swung at the ball. The bat was held in a steady horizontal position by holding onto the handle with one hand, and the ball was dropped vertically through a height of about 0.8 m onto a point 6 in. from the end of the barrel. Sometimes the ball landed exactly on the top of the bat, and sometimes it landed an inch or two to the left or right, but every drop was recorded on film. About 50 such drops were filmed, first by dropping the ball without spinning it. Another 50 drops were recorded with the ball spinning clockwise at an average rotation speed of 12.6 rev s^{-1}, and then another 50 drops were recorded with the ball spinning counter-clockwise at an average rotation speed of 11.5 revolutions s^{-1}. To spin the ball, a strip of felt was wound around the ball three or four times and the ball was allowed to fall vertically while holding onto the top end of the felt strip.

There was no need to swing the bat at the ball in this experiment. The physics of the scattering and collision process is the same regardless of whether the bat is swung or not. The rebound speed and angle of the ball will be different, but the point is that it is easy to take the results of this experiment and then calculate the effect of swinging the bat at the ball. The experiment that was done is equivalent to

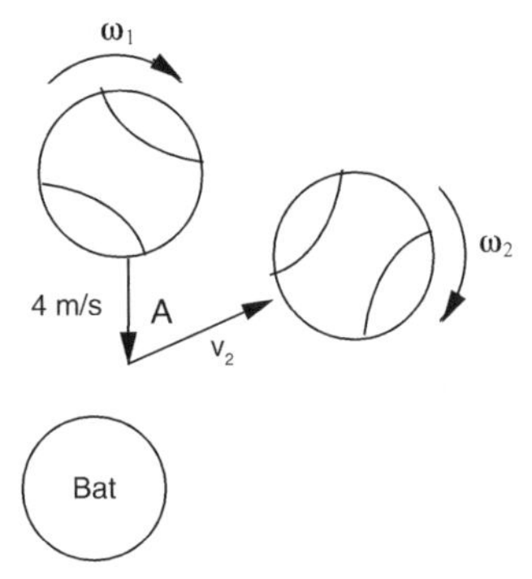

Fig. 17.2 The ball was dropped vertically onto a stationary bat at speed $v_1 = 4.0\,\mathrm{m\,s^{-1}}$ (9 mph) and bounced off the bat at speed v_2 at an angle A, with spin ω_2. The initial spin of the ball is denoted by ω_1, which was either zero or $+12.6$ rev s^{-1} or -11.5 rev s^{-1}

swinging a bat at the ball and attaching a camera to the end of the bat so that the bat appears to remain at rest before the collision. Alternatively, the camera could have been attached to a trolley that followed the bat as it was swung. It is a simple matter to then calculate the speed of the bat and the ball in a reference frame where the camera remains at rest and the bat is swung at the ball. All that needs to be done is to subtract the speed of the camera from the speeds of the bat and the ball.

Experimental Results

Some typical experimental results are shown in Fig. 17.3 where the ball was dropped either onto the top (center) of the bat or 1 in. to the right. There is a lot of information contained in the six separate results shown in Fig. 17.3. The actual numbers can be ignored for the moment to concentrate on the physics of what happened in each case. In summary, the following things happened:

Drop A. The ball was dropped without spin onto the top of the bat. The ball bounced vertically without spin, at low speed. This is similar to dropping a ball on a hard floor, but the ball does not bounce as well off a bat since the ball gives some of its energy to the bat. The bat recoils away from the ball, robbing the ball of some of its kinetic energy.

Drop B. The ball was dropped without spin to land on the bat 1 in. to the right. This is similar to dropping a ball onto an inclined platform since the surface of the bat slopes down to the right at the impact point. As a result, the ball is deflected to the right and it spins clockwise when it bounces due to the friction force acting on the bottom of the ball. Surprisingly, the bottom of the ball does not slide down the sloping surface of the bat. Rather, the bottom of the ball grips the bat and the top of the ball rolls forward during the bounce, causing it to spin clockwise when it bounces.

Drop C. The ball was spun clockwise at 12.6 rev s^{-1} and then dropped onto the top of the bat. Instead of bouncing vertically, as in drop A, the ball bounced to the right at an angle of 30°. Again, a similar thing happens when a spinning ball is dropped vertically onto a hard floor. The bottom of the ball starts sliding to the left when it hits the floor (or the bat), resulting in a friction force acting to the right on the bottom of the ball. The friction force does two things to the ball. It pushes the ball to the right, and it decreases its rate of spin, in this case from 12.6 to 3.6 rev s^{-1}.

Drop D. The ball was spun clockwise at 12.6 rev s^{-1} then dropped onto the bat, landing 1 in. to the right. The ball deflected through a large angle (75°) to the right, due to the combined effect of the downward slope of the bat and the clockwise spin of the ball. The ball bounced with a large clockwise spin, again due to the combined effects of the initial clockwise spin and the effect of the slope.

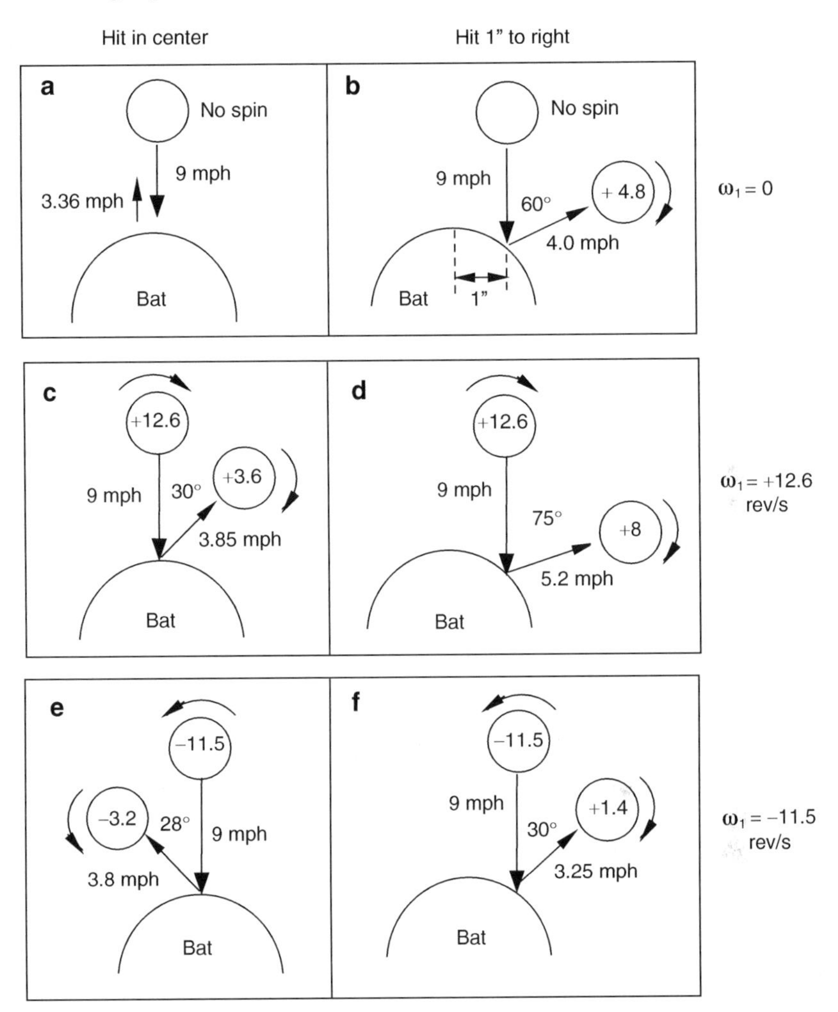

Fig. 17.3 Experimental results obtained by dropping a baseball onto a hand-held bat at a point 6 in. from the end of the barrel, either in the *middle* of the bat or 1 in. to the *right*. The numbers inside each ball refer to the ball spin, in rev s^{-1}, and the angles are the scattering angles

Drop E. The ball was spun counter-clockwise at 11.5 rev s^{-1} and dropped onto the top of the bat. The ball deflected to the left because the friction force pushed the ball to the left, and it bounced with reduced spin.

Drop F. The ball was spun counter-clockwise at 11.5 rev s^{-1} and dropped onto the bat, landing 1 in. to the right. The ball deflected to the right by only 30°, despite the slope of the bat, since the initial spin caused the ball to deflect back to the left, thereby reducing the deflection caused by the slope of the bat.

17.3 Swinging the Bat at the Ball

The results in Fig. 17.3 were obtained when the bat was held at rest in the hand. Of greater interest is the result when the bat is swung at the incoming ball. That is a little more difficult to determine experimentally, but it is very easy to work out theoretically. We simply add a constant vertical speed to all the results in Fig. 17.3, assuming that the camera was fixed in position and the bat was moving vertically upwards. The results of adding 4 mph to all the vertical speeds in Fig. 17.3 are shown in Fig. 17.4. In that case, the bat moves vertically up at 4 mph and the ball falls onto

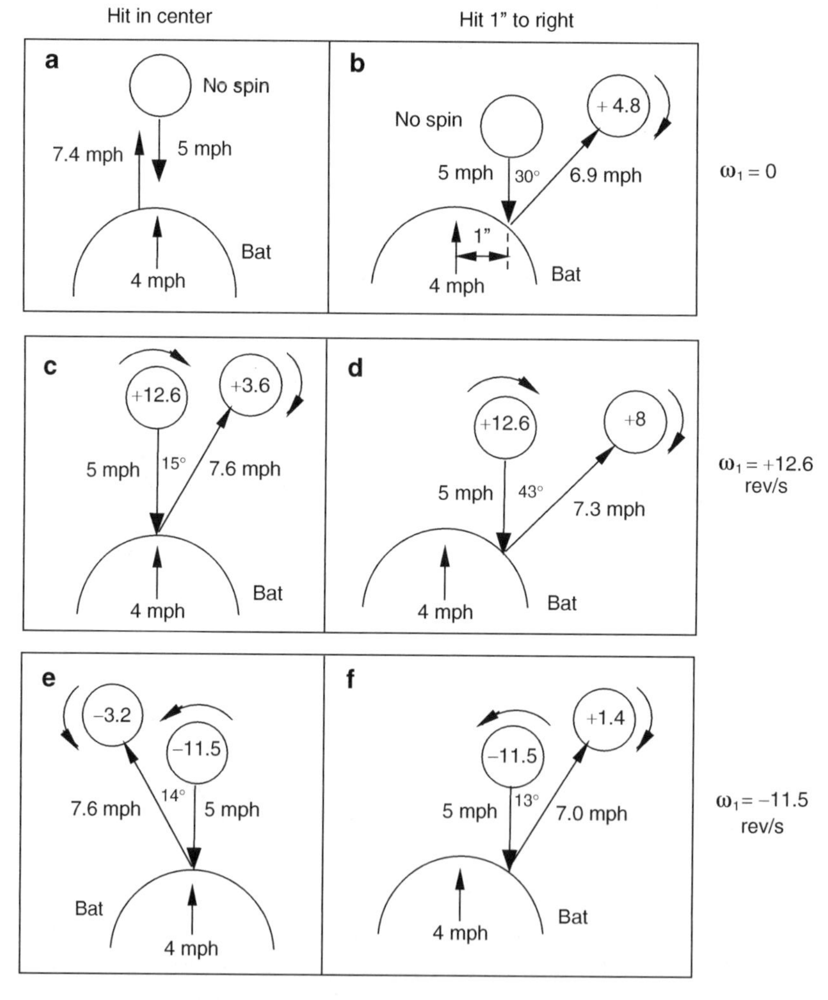

Fig. 17.4 The same results as those in Fig. 17.3 adjusted for a case where the bat is swung vertically at 4 mph and the ball falls onto the bat at 5 mph. The spin remains the same but the outgoing speeds are larger and the scattering angles are smaller

the bat at 5 mph. The relative speed before the collision is unchanged and is still 9 mph. The horizontal speeds remain the same, and so do the various ball spins. The ball rebounds at a greater speed in the vertical direction when we add 4 mph to all the vertical speeds, so it rebounds at an angle closer to the vertical.

Even though the experiment was done at a ball speed of only 9 mph, similar rebound angles can be expected if the ball impacts the bat at greater speeds. The angles won't necessarily be exactly the same since the ball distorts more in a high speed collision. However, the bounce speeds and spins will be proportionally larger. To scale the results to a real game of baseball, the speeds would need to be increased by a factor of nearly 20, and the incoming ball spins would need to be doubled or tripled. Before we attempt to do that, we need to examine the experimental data in more detail to see how the various quantities scale.

17.4 Additional Experimental Results

All results obtained when the ball was dropped with zero spin onto a stationary bat are shown in Figs. 17.5 and 17.6. When the ball landed in the middle of the bat it bounced vertically (i.e., A = 0) at a speed $v_2 = 1.5\,\mathrm{m\,s^{-1}}$ on average, and it bounced without any spin. If the ball struck the bat at a point to the left or right of center, then the ball bounced to the left or right, at a speed greater than $1.5\,\mathrm{m\,s^{-1}}$, spinning clockwise when it bounced to the right (as shown in Fig. 17.2) or counter-clockwise when it bounced to the left. These results are as expected, and also agree with theoretical calculations based on the assumption that the ball grips the bat when it bounces. That is, the ball does not slide or roll across the bat as it bounces, but it grips the bat. The bat and the ball grip together in a manner that is similar to the engagement of two gears. If one gear wheel rotates, it causes the other gear wheel to rotate in the opposite direction. The gear effect is important in golf [2] and is also

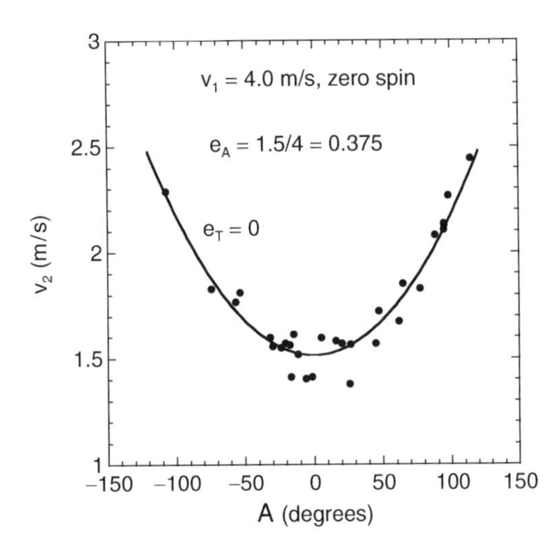

Fig. 17.5 Measured values of the bounce speed v_2 vs. the scattering angle A when the incident ball was dropped to impact at $4\,\mathrm{m\,s^{-1}}$ without spin. The curved line labeled $e_T = 0$ is the theoretical calculation described in Appendix 17.1

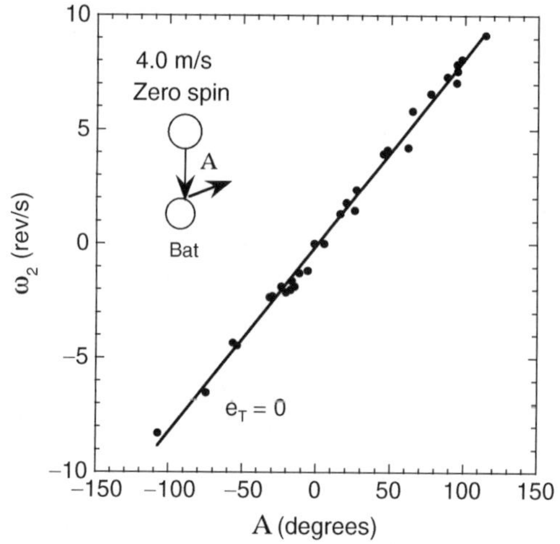

Fig. 17.6 Measured values of the ball spin ω_2 vs. the scattering angle A when the incident ball was dropped to impact at $4\,\mathrm{m\,s^{-1}}$ without spin. The line labeled $e_T = 0$ is a theoretical calculation

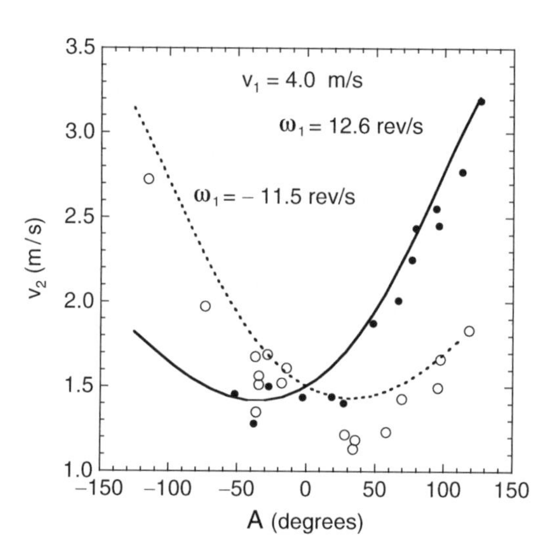

Fig. 17.7 Measured values of the bounce speed v_2 vs. the scattering angle A when the incident ball was dropped to impact at $4\,\mathrm{m\,s^{-1}}$ with spin. The curved lines are theoretical calculations with $e_T = 0$

important in billiards [3]. In a bat and ball collision, if the ball grips the bat while the ball is rotating, it will cause the bat to rotate in the opposite direction. However, we focus here on the resulting spin of the ball.

Results obtained when a spinning ball was dropped on the bat are shown in Figs. 17.7 and 17.8. When the ball landed in the middle of the bat, it bounced about 30° away from the vertical, either to the right in Fig. 17.2 if the initial spin was clockwise ($\omega_1 = 12.6\,\mathrm{rev\,s^{-1}}$), or to the left if the initial spin was counter-clockwise ($\omega_1 = -11.5\,\mathrm{rev\,s^{-1}}$) and it bounced with a reduction in spin (from 12.6 to $3.6\,\mathrm{rev\,s^{-1}}$ or from -11.5 to $-3.5\,\mathrm{rev\,s^{-1}}$). A similar result is obtained when a spinning baseball is dropped vertically onto a horizontal wood floor. The ball bounces off toward one side, spinning in the same direction as the incident ball but spinning more slowly.

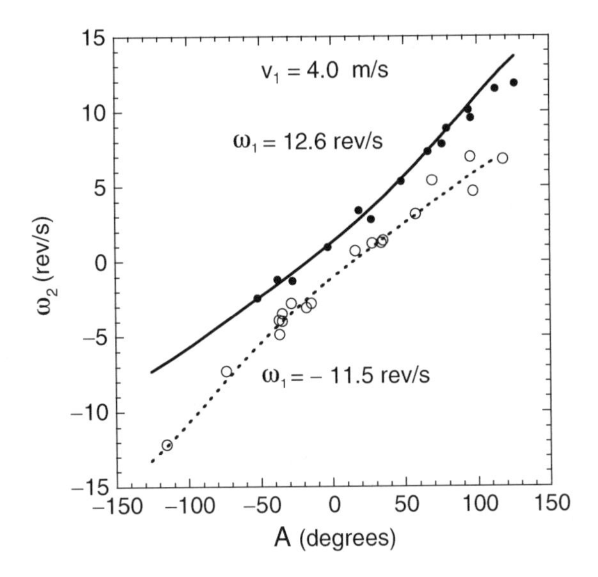

Fig. 17.8 Measured values of the ball spin ω_2 vs. the scattering angle A when the incident ball was dropped to impact at 4 m s^{-1} with spin. The curved lines are theoretical calculations with $e_T = 0$

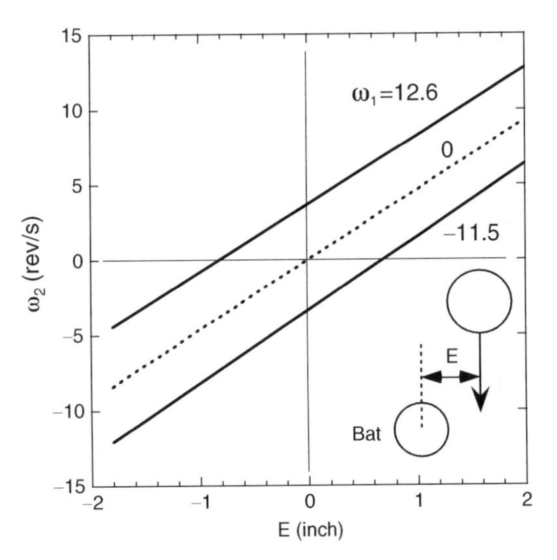

Fig. 17.9 Outgoing ball spin ω_2 vs. the impact distance E when the incident ball was dropped to impact at 4 m s^{-1} with $\omega_1 = 0$, or 12.6 rev s^{-1} or -11.5 rev s^{-1}

The most important results are those shown in Fig. 17.9. Given that a clockwise spinning ball bounces to the right when it lands in the middle of the bat, it bounces even further to the right when it impacts to the right of center in Fig. 17.2, and it bounces with a clockwise spin greater than for a ball incident without spin. If a clockwise spinning ball impacts to the left of center then the incident spin of the ball tends to deflect the ball to the right, but the slope of the bat tends to deflect it to the left. As a result, when the ball impacts about 3/4 in. left of center the ball bounces almost vertically and with almost zero spin. The opposite effect occurs when the incident ball is spinning counter-clockwise. That is, the incident spin tends to deflect the ball to the left. So, a ball landing to the left of center deflects a long way to the left and with relatively large spin, while a ball landing the same distance to the right of center does not spin as fast and is scattered through a smaller angle.

17.5 High Speed Results

The scattering process observed at low ball speeds can also be expected at high ball speeds. Consider the collision shown in Fig. 17.10 where a ball is incident from the right in a horizontal direction at 85 mph and the bat is incident horizontally at 55 mph. If the center of the bat is located below the center of the ball then the ball will be deflected upward, with backspin. That is exactly what is needed to hit a ball to the outfield or beyond, although the same or at least a similar result could be achieved by swinging the bat up at the ball and striking the ball in the middle to project the ball high up into the air. If the ball is falling downward at the same angle as the bat is rising upward, then that is equivalent to the situation shown in Fig. 17.10 (assuming that Fig. 17.10 is rotated through an appropriate angle).

To calculate the outgoing speed, spin and launch angle of the ball, it helps to change the reference frame so that the bat is at rest and the ball is incident at 140 mph. The equations describing such a collision are given in Appendix 17.1. Using the results of those calculations, we can then change back to the original frame of reference where the bat approaches the ball at 55 mph. The results are shown in Fig. 17.11 for a case where the bounce factor $q = e_A = 0.2$, $e_T = 0$, and the ball is incident with backspin and is spinning at 1,200 rpm. The value of q for such a collision depends on whether the bat strikes the ball near the tip of the bat or further along the barrel, but $q = 0.2$ is typical for a high speed collision. The results shown in Fig. 17.11 are very similar to those quoted in [4] and show that (a) the outgoing ball speed drops by 6% if $E = 1$ in., and by 24% if $E = 2$ in., (b) the launch angle increases up to about 76° above the horizontal when $E = 2$ in., and (c) the outgoing ball spins at about 8,000 rpm when $E = 2$ in.

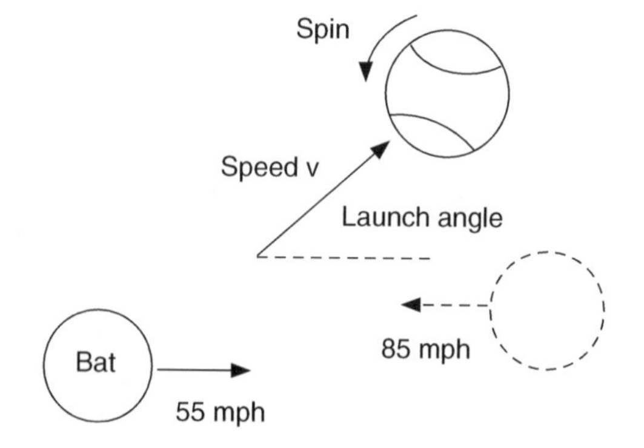

Fig. 17.10 A ball incident horizontally from the right at 85 mph is struck by a bat moving horizontally to the right at 55 mph. If the center of the bat is below the center of the incoming ball then the ball will be launched at an angle above the horizontal with backspin. The result is shown in Fig. 17.11

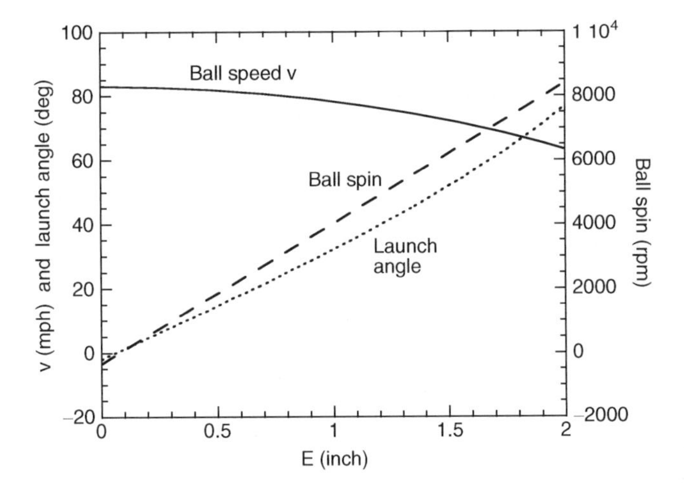

Fig. 17.11 Calculated results for the collision shown in Fig. 17.10. E is the vertical distance between the centers of the bat and incoming ball in Fig. 17.10. The outgoing ball spin is shown on the vertical axis on the *right* side of the graph

An interesting question is whether it is better to strike the ball head-on, with $E = 0$, or whether the ball will travel farther when E is about 1 in. or so. For the results shown in Fig. 17.11, the ball travels farthest, about 320 ft, when $E = 0.8$ in. Despite the slight drop in launch speed when $E = 0.8$ in., the increase in launch angle and backspin helps to carry the ball farther, as noted in [4]. When E is about 2 in. or so, the ball pops up into the air and lands close to the batter, as shown in Chap. 6.

17.6 How to Hit Home Runs

In 2003, Sawicki et al. [5] published a detailed paper, based on observations of lift and drag coefficients measured over the previous 44 years, indicating that a curve ball can be hit farther than a fast ball or a knuckleball. A curve ball is pitched with topspin at a relatively low speed, a fast ball is pitched with backspin at a higher speed, and a knuckleball is pitched without spin at relatively low speed. Typical pitched ball spin values are $-200\,\mathrm{rad\,s^{-1}}$ for a fast ball, $+200\,\mathrm{rad\,s^{-1}}$ for a curveball and zero spin for a knuckleball. 200 rad/s is about 32 rev/s or about 1,900 rpm. If all three pitches were struck with the same outgoing launch angle and spin, then a fast ball would travel farther since the outgoing ball speed increases when the incoming ball speed increases (given that a faster incoming ball bounces at higher speed off a bat, regardless of whether the bat is swung or not).

If all three balls were struck at the same speed and launched at the same angle, then the ball traveling farthest would be the one with the most backspin. The Magnus force acts upward on a ball spinning backward, is approximately proportional to the spin, and will therefore carry the ball farther before it lands.

Given that a fast ball is incident with backspin, the spin direction of the ball needs to be reversed if the outgoing ball is to be struck with backspin. A curve ball is already spinning in the correct direction to be struck with backspin. Consequently, if the same bat swing is used to strike each ball, and if each ball was incident at the same speed, then the curve ball would gain more backspin than the fast ball. Sawicki et al. found that a curveball indeed acquires more spin than a fast ball, despite the fact that the curve ball is incident at lower speed, but it exits from the bat at a slightly lower speed. However, the additional spin more than makes up for the lower batted speed, with the result that curve balls can be hit slightly farther.

Adair [6] criticized the study on several grounds, and the interested reader can follow the arguments in the original American Journal of Physics articles. Part of the problem is that accurate measures of lift and drag coefficients, and accurate measures of e_T or e_x have still not been obtained at the high ball speeds and spin rates found in professional or even in amateur baseball or softball.

Appendix 17.1 Scattering Model

The experimental data shown in Figs. 17.5–17.8 agree very well with a theoretical bounce model where it is assumed that the ball grips the bat and then bounces with a value of $e_T = 0$, e_T being closely related to the tangential coefficient of restitution, e_x. The tangential COR was defined in the previous chapter for a ball bouncing off a very heavy surface. A stationary bat is not a particularly heavy surface and it recoils when a ball strikes the bat. In that case, e_x and e_y are both defined in terms of the relative speed of the bat and the ball after and before the collision. However, if we ignore the recoil speed of the bat after the collision then we can define simpler versions of e_x and e_y that we can describe as "apparent" values of the COR, given by the ratio of the ball speed after the collision to the ball speed before the collision. The apparent value of e_x is then labelled e_T and the apparent value of e_y is labelled as e_A or q.

If $e_T = 0$ then the ball bounces off the bat with a tangential speed, $v_{x2} = R\omega_2$ where R is the ball radius and ω_2 is the angular velocity of the ball after the collision. This result is similar to that for a rolling ball but it is not exactly the same since the bat itself rotates and translates after the collision. A ball that rolls along a stationary surface does so with $v_x = R\omega$ but if it continues to do so while the surface itself is moving then the bottom of the ball actually slides along the surface. That is what happens when a ball bounces off a bat. The ball grips the bat, both are set in motion and then the ball slides backwards along the surface before finally bouncing off the bat with $e_T = 0$.

A model of the scattering process is shown in Fig. 17.12. A ball of radius r falls vertically onto a stationary bat of radius R. At least, this was the situation in the experiment described above. Once we determine the basic physics of the scattering

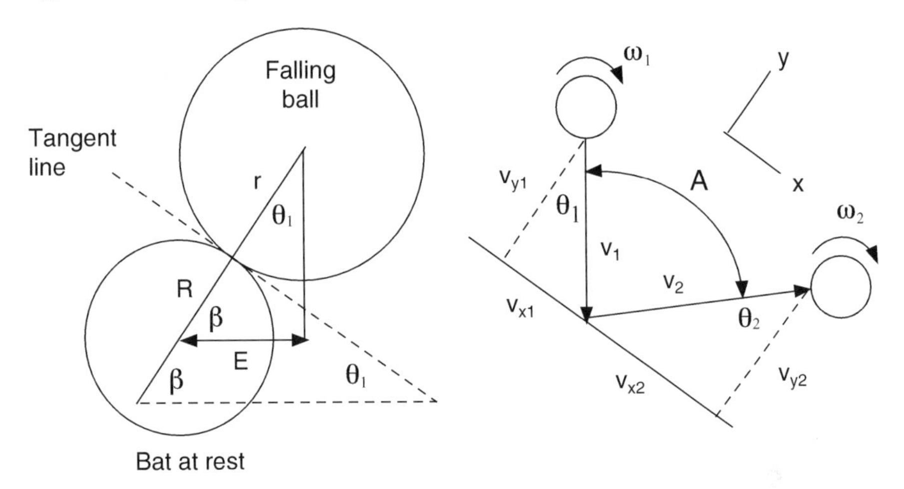

Fig. 17.12 A ball of radius r falls vertically at speed v_1 onto a stationary bat of radius R, and bounces at speed v_2. The ball is scattered through an angle A. The horizontal distance between the ball center and the bat center is E. We can model the collision as an oblique bounce off a flat surface that is inclined at an angle parallel to the *dashed tangent line*

process in this case, then it is relatively easy to consider a more realistic case where a bat and ball approach each other in a horizontal direction or in some other direction.

As shown in Fig. 17.12, the horizontal distance between the bat and ball centers is denoted by E. That distance is commonly called the impact parameter in scattering experiments. The line joining the ball centers is inclined at an angle β to the horizontal, where $\cos\beta = E/(r + R)$. The ball is incident an an angle $\theta_1 = 90 - \beta$ to the line joining the ball centers and rebounds at angle θ_2. The ball is, therefore, scattered through an angle A $= \theta_1 + \theta_2$.

Despite the curvature of the bat, we can model the collision as one where the ball bounces off a flat surface inclined at an angle parallel to the tangent line shown in Fig. 17.12. We can even choose to ignore motion of the bat. Suppose that the bat recoils at speed V_y in a direction perpendicular to the surface. Then the coefficient of restitution (COR) in the y direction is defined by

$$e_y = \frac{(v_{y2} + V_y)}{v_{y1}} \tag{17.1}$$

We can ignore motion of the bat in the y direction by defining an apparent coefficient of restitution, $e_A = v_{y2}/v_{y1}$, which is defined purely in terms of the ball speed in the y direction. Similarly, we can ignore motion of the bat in the x direction by defining an apparent tangential coefficient of restitution, e_T, considering only the tangential velocity of the contact point on the ball, given by $v_x - r\omega$. We define e_T by the ratio

$$e_T = -\frac{(v_{x2} - r\omega_2)}{(v_{x1} - r\omega_1)} \tag{17.2}$$

The equations for the bounce are then identical to those derived in Appendix 16.1, except for the fact that e_y is replaced by e_A and e_x is replaced by e_T. The result of the bounce in Fig. 17.10 is therefore described by the equations

$$\frac{v_{y2}}{v_{y1}} = e_A \tag{17.3}$$

$$\frac{v_{x2}}{v_{x1}} = \frac{(1-\alpha e_T)}{(1+\alpha)} + \frac{\alpha(1+e_T)}{(1+\alpha)}\left(\frac{r\omega_1}{v_{x1}}\right) \tag{17.4}$$

and

$$\frac{\omega_2}{\omega_1} = \frac{(\alpha - e_T)}{(1+\alpha)} + \frac{(1+e_T)}{(1+\alpha)}\left(\frac{v_{x1}}{r\omega_1}\right) \tag{17.5}$$

The theoretical curves in Figs. 17.5–17.9 were obtained using $e_A = 0.375$, $e_T = 0$ and $\alpha = 0.4$, in which case $v_{y2}/v_{y1} = 0.375$,

$$\frac{v_{x2}}{v_{x1}} = 0.714 + 0.286\left(\frac{r\omega_1}{v_{x1}}\right) \tag{17.6}$$

and

$$\frac{\omega_2}{\omega_1} = 0.286 + 0.714\left(\frac{v_{x1}}{r\omega_1}\right) \tag{17.7}$$

References

1. R. Cross, A. Nathan, Scattering of a baseball by a bat. Am. J. Phys. **74**, 896–904 (2006)
2. R. Cross, A. Nathan, Experimental study of the gear effect in ball collisions. Am. J. Phys. **75**, 658–664 (2007)
3. R. Cross, Cue and ball deflection (or "squirt") in billiards. Am. J. Phys. **76**, 205–212 (2008)
4. M.K. McBeath, A.M. Nathan, A.T. Bahill, D.G. Baldwin, Paradoxical pop–ups: Why are they difficult to catch?. Am. J. Phys. **76**, 723–729 (2008)
5. G.S. Sawicki, M. Hubbard, W.J. Stronge, How to hit home runs: Optimum baseball swing parameters for maximum range trajectories, Am. J. Phys. **71**, 1152–1162 (2003)
6. R.K. Adair, Comment on How to hit home runs: Optimum baseball bat swing parameters for maximum range trajectories. Am. J. Phys. **73**, 184–185 (2005)

Chapter 18
Bat and Ball Projects

In the matter of physics, the first lessons should contain nothing but what is experimental and interesting to see. A pretty experiment is in itself often more valuable than twenty formulae extracted from our minds.

– Albert Einstein

18.1 Introduction

Twelve simple projects are described in this chapter that you can undertake at home or elsewhere. The projects deal with the flight of balls through the air and the properties of bats and balls. The object of each project is to get you thinking about the physics of baseball and softball. The best way to do that is to undertake a few experiments yourself. You can play or watch baseball or softball for your whole life without ever thinking about the physics behind it. But if you do start thinking about it, then you will probably be surprised that there is much more physics in the game than you ever thought possible. Depending on your background and interests, a baseball or softball game can be viewed as a whole series of physics experiments, an exercise in psychology or biomechanics, or just simply a fun sport to play and watch. To others, it is more important than any of those things. It is matter of life and death.

The project experiments described below are meant to serve several purposes. First, the phenomena described in this book are all based on experimental evidence. The experimental results are just as important, if not more important, than our attempts to explain them. The second purpose is just as important as the first. That is, the projects were chosen so that you can do them yourself without the need for expensive equipment. The experiments are not only relevant to the physics of baseball and softball, but are simple enough to be conducted in a meaningful way by almost anyone. The experiments are interesting and fun in themselves, and will hopefully encourage and motivate you to explore some of these or your own experiments in more detail, and to seek out the explanations provided in this book and in the additional references. A third purpose is to demonstrate how physicists go about their work. Some people think we just calculate everything. More often than not, we do an experiment, try to explain it, discover that we don't have enough information to explain it properly, then plan a better experiment. This process can easily go on for a year or more before we figure out what we want to know. Only rarely does the process take less than a week, although each project described below should take only a few hours or less.

R. Cross, *Physics of Baseball & Softball*, DOI 10.1007/978-1-4419-8113-4_18,

As any physics student knows, physics can be dry and uninteresting if presented as a series of laws and equations that need to be memorized for the purpose of passing an exam. The most interesting part of any physics lesson or lecture is usually the demonstration of the effects being described, especially if the demonstration ends in a spectacular explosion, or at least a loud noise or something catching fire. None of the project experiments lead to such a result, but they are still interesting since they concern the physics of baseball and softball, and since ball games are something that many people are passionate about. It is possible to be just as passionate about experimental physics, not just because it is fun but because it is the best way to determine what really happens to a bat or ball when someone throws or hits a baseball or softball. By that we mean the physics of the process. If you want to know what happens to your body when you throw or hit a ball, then it is best to ask a biomechanist or physiologist.

18.2 Flight of the Ball

The first four projects are concerned with the flight of the ball through the air. Everyone knows that gravity pulls the ball downward, but what is the effect of the air? Does it make any difference if the ball is spinning? How much difference?

Project 1: Effect of Gravity on the Flight of a Ball

The effect of gravity on a ball can be demonstrated and understood by a simple experiment. Get hold of two balls, one in each hand, and hold them both at about head height. Throw one of them at low speed in a horizontal direction so that it lands about 6–10 ft in front of you. At the instant you release the ball, release the other ball so that it falls straight down. The question here is: which ball will land first? A common answer is that the thrown ball takes longer to land because it has to travel farther. The correct answer is that both balls land at the same time. If you find that they *don't* land at the same time then you made a mistake. Either you didn't release both balls from the same height, or at the same time, or you threw one of them slightly upward or downward rather than horizontally.

If you were to take video film of the two balls, then you would get a result similar to that shown in Fig. 18.1. If the ball is dropped vertically from a height of 6 ft then it falls in a straight line and takes just over 0.6 s to hit the ground. As it falls, it accelerates, so it travels a greater distance in any 0.1-s period than in the previous 0.1-s period. If the ball is thrown horizontally at 15 ft s^{-1}, then it will land about 9 ft away but it still takes just over 0.6 s to land. The thrown ball travels farther but it doesn't take longer to get there since it is launched at a higher speed. At any given time, both balls will have fallen exactly the same vertical distance, because the acceleration in the vertical direction is exactly the same.

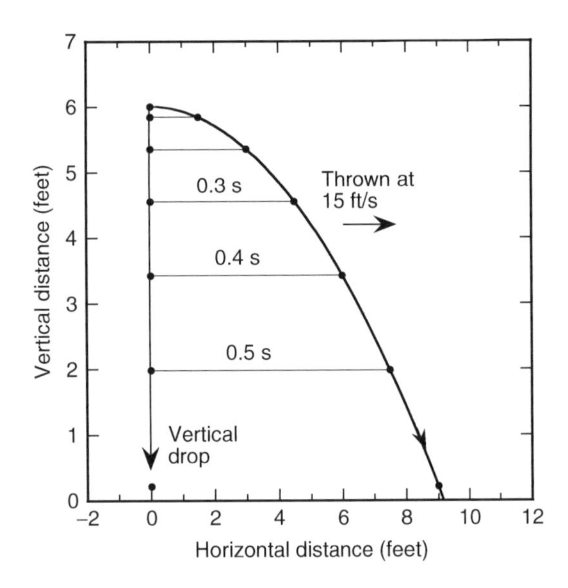

Fig. 18.1 A ball dropped from a height of 6 ft takes just over 0.6 s to hit the ground. If the ball is thrown horizontally at 15 ft s^{-1}, it takes exactly the same time to hit the ground and lands about 9 ft away. The position of the ball is shown every 0.1 s

There is no acceleration in the horizontal direction, so the ball continues to travels at 15 ft s^{-1} in the horizontal direction until it hits the ground. After 0.1 s it travels 1.5 ft horizontally, and after 0.5 s it is 7.5 ft from the starting point.

If a ball is traveling at high speed and travels a long distance then the effects of ball spin and air resistance become important. Projects 2, 3 and 4 illustrate the effects of spin and the air on relatively light objects since the effects are more easily observed, and since the basic physics is exactly the same.

Project 2: Magnus Force on a Spinning Ball

All objects curve down toward the ground during their flight through the air, due to the vertical gravitational pull of the earth. If the object is spinning, there is an additional force on the object, known as the Magnus force. The additional force arises from the fact that air streaming past the object travels at a different speed on either side of a spinning object. The resulting Magnus force acts in a direction perpendicular to both the motion of the object and to the direction of the spin axis. For example, if a golf ball is struck with backspin and is traveling horizontally, the Magnus force acts vertically upward on the ball and helps to keep the ball in the air for a long time. If a tennis ball is hit with topsin, then the Magnus force acts vertically down on the ball and causes the ball to drop down onto the court faster than it would if gravity alone was the only vertical force on the ball. A baseball or softball pitched with spin about its vertical axis curves either to the left or right, depending on the direction of spin, while a baseball or softball spinning rapidly about a horizontal axis can curve up like a golf ball or down like a tennis ball, depending on the amount and direction of spin. If a ball is spinning slowly about a

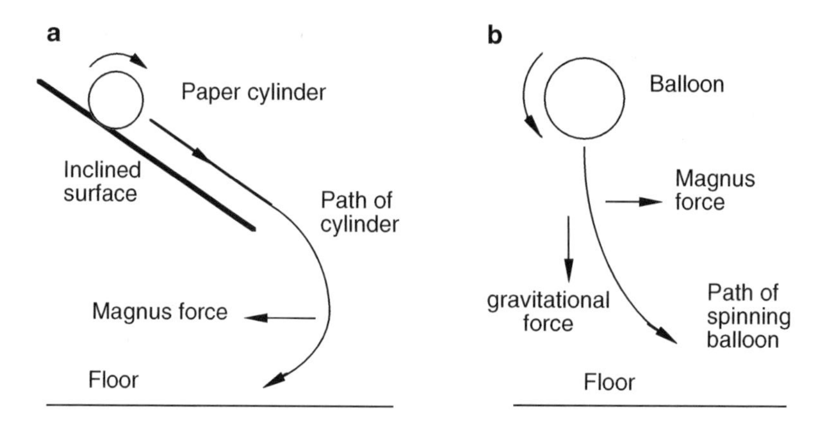

Fig. 18.2 Examples of the Magnus force acting on (**a**) a spinning paper cylinder and (**b**) a spinning balloon

horizontal axis, then it will curve down due to gravity regardless of the spin direction since the Magnus force is then much smaller than the gravitational force.

A simple demonstration of the Magnus effect is shown in Fig. 18.2a. The object here is a paper cylinder constructed from half an ordinary sheet of writing paper with one edge joined to the opposite edge with adhesive tape. If the cylinder is rolled down a flat, inclined sheet of cardboard or a plank of wood, then the cylinder will spin through the air after it rolls off the bottom end. One would normally expect that the cylinder would fall to the floor and land a few feet beyond the bottom end of the incline. Instead, the cylinder curves back underneath the incline due to the Magnus force. In this case, the cylinder falls vertically through the air and the Magnus force acts in a horizontal direction, back toward the original launch point. The Magnus force is not very large but it is large enough to curve a light paper cylinder by 1 or 2 ft.

The Magnus effect also works well with a light, spinning balloon although a balloon on its own tends to wobble as it spins and falls through the air. The wobbling effect can be reduced considerably by tying a length of string a few times around the balloon's equator and then taping it in place. If the string lies along the equator, then the balloon should be spun so that its axis passes through the north and south poles. However the balloon first needs to be tilted through 90° so that the spin axis is horizontal and the string rotates in a vertical plane. That way, as the balloon falls vertically to the floor, the Magnus force pushes the balloon off to one side. Further details are given in [1].

Project 3: Lift Forces

A baseball or a softball that is spinning backwards as it flies through the air experiences an upward force that holds it in the air for a relatively long time, and can

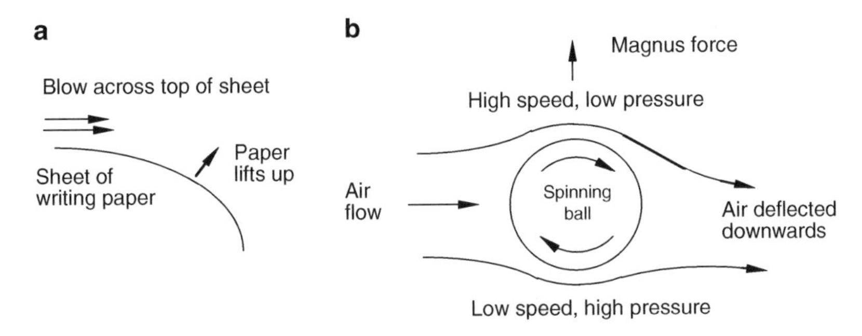

Fig. 18.3 Examples of lift forces acting on (**a**) a sheet of paper and (**b**) a spinning ball

even cause the ball to rise up against the force of gravity. Such a force is called a lift force in aerodynamics. Lift forces are needed to keep aeroplanes in the air, but aeroplanes don't need to spin backwards to experience a lift force. Rather, the wings of an aeroplane act to deflect the incoming air downwards, and the resulting force of the air on the wings is an upwards lift force. You can demonstrate this for yourself in a simple way, as illustrated in Fig. 18.3a.

Take an ordinary sheet of writing paper and hold it by one edge in an approximately horizontal position. The far end will drop down. Now blow air over the horizontal part. You might expect that blowing onto the upper side of the sheet of paper will make it curve down even more. Instead, the paper lifts up, since the air pressure on the upper side drops. A similar result is obtained if you hold two sheets of paper in a vertical position so that they are about 1/2 in. apart. If you blow air between the two sheets then instead of blowing apart, the two sheets move towards each other since the air pressure is reduced between the two sheets.

Now consider the situation shown in Fig. 18.3b where air is approaching from the left and passing around a spinning ball. Normally, the air is at rest in front of an approaching ball, but the situation shown in Fig. 18.3b is equivalent in terms of the physics of the problem. If a ball is spinning and approaching from the right, the resulting forces on the ball are the same as when the ball is at rest, but spinning on a fixed axis, and the air is approaching from the left. That is how objects are tested aerodynamically in a wind tunnel. The Magnus force on the ball acts upwards, as a positive lift force, and can be explained in two ways. The air is deflected downwards by the spinning ball, a result that can be seen clearly if smoke is added to the air stream. If the air is deflected downwards by the ball, then Newton's third law says that there is an equal an opposite force exerted by the air on the ball. The alternative explanation is that the air speeds up as it passes over the top of the ball, since it is accelerated by the spinning ball. Consequently, there is a drop in pressure above the ball so the net force on the ball is upwards.

Project 4: Drag Forces

One of the most famous physics experiments of all time was reported to have been conducted by Galileo in the early 1600s at the Leaning Tower of Pisa in Italy. It seems more likely that he did not actually conduct such an experiment himself, but it is claimed that he dropped two balls from the top of the tower to see if they would land at the same time. The point of the exercise was to see whether a heavy ball would fall faster than a light ball. It is obvious that the force on a heavy ball is greater than the force on a light ball because it is harder to lift a heavy ball. However, it was not obvious at the time whether the heavy ball would fall faster or not, given that a heavy ball is harder to accelerate than a light ball. The result of the experiment was that both balls landed at the same time. We "explain" that result today by saying that the acceleration due to gravity is the same for all objects, regardless of their weight.

If a baseball and a softball are dropped from a certain height, and released at the same time, then they will both hit the ground at the same time (at least, within a tiny fraction of a second if dropped from a height of say 6 ft). However, if you drop a baseball and a sheet of paper of the same diameter as the baseball, then the baseball will hit the ground well before the paper lands. The acceleration due to gravity is the same for the ball and the sheet of paper, and even the drag force due to the air is about the same, at least at the start of the fall. The drag force is the force of the air pushing against the ball or the sheet of paper, opposing its motion. The drag force is much smaller than the force of gravity acting on the ball, so the effect of the air on the ball is very small. However, the drag force acting on the sheet of paper is about equal to the force of gravity on the sheet of paper. As a result, the sheet of paper falls much more slowly than the baseball.

We can measure the acceleration due to gravity by measuring the time taken for a baseball to hit the ground. Similarly, we can measure the drag force on a sheet of paper by measuring the time it takes for the paper to hit the ground. If we use a circular sheet of paper of the same diameter as a baseball, then the drag force on the paper will be similar to that on the ball, but not exactly the same since the drag force depends on two factors. One factor is the area of the ball or the sheet of paper and the other factor is the speed of the ball or the sheet of paper. Since the ball falls faster, the drag force on the ball will be larger than the drag force on the sheet of paper.

Suppose we drop a baseball from a height of 6 ft and measure the fall time with a stop watch to be 0.63 s. The formula for the vertical distance traveled, s, in a time t, is $s = 0.5at^2$ where a is the vertical acceleration. Since $s = 6$ ft and $t = 0.63$ s, we find that $a = 30\,\mathrm{ft\,s^{-2}}$, which is reasonably close to the accepted value for the acceleration due to gravity, which is $32\,\mathrm{ft\,s^{-2}}$. A more accurate result would be obtained if we filmed the fall with a video camera.

If a sheet of paper is dropped from a height of 6 ft then it will take about 3 or 4 s to fall to the ground depending on how it falls. A flat sheet of paper twists and turns as it falls, and can even rise upward for a short time during the fall. The fall is too erratic to measure the drag force properly (Fig. 18.4). A better technique is to make a pointy cone out of the sheet of the paper, and let it fall with the pointy end down. A cone

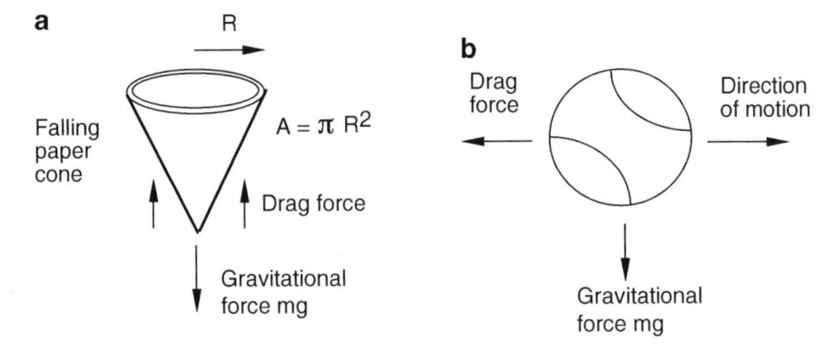

Fig. 18.4 Drag forces on a paper cone and on a baseball. The drag force always acts backwards, opposing the motion of an object. The drag force acts vertically up on the falling paper cone, but it acts to the left on the ball if the ball is traveling to the right

is more streamlined and it will fall straight down, taking about 1.5–2 s to fall. If it takes 2 s to fall 6 ft, then $s = 6$ ft and $t = 2$ s so $a = 2s/t^2 = 2 \times 6/2^2 = 3\,\text{ft}\,\text{s}^{-2}$, which is about ten times smaller than the acceleration due to gravity. The total force on the cone is therefore about ten times smaller than the gravitational force. This result implies that the drag force acting upward on a paper cone is almost equal to the force of gravity acting downward on the cone. In fact, the drag force increases as the speed of the cone increases, until the drag force is equal to the force of gravity on the cone. At that point, the total force on the cone drops to zero and the cone then coasts down to the ground at a constant "terminal" speed. The drag force is then equal to the weight of the cone. By changing the size, shape and weight of the cone, it is easy to show that the drag force is proportional to the cross-sectional area of the cone and proportional to its speed squared, as explained in Sect. 3.4.

18.3 Physical Properties of a Bat

The four most important properties of a bat are its weight, length, balance point and swing weight. The following projects describe how you can measure the balance point and the swing weight, and how you can locate the sweet spot of a bat.

Project 5: The Sweet Spot of a Bat

The origin of the sweet spot of a baseball bat has been debated for many years [2,3]. The following experiment shows conclusively that the sweet spot is located at a region of the bat where bat vibrations are minimized. However, to show that this is the case, it helps to experiment with a straight beam of wood with a rectangular cross section rather than a normal bat. The problem with a normal bat is that the two main contenders for the sweet spot are so close together that it is almost impossible to choose between them. For a rectangular cross-section beam, these two points are well separated, making a positive identification of the sweet spot much easier.

In terms of the feel of a bat, there are two possible impact locations along the barrel where the bat can potentially feel good. By "good" we mean that there is very little shock or vibration transmitted to the handle, despite the extremely large force exerted by the bat on the ball, and vice-versa. At other impact points on the barrel, the batter feels a greater amount of shock and vibration. One of the two potentially good spots is known as a vibration node. The other potentially good spot is known as the center of percussion (COP). Many articles have been written about the sweet spot. Roughly half the articles claim that the sweet spot is the vibration node, while the other half claim that the sweet spot is the COP. These two spots are described in more detail in Chap. 14. Here, we describe a simple experiment you can do yourself to distinguish between the two spots.

The vibration node of a bat is easily located by listening to the bat. To do this, you need to hold the bat in a special way, with the knob end close to one ear so that you can hear the knob vibrate. The bat must be held lightly, between the thumb and one finger, at a point about 6 in. from the knob, as shown in Fig. 18.5. Then tap the barrel with the ball of one finger of the other hand, and listen for a gentle humming sound. If you tap the barrel with a fingernail, you will hear a click sound, and the gentle hum will be harder to hear. If you have trouble hearing the hum, try holding the bat 7 in. from the knob, rather than 6 in., or try hitting the barrel near the far end of the barrel. If you still have trouble, then hit the barrel gently with a baseball, and the hum will then be a bit louder. Once you can hear the hum, then tap the barrel at various points along the whole length of the barrel. You will be able to locate a spot where the hum disappears. That spot is the vibration node. Actually, it is one of several different nodes along the barrel, but it is the easiest one to find. Technically, it is known as the node of the fundamental mode. The node point is typically about 6 or 7 in. from the end of the barrel, and it is the sweet spot of the bat. However, to prove that it is the sweet spot, we also need to prove that the COP is not the sweet spot.

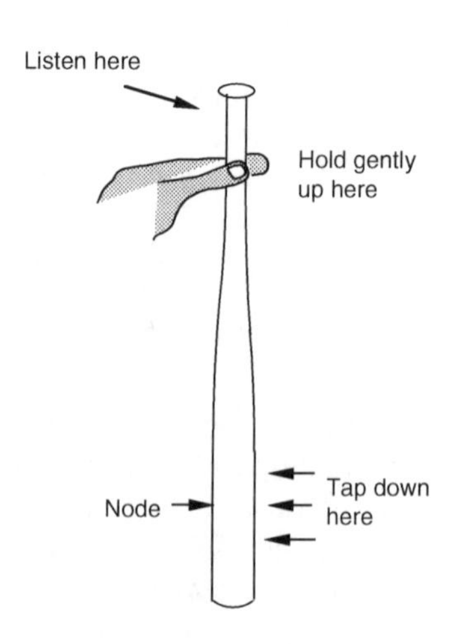

Fig. 18.5 Listening for the vibration node of a bat

For the second part of the experiment, you will need to find a straight beam of wood roughly the same length and width as a wood bat. A suitable beam would be 2 or 3 ft long, roughly 1 in. thick and 1 or 2 in. wide. In that case, the node point is located at a distance $0.22L$ from the end of the beam, where L is the length of the beam. For example, if $L = 33$ in. then the node point is 7.3 in. from the end. By listening to the beam, you will be able to confirm that the node point is indeed located at this spot.

The COP for the beam is not as well defined as the node point, since it is defined in terms of a rotation axis somewhere near the opposite end of the beam. For example, if the rotation axis is at one end of the beam, then the corresponding COP is located at a distance $L/3$ from the other end. So, if $L = 33$ in., the COP is located 11 in. from the end of the beam. This means that if the beam is freely suspended and is struck 11 in. from one end, the beam will rotate about an axis through the other end. If that is where a batter holds the beam, then the impact will feel good because the handle will not jerk forward or backward out of the hand. However, if the batter holds the beam further along, say 3 in. from the end of the beam, then the COP for an axis 3 in. from one end is located at a distance x from the other end where $x/L = (L-9)/(3L-18)$. For example, if $L = 33$ in., then $x = 9.78$ in. The COP of a 33 in. beam is therefore somewhere in the region of 9.78–11 in. from one end, whereas the node point is 7.3 in. from the end, at least 2 in. away (Fig. 18.6).

Armed with this information, you can locate and identify the sweet spot of your uniform beam of wood. To do so, simply hold the beam at one end in one hand and drop a baseball (or a golf ball) at various points along the other end. You will easily locate an impact point that feels best. It will be at the node point, not at the COP.

Holding onto a bat with one or two hands shifts the COP even further away from the tip of the barrel, as described in Chap. 14. That leaves the node point as the only real contender for the sweet spot. There is, however, a slight complication. That is, a bat has several other node points close to the fundamental node point. A bat

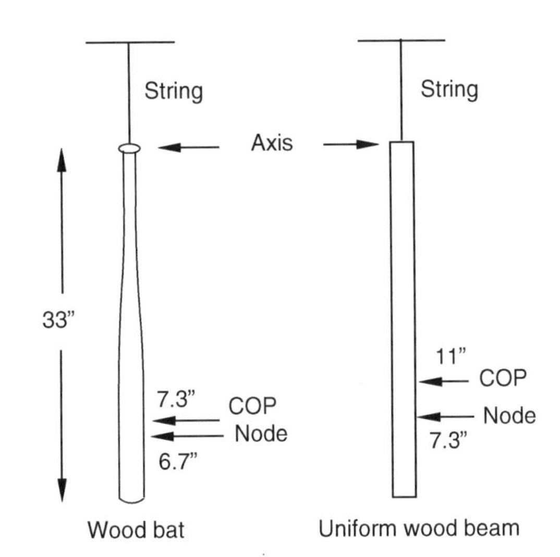

Fig. 18.6 Locations of the node point and the COP for a wood bat and a uniform wood beam each 33 in. long. If the bat and the beam are freely suspended by a length of string then an impact at the COP will cause each of them to rotate about an axis at the top end. For a real bat, the COP is close to the node point, making it difficult to decide which one feels best

can vibrate at several different frequencies simultaneously. The lowest frequency vibration, or the fundamental mode, occurs at about 170 Hz, and is responsible for the humming sound that you can hear. The bat can also vibrate at a higher pitch, around 530 Hz, and its node is about 2 in. closer to the end of the barrel. If you are musically sensitive to pitch, you might also be able to locate the node of the 530 Hz mode.

If you hold the bat close to the knob itself, then the humming sound disappears. The high pitched sound you hear then, with an aluminum bat, corresponds to a hoop mode vibration of the barrel at a frequency between about 1,000 Hz and 2,000 Hz, depending on the stiffness of the barrel.

Project 6: Balance Point of a Bat

A bat has a heavy end and a light end. It can be balanced on one finger at a point about 12 in. from the heavy end. The balance point of a bat is the point where the whole mass of the bat is effectively located. Technically, we refer to such a point as the center of mass, and it is located along the central axis of the bat, immediately above the point at which it balances on one finger (Fig. 18.7).

For any given bat weight, the location of the balance point provides a good indication of how easy it is to swing the bat. A bat with its balance point closer to the knob will be easier to swing. A bat with its balance point farther from the knob will be harder to swing. However, an even better measure of the speed at which a bat can be swung is a quantity commonly known as the swing weight of the bat. We will explain how the swing weight can be measured in Project 7. Here we describe a convenient method of measuring the distance between the balance point and the knob.

If you were to cut the bat through its center of mass, and weigh each of the two separate parts, you might be surprised by the result. The barrel part is heavier than the handle part. The centre of mass is not the point where half the mass is on one side and half on the other. In the case of a perfectly symmetrical object such as a straight, uniform rod or a solid sphere, the center of mass is in the center of the object, with half the mass on one side and half on the other side. However, this is not the case with unsymmetrical objects.

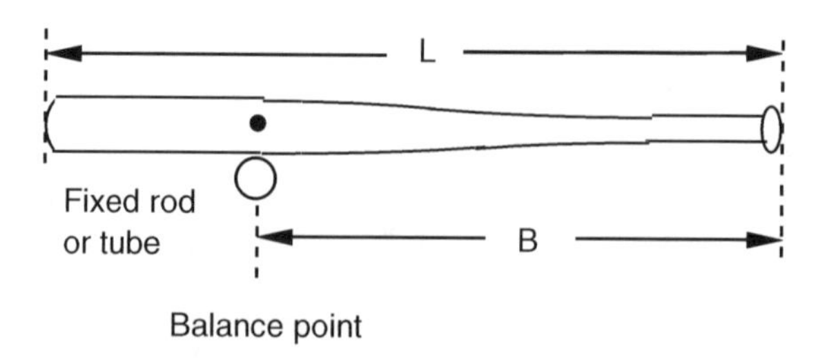

Fig. 18.7 The balance point of a bat can be measured by balancing the bat on a rod or tube. For wood bats, B/L is typically about 0.66. For an aluminum or composite bat, B/L is typically about 0.60 and most are in the range 0.57–0.63

Suppose we balance a bat on a rectangular block of wood or metal about 4 or 5 mm wide. That is a lot easier than trying to balance the bat on a sharp edge. If we add a coin at the end of the barrel and another coin of the same weight near the end of the handle, then the bat will be unbalanced and the handle end will rotate downward. To keep the bat balanced, the two coins must be located at the same distance from the balance point. Alternatively, the bat could be balanced with two coins at the barrel end and one coin near the handle end. To keep the bat balanced, the torque (weight times distance from balance point) acting on one side of the balance point must be equal to the torque acting on the other side.

The balance point represents a point where the clockwise torque due to all the mass on one side of the balance point is equal to the counter-clockwise torque due to all the mass on the other side. The significance of the balance point is that if a force is applied in line with the balance point then the bat will not rotate. If a force is applied anywhere else, then the bat will rotate unless an equal and opposite torque is applied on the other side of the balance point.

To locate the balance point of a bat, a simple but not very accurate method is to balance the bat in a horizontal position on one finger. In theory, a more accurate method is to try to balance the bat on a sharp edge. In practice, that is almost impossible since the bat will tend to rotate one way or the other no matter where you position the bat. A better method is to balance the bat across a circular tube about 3/4 in. in diameter. To stop the tube itself rotating, it needs to be mounted in such a way that it can't rotate, and it needs to be mounted an inch or so above a horizontal surface so that the bat is free to rotate up or down when it is unbalanced. For example, the tube could be inserted through holes drilled through two blocks of wood each about 2 in.^{2}. A ruler underneath the bat can then be used to measure the distance from the balance point to the end of the knob to within about 1 mm. You can then compare different bats to find out whether you prefer the balance point to be closer to or further from the knob.

An alternative (and the official) method of measuring the balance point is to place the bat horizontally on two scales, one near each end of the bat. The scale near the barrel end will have a higher reading than the scale near the handle end. Adding the two weights gives the total weight of the bat. Multiplying each weight by the distance to the (unknown) balance point gives the two torques. Since the torques about the balance point are equal and opposite, you can calculate the distances from the middle of each scale to the balance point. Accurate distances can be calculated if the bat rests in a raised V-shape edge mounted in the middle of each scale.

Project 7: Swing Weight of a Bat

One of the most important physical properties of a bat is its swing weight, because it determines the ease at which it can be swung, and because it is one of two main factors that determines the inbuilt power of the bat. For a wood bat, it is essentially the only factor. Most people judge a bat by its weight, since that is relatively easy to determine. Swing weight is harder to judge, and harder to measure, but an experienced batter should be able to distinguish small differences in swing weight between different bats if he or she knows what to look for.

Ordinary weight determines whether a bat is heavy or light when it is lifted up. Swing weight determines whether a bat feels heavy or light when it is swung. At first sight, it might seem that there is no real difference, but there is. If you pick up a bat by holding onto the barrel then it will feel just as heavy as when you pick it up by the handle. But if you swing a bat by holding onto the barrel end instead of the handle end, then the bat will feel much lighter since it is then much easier to swing and you can swing the bat a lot faster.

It is not difficult to measure the swing weight of a bat, but it takes a few minutes and involves a few calculations. The bat needs to be mounted like a pendulum so that it can swing back and forth, and the time for each back and forth swing needs to be measured with a stopwatch or some other electronic timing device. That is the officially approved method and it is the method that we will now describe. There is an even better method, commonly used to measure the swing weight of tennis racquets and golf clubs, but it is not widely used to measure the swing weight of bats. Racquets and clubs are measured by swinging them back and forth in a horizontal plane against a fixed spring. Bats are usually measured by swinging them in a vertical plane, using gravity to swing them back and forth as a pendulum. By swinging a racquet in a horizontal plane, gravity does not play any role, so it is not necessary to know the mass or the balance point of the racquet to work out its swing weight. The end result is a faster and more accurate measurement of the swing weight.

The swing weight of a bat can be estimated to within about 3% accuracy using the formula

$$I_6 = (0.76B/L - 0.2)ML^2,$$

where B is the balance distance in inches, M is the bat mass in oz and L is the bat length in inches. The formula is not a theoretically derived formula. It just happens to give a good fit to experimentally measured values of swing weight for a wide variety of bats. I_6 is the swing weight (in oz in.^{2}), defined as the moment of inertia of the bat when the bat rotates about an axis in the handle located 6 in. from the knob. For a bat of any given mass and length, the formula shows that swing weight increases as the balance distance increases. To determine the swing weight more accurately, it needs to be measured. The formula might actually be correct to within 1% for some bats, but it might give an answer for other bats that is 2 or 3% too high or 2 or 3% too low.

To measure bat swing weight by the pendulum method, the bat needs to be suspended near the knob end so it can swing like a pendulum. The easiest way is to tie string around the knob and loop it over a horizontal rod. The bat will then swing about an axis an inch or two beyond the knob. If you are measuring an old bat, an alternative method of mounting the bat is to drill a hole through the handle and then insert the rod through the hole. Ideally, a collar should be constructed to clamp around the handle at a point 6 in. from the knob, since that is the official method. The collar needs to rest on two sharp edges so that the bat can swing

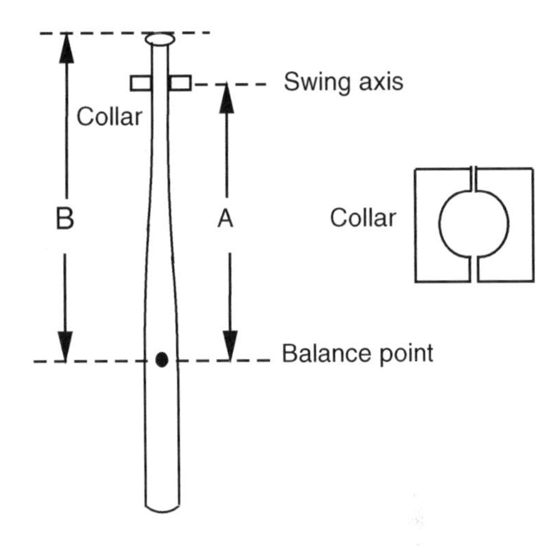

Fig. 18.8 The swing weight of a bat can be measured by swinging it as a pendulum about an axis located a distance A from the balance point

freely (Fig. 18.8). A simpler method is to hang the knob on two parallel rods spaced about 1 in. apart and adjusted so that the bat can swing freely in a direction parallel to the two rods.

If the bat swings freely then the bottom end can be pulled aside about 6 in. then released. After a few swings, start timing the oscillations and measure the time for say ten complete (back and forth) cycles. One complete cycle means that the end of the barrel starts at a point say 6 in. to the right, moves 6 in. to the left and then returns to the starting point. Repeat several times and take an average time for ten cycles, then divide the answer by ten to get the time, T, for one cycle. To calculate the swing weight, it is also necessary to measure the mass M of the bat, the distance B from the end of the knob to the balance point, and the distance A between the balance point and the axis of the pendulum.

There are three different swing weights that can be calculated from this information, depending on which of three different axes we choose. We can denote these swing weights as follows:

I_A = Swing weight for the actual axis used to measure the time T.
I_B = Swing weight for an axis passing through the balance point.
I_6 = Swing weight for an axis 6 in. from the end of the knob.

The official method is to swing the bat about an axis 6 in. from the end of the knob. However, if we measure the swing weight I_A about a different axis then it is easy to calculate I_6, as follows:

If M is measured in oz, and if A and B are measured in inch, and we take the acceleration due to gravity to be $9.81\,\text{m}\,\text{s}^{-2} = 386.22\,\text{in}\,\text{s}^{-2}$ then the three swing swing weights, in oz in.^{2} are given by

$$I_A = 9.783MAT^2$$
$$I_B = I_A - MA^2$$
$$I_6 = I_B + M(B-6)^2 = 9.783MAT^2 + M[(B-6)^2 - A^2]$$

For example, suppose that $M = 31$ oz, $B = 20$ in., $A = 19$ in., and $T = 1.602$ s. Then

$$I_A = 9.783 \times 31 \times 19 \times 1.602^2 = 14,788 \text{ oz in.}^2$$
$$I_B = 14{,}788 - 31 \times 19^2 = 3597 \text{ oz in.}^2$$
$$I_6 = 3597 + 31 \times 14^2 = 9673 \text{ oz in.}^2$$

These formulas were derived using the parallel axis theorem, described in Chap. 1.

18.4 Impact of a Bat and a Ball

What happens when a bat collides with a ball? The following four projects shed light on the most important effects.

Project 8: The Trampoline Effect

For this experiment, you need a baseball or a softball and a tennis racquet. Drop the ball on a hard surface, such as a concrete driveway or sidewalk, and measure the drop height h_1 and the bounce height h_2 (from the concrete to the bottom of the ball, not the top of the ball). Repeat the experiment by dropping the ball onto the strings of the racquet. The racquet needs to be clamped firmly under foot onto the hard surface so that the racquet itself does not bounce or vibrate. The ball bounces to a greater height off the strings than it does off the concrete.

The strings act as a trampoline and store some of the elastic energy during the collision. The ball squashes slightly during the collision and also stores some elastic energy. More than half of the elastic energy stored in the ball is lost during the collision, but almost all of the elastic energy stored in the strings is recovered and given back to the ball. If the ball loses say 60% of its stored elastic energy, then it will bounce to only 40% of its drop height when dropped on a hard surface.

When the ball is dropped on the strings, the ball will store about 40% of the total elastic energy and the strings will store about 60% of the total elastic energy, since the strings are softer than the ball. If the ball loses 60% of its stored elastic energy, and if the strings lose only 10% of their stored elastic energy, then the fraction of the total elastic energy lost is $0.6 \times 0.4 + 0.1 \times 0.6 = 0.3$, in which case the ball will bounce to 70% of its drop height. You might find that the ball bounces even higher than this, in which case the strings will have stored more of the elastic energy than we assumed.

The energy loss in the strings can be measured separately by dropping a hard steel ball or a hard billiard ball onto the strings, in which case the strings will store all of the elastic energy and the ball will store none.

You could try bouncing the ball off an aluminum bat in this way, to see if the bat acts as a trampoline, but a better method is described in Project 10.

Project 9: Impact Duration

The collision time between a bat and a ball is about 0.001 s. The time is so short that it cannot possibly be seen by eye, and it cannot be measured with a stop watch. It can be measured with a very high speed video camera capable of recording at least 5,000 frames every second, but such cameras are extremely expensive and not particularly accurate if there are only 5 or 10 frames recorded during the collision. A much cheaper and much more accurate method of measuring both the collision time and the force on the ball is to use a piezoelectric disk [4]. Such a disk costs about $3 and can be obtained in the form of a piezo buzzer from an electronics store or in the form of a buzzer in a musical greeting card. An even better and cheaper disk can be obtained from companies that supply piezo disks in their raw state, without the plastic cover that normally surrounds piezo buzzers. They can be obtained from Edmund Scientific or at www.digikey.com. They sell a wide range of suitable piezo disks for about $1 each, listed under audio products or buzzer elements or piezo benders. Some have leads already attached. If not, you can solder a fine wire lead to each side of the piezo disk yourself (or get someone to do it for you). A disk about 20 or 30 mm in diameter will be suitable.

Piezoelectric devices can be used in two ways. The most common application is to generate a buzzing sound, either as a warning sound for a burglar alarm or as an indication of the time in a clock alarm or as a series of musical notes, as in a musical card sent as a birthday or Christmas card. An alternating voltage applied to the buzzer causes it to vibrate and emit a sound. A piezo disk can also be used in reverse. When a force is applied to the disk, a voltage signal is generated. The voltage can be so large that it causes an electrical spark which can be used to ignite a gas appliance. The voltage is proportional to the force on the disk and drops to zero when the force decreases to zero. It is just what we need to measure the force on a baseball.

A piezo disk is typically about 20 mm in diameter, about 0.3 mm thick and is made from a special type of ceramic. It is normally attached on one side to a thin brass disk, and has a silver electrode on the other side. A thin wire lead is attached to the brass disk and another thin wire lead is soldered to the silver electrode. When an alternating voltage is applied across the two leads, the disk vibrates. When a force is applied to the disk, a voltage is generated between the two leads. The disk is normally housed in a plastic case when used as a buzzer and needs to be removed carefully with a pair of pliers to cut away the plastic housing. When that is done, the disk can be attached to a flat, solid surface with adhesive tape and a baseball can be safely dropped onto the disk. The voltage generated is proportional to the impact force, is typically a few volts and it lasts as long as the collision time, about 0.003 s at very low ball speeds.

The catch in this experiment is that you need access to an oscilloscope to measure the voltage. All physics laboratories have suitable oscilloscopes, which need to be a storage type. That is, it needs to capture the voltage signal and store it in memory so you can examine the signal after the collision is over. In addition, you need a voltage probe. One end of the probe connects to the two piezo leads, and the other end plugs into the oscilloscope.

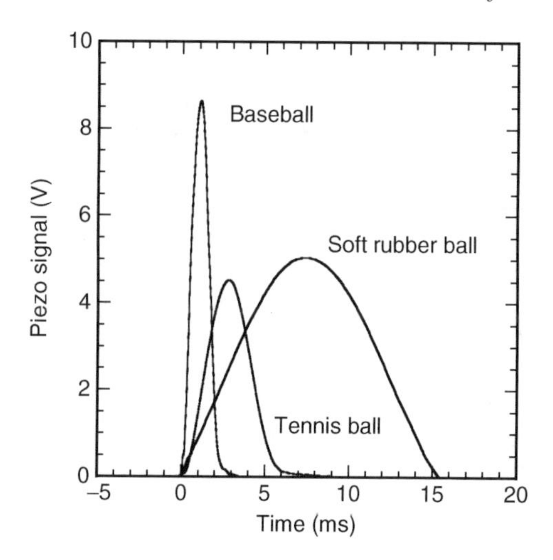

Fig. 18.9 Signals recorded with a 15 mm diameter, 0.3 mm thick piezo disk, for three different low speed balls. The impact time of the baseball was about 3 ms. At high speed, the impact time of a baseball is typically less than 1 ms. Negative time here is time before the ball hit the piezo, showing that there was no force on the piezo until the ball landed

Typical results are shown in Fig. 18.9. The results in Fig. 18.9 can be used to determine the stiffness of each ball. The collision time T of a ball of mass m and stiffness k is given to a good approximation by $T = \pi\sqrt{m/k}$. For the baseball, $m = 0.145\,\text{kg}$, and $T = 0.003\,\text{s}$, so $k = 1.6 \times 10^5\,\text{N m}^{-1}$. However, at high ball speeds where T is about 1 ms, the ball stiffness increases to about $1.4 \times 10^6\,\text{N m}^{-1}$. The collision time of a tennis ball is longer than that of a baseball, despite its smaller mass, because a tennis ball is much softer than a baseball.

To obtain the results in Fig. 18.9, a 1-nF capacitor was soldered across the two piezo leads. This results in a smaller but a more reliable voltage signal. The way a piezo works is that an electrical charge is generated when a force is applied to the disk, but that charge flows off the piezo disk and through the voltage probe and the oscilloscope. The time taken for the charge to drop to zero is increased when an additional capacitor is connected across the piezo disk. If that time is increased to say 0.5 s, then the voltage signal generated by the piezo is reliably proportional to the force on the piezo for at least 0.05 s, more than enough to record the force exerted by a baseball or a tennis ball. If you push on the piezo with your finger, and hold your finger on the piezo for say 1 s, then the push force is recorded correctly for the first 0.05 s, but the signal drops to zero after about 0.5 s even though you are still pushing on the piezo. If you then lift your finger off the piezo, the signal suddenly decreases to a negative value and then returns to zero after an additional 0.5 s.

A problem with very thin piezo disks is that the output voltage cannot exceed about 20 V. Another problem, with small diameter disks, is that the contact area of the ball can be larger than the diameter of the disk. As a result, large forces on the disk are not recorded reliably. To overcome this problem, it is necessary to use disks that are about 5 or 6 mm thick and about 50 mm in diameter. Such disks are more expensive, around \$60 or more, but they can be used to measure the force on a baseball incident at relatively high speed. They need to be mounted on a flat, heavy surface so that the ball does not smash the ceramic disk. If one side of the disk is attached to say the flat end of a metal cylinder with superglue (or a conducting paste) then one lead can be attached to the metal cylinder and the other lead needs to be soldered to the other side of the disk.

Project 10: Bat and Ball Collisions

A very nice experiment illustrating the physics of bat and ball collisions can be performed in the following way. A bat is mounted so that it can swing like a pendulum about an axis a few inches from the knob, as indicated in Fig. 18.10. A ball is suspended by a length of string so that it can also swing like a pendulum. That way, the bat can be swung so that it strikes a ball at rest, or the ball can be swung so it strikes a bat at rest, or the bat and the ball can both be swung so that they collide with each other. The results are very interesting and absolutely fundamental in terms of understanding the collision between a bat and a ball.

In this experiment, it is not necessary that the bat or the ball be swung at high speed. The basic physics of the collision is the same, regardless of the actual speeds of the bat and the ball. Low speed collisions are much more convenient in terms of setting up and taking measurements. In fact, most of the measurements can be done with a ruler, but greater accuracy can be obtained by filming the collision with a video camera. The beauty of the experiment is that (a) it can be done by almost anyone, (b) it illustrates the basic physics behind bat and ball collisions in an enlightening way, and (c) it contains many hidden and subtle features that have

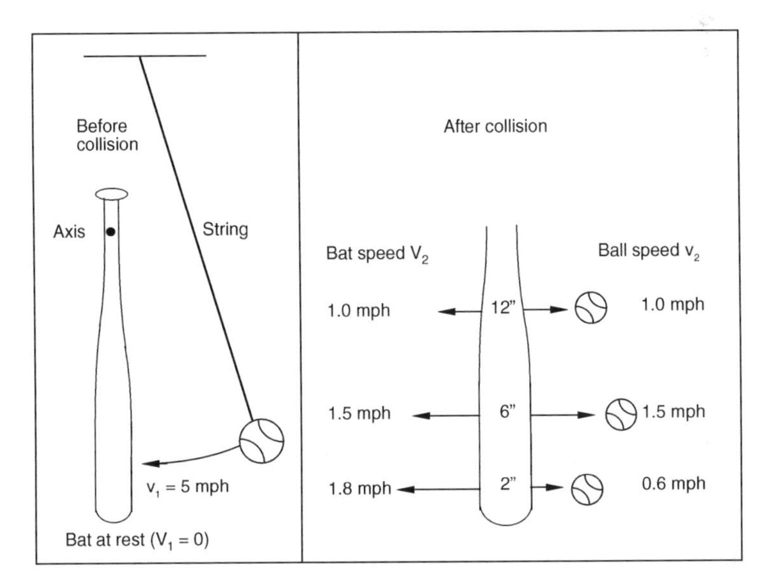

Fig. 18.10 A ball incident at 5 mph on a bat at rest. In this experiment, the ball bounced best about 6 in. from the end of the barrel

still not been properly explained. For example, you could ask an innocuous question such as "what happens if the ball is spinning when it hits the bat or if it hits the bat toward one edge?" and then you can find out the answers yourself by filming the results with a video camera. For the moment, we will put these other issues aside and concentrate on the straightforward problem where the ball is not spinning and where it strikes the bat at right angles and in the middle of the bat.

The bat and the ball need to be mounted in a suitable manner. The ball is a bit easier, although it helps to use two equal lengths of string in a V-shaped support to prevent the ball twisting around before or after the collision. One way to tie string to the ball is to hammer a few small nails into the ball and then bend the top end of each nail into a hook. Another way is to screw two small eye screws into the ball. Another way is to glue metal hooks onto the ball or attach them with double-sided tape. Mounting the bat is easy if you are prepared to drill a hole straight through the handle. Otherwise, the bat can hang freely with the knob supported on two parallel rods spaced about 1 in. apart. Alternatively, you can tie string around the knob and swing the bat about an axis an inch or two beyond the end of the handle.

Results of such an experiment are shown in Figs. 18.10 and 18.11 for an Easton BK7 aluminum bat and a standard 145 g baseball. In Fig. 18.10 the bat was at rest before the collision and the ball was swung into the bat to impact at various points along the barrel. The ball was incident at speeds varying from from about 4–5 mph, but the results in Fig. 18.10 were scaled as if the ball was incident at exactly 5 mph at every spot. For example, if the ball was actually incident at 4 mph at one particular spot, the measured bat and ball speeds were multiplied by 5/4 before including them in Fig. 18.10.

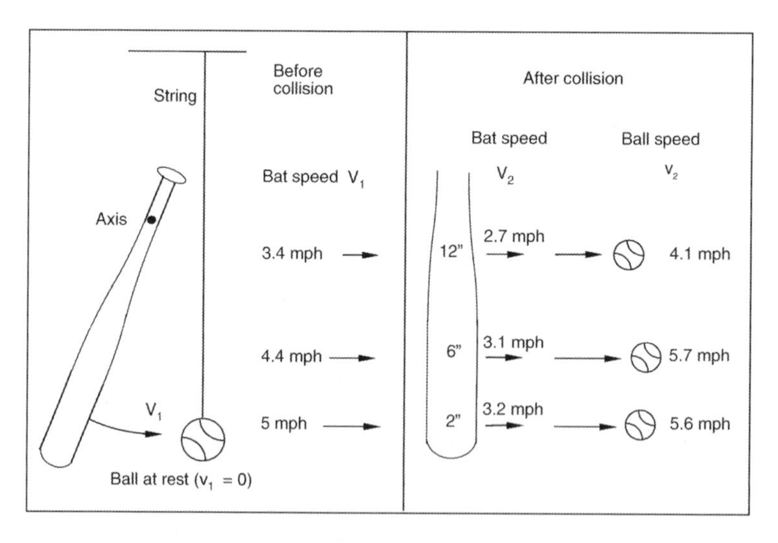

Fig. 18.11 When the bat was swung through an angle of 50° onto a stationary ball, the tip of the bat struck the ball at a higher speed than points further along the barrel. The ball exit speed was largest about 6 in. from the tip, but the exit speed near the tip was almost as large

An additional bounce experiment was performed to calibrate the behavior of the ball. Using the arrangement shown in Fig. 18.10, the bat was replaced with a heavy block of polished granite. A smooth, heavy block of concrete would also work, such as those sold for making walkways or driveways. A ball incident at 5 mph bounced off the granite at 2.95 mph on average, giving an average value of the COR of $2.95/5 = 0.59 \pm 0.01$.

The coefficient of restitution, COR, for a collision between any two objects is defined as the relative speed after the collision divided by the relative speed before the collision. In the case of a ball bouncing off a heavy block of granite, the block is at rest before and after the collision, so the COR is just the exit speed of the ball divided by the incident speed of the ball. However, in the case of a bat and ball collision, we define the latter ratio as the bounce factor, q, for the collision.

To explain the difference between the COR and the bounce factor, it helps to consider the results in Fig. 18.10. The ball was incident at 5 mph at each spot, and the ball bounced best at a point about 6 in. from the end of the barrel, at a speed of 1.5 mph. The bounce off a bat is weaker than the bounce off a granite block since some of the energy of the ball is given to the bat. The ratio of the exit speed of the ball (1.5 mph) to the incident speed (5 mph) in this case was $q = 1.5/5 = 0.3$. The bat itself recoiled at 1.5 mph at the impact point, so the relative speed of the bat and the ball after the collision was $1.5 + 1.5 = 3.0$ mph. The relative speed before the collision was 5.0 mph, so the COR in this case was 3.0/5.0 = 0.60. That is, the COR was slightly larger than that for the collision between the ball and the block of granite. This is a significant result and we will return to it shortly.

At the other two impact points in Fig. 18.10, the following results were obtained:

(a) At a point 2 in. from the tip of the barrel, $q = 0.6/5 = 0.12$ and $\text{COR} = 2.4/5.0 = 0.48$.
(b) At a point 12 in. from the tip of the barrel, $q = 1/5 = 0.2$ and $\text{COR} = 2/5 = 0.40$.

The COR values at the 2-in. and 12-in. points were both less that the corresponding COR for an impact on granite. The reason is slightly complicated and is explained in more detail in Chap. 12, but it is due to loss of energy caused by vibrations of the bat. The increase in the COR at the 6 in. point is due to two factors. One is that bat vibrations are much weaker at this point. The other is that a trampoline effect exits for hollow bats, and it works best at a point about 6 in. from the end of the barrel.

The results of swinging the bat at a stationary ball are shown in Fig. 18.11. The bat was pulled aside and allowed to swing through an angle of 50° before it collided with the ball. As a result, the impact speed varied along the bat, being zero at the rotation axis and a maximum at the tip of the bat. The bat collided with the ball at 5 mph at a point 2 in. from the tip of the barrel, but at only 3.4 mph at a point 12 in. from the tip.

The exit speed of the ball in this case is proportional to the speed of the bat at the impact point, with the result that the exit speed of the ball was almost as large near the tip of the bat as it was at the 6 in. position. However, the exit speed of the ball also depends on the bounce factor, q. The formula for the exit speed, v_2, when the ball is initially at rest, is

$$v_2 = (1 + q)V_1,$$

where V_1 is the speed of the bat, just before the collision, at the impact point. We can use this formula, together with the results shown in Fig. 18.11, to calculate the value of q at each of the three impact points. For example, at the 6 in. point, $v_2 = 5.7$ mph and $V_1 = 4.4$ mph, so $5.7 = (1 + q) \times 4.4$, giving $q = 5.7/4.4 - 1 = 0.3$, which is the same as the result found using the results in Fig. 18.10. Similarly, $q = 5.6/5 - 1 = 0.12$ at the 2 in. position, and $q = 4.1/3.4 - 1 = 0.2$ at the 12 in. position, as we found before.

It is also interesting to calculate the COR values from the experimental results shown in Fig. 18.11. At the 2 in. position, COR $= (5.6 - 3.2)/5.0 = 0.48$. At 6 in., COR $= (5.7 - 3.1/4.4) = 0.59$. At 12 in., COR $= (4.1 - 2.7)/3.4 = 0.41$. These values are the same as those obtained when the bat was initially at rest, within experimental error. The q and COR values are therefore the same, regardless of whether the bat is at rest or the ball is at rest before the collision. Furthermore, the q and COR values are the same even if the bat and ball are both approaching each other before the collision. In the latter case, the formula for the ball exit speed is

$$v_2 = (1 + q)V_1 + qv_1$$

When the incident ball speed is zero (that is, $v_1 = 0$) then $v_2 = (1 + q)V_1$. When the incident bat speed is zero (that is, $V_1 = 0$) then $v_2 = qv_1$. Consequently, the ball exit speed is just the sum of the two separate parts – the speed when the ball is at rest plus the speed when the bat is at rest. The results of adding the speeds in Figs. 18.10 and 18.11 are shown in Fig. 18.12.

The results in Fig. 18.12 are of more practical interest than those in Figs. 18.10 and 18.11, but we can see how they arise when we examine the separate results in Figs. 18.10 and 18.11. In particular, we see that the bounce factor, q, plays a very important role in determining the performance of a bat. Different bats will have different values of q, but most bats are fairly similar in that q is relatively small near the tip of a bat and q is typically about 0.3 at a point about 6 in. from the tip of the barrel. The significance of q is that if one bat has a larger value of q than another, then the ball exit speed will be greatest off the bat with the larger q value, assuming that both bats are swung at the same speed. In general, the ball will bounce better off a heavy bat than a light bat, and will therefore have a larger q value, but heavy bats can't be swung as fast as light bats. Consequently, it is not immediately obvious whether heavy bats are better than light bats in this respect. The problem is examined in more detail in Chap. 11.

A more complete set of q and COR values for the Easton bat is shown in Fig. 18.13, where it can be seen that q is a maximum about 6.5 in. from the tip of the barrel, while the COR is a maximum about 5 in. from the tip of the barrel. The main feature of interest in the COR profile is that the COR can be greater than that

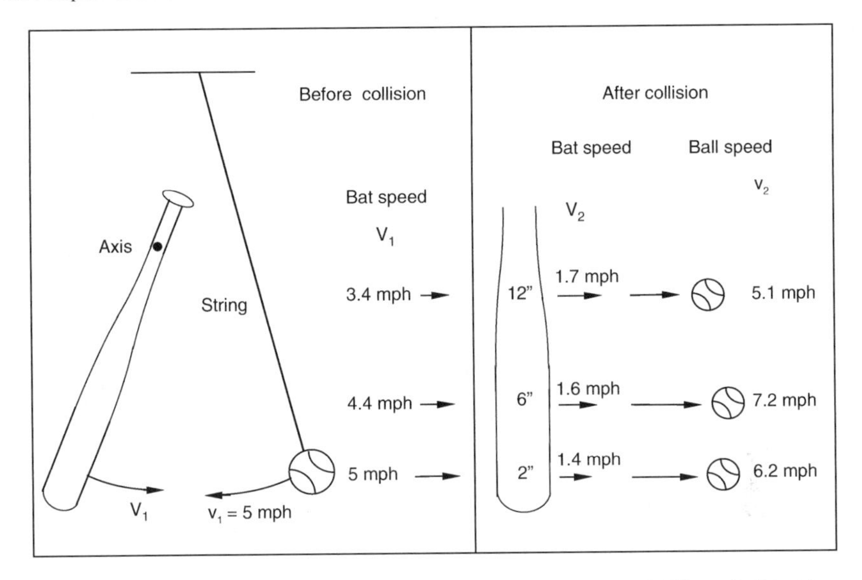

Fig. 18.12 Calculated results when the bat is swung through an angle of 50° at a ball incident at 5 mph. The bat and ball speeds after the collision are equal to the results in Figs. 18.10 and 18.11 added together

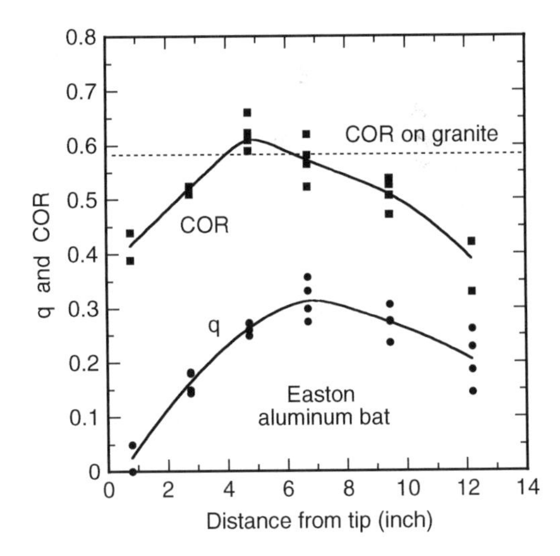

Fig. 18.13 Measured values of the bounce factor q and the COR vs. the impact distance from the tip along the barrel. Each data point represents a different collision, and the *solid lines* are best fit curves to guide the eye

for a collision between a ball and a heavy, rigid surface, as a result of the trampoline effect. Of greater practical interest is the q profile for a given bat, since q is easier to measure and since it determines the exit speed of the ball for any given swing speed of the bat, at any given impact point.

Measurement of Bat and Ball Speeds

The experimental results in Figs. 18.10–18.13 were obtained by using a video camera to film all collisions, and then measuring the distanced traveled by the bat and the ball from one frame to the next. At 30 frames s^{-1}, the time between successive frames is 1/30 s. A simpler technique can be used to measure the ball speed, since the speed is proportional to the horizontal distanced traveled by the ball. Suppose that the ball is raised to a height h above the impact point by pulling the ball aside though a horizontal distance x, as shown in Fig. 18.14. If the length of the pendulum is R, then $x^2 + y^2 = R^2$ where $y = R - h$. Hence, $x^2 = 2Rh - h^2$. Provided that $h \ll R$, we can ignore the term h^2 here and then $h = x^2/(2R)$.

After the ball drops through a height h, its kinetic energy will be given by $mv^2/2 = mgh$, so $v^2 = 2gh = gx^2/R$. Hence, $v = x\sqrt{g/R}$. The ratio of the ball exit speed to the incident speed is therefore the same as the ratio of the horizontal displacements in each case, so the q value can be measured with a ruler to within about 5% accuracy if desired. For example, if the ball is released when $x = 100$ cm and rebounds to $x = 25$ cm, then $q = 0.25$. Given that it is difficult to judge the rebound distance by eye, it is better to film each impact with a video camera to determine the rebound distance. The horizontal displacement of the bat can also be found by inspection of the video film.

Project 11: Is the Handle End of a Bat Important?

Many people believe that it is important to grip the handle end of a bat firmly so that the whole weight of the player's body is transferred through to the ball. Conversely, if a player relaxes his or her grip as the bat strikes the ball, then it is only the weight of the bat that is transferred to the ball. Another way of saying the same thing is this. A youngster swinging a bat will not be able to hit the ball as far as a heavy adult, even if they both swing the same bat at the same speed.

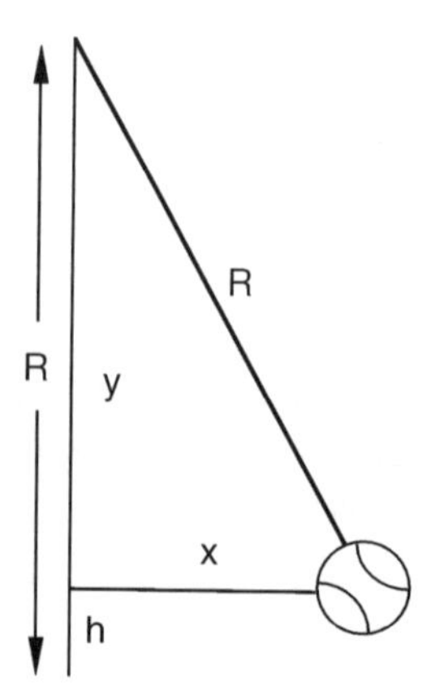

Fig. 18.14 The impact or bounce speed of the ball is proportional to the horizontal displacement x when $h \ll R$

It turns out that all of these statements are incorrect. A player could actually let go of the handle just before striking the ball and the result will be the same as if the player maintained an iron grip. The physics of the situation is that when the bat strikes the ball, a bending wave is transmitted from the impact point toward the handle end. When the wave gets to the end of the handle, it reflects off that end and travels back toward the impact point. The size (or amplitude) of the reflected wave depends on the weight connected to the handle and whether the handle is clamped rigidly or relatively free. However, by the time the wave gets back to the impact point, the ball has already left the bat and is flying through the air. So, the reflected wave has no effect on the speed of the ball coming off the bat. The ball has no way of knowing whether the handle was held firmly or loosely or even if it was held in a vice, or even if the bat was 6 ft long.

The effect of wave propagation along the bat leads to three very surprising conclusions. The first is that only part of the weight of the bat is effective in sending the ball on its way. The second is that the effective weight of the bat depends on which part of the bat collides with the ball. And the third is that the weight of the batter does not add anything at all to the effective weight of the bat. In general, a heavy batter can swing the bat faster than a light batter, but the actual weight of the batter has got nothing to do with the final collision between the bat and the ball.

Various experiments can be (and have been) done to prove that this is indeed the case. The arrangement shown in Fig. 18.10 is suitable. The ball will bounce off the bat at the same speed, regardless of whether the handle end is free to pivot about an axis through the handle, or whether the handle end is rigidly clamped, say in a vice. There may, however, be a difference in the bounce speed if the ball collides with the bat near the tip of the barrel. In that case, the ball is likely to come to a complete stop when testing light bats, and will not bounce at all if the bat is pivoted at the handle end. If the bat is then clamped at the handle end, the ball will come to a complete stop as before, the bat will start vibrating, and the vibrating bat will strike the stationary ball, ejecting it as a result of a second collision.

An alternative experiment is to collide a ball on a flat rather than a curved surface. A tennis ball and a tennis racquet will work. Compare the bounce height of the ball, off the middle of the strings, when the handle is (a) held by hand and (b) clamped firmly by hand on the edge of a table. The ball will bounce to the same height in both cases, although there may be a difference at the far end of the racquet due to the double bounce effect described in the previous paragraph.

Another experiment would be to measure the bounce of a light superball off a beam of wood or aluminum about 3 ft long [5]. A suitable arrangement is to mount the ball as a pendulum bob on the end of a length of string and let it collide with the beam when the beam is supported horizontally. The bounce speed (or distance) will be the same, regardless of the length of the beam or the method of supporting the "handle" end of the beam. However, if the beam is only about one foot long, or if the ball collides about one foot from the handle end, then the bounce speed will indeed depend on the support method since the reflected wave will then arrive back at the impact point before the ball bounces.

Regardless of how the experiment is done, you will find that the ball bounces best at points about 6 in. or more from the impact end, and that the bounce speed progressively reduces as the impact point moves closer to the tip of the bat or the beam.

Project 12: Measure the Spin of a Pitched Ball

Attach one end of long, flat tape to a ball. Allow the rest of the tape to lie flat on the ground between the rubber and the plate. Throw the ball to the catcher. After the ball is caught, count the number of turns of tape wrapped around the ball, typically between 6 and 15. By estimating or measuring the time taken for the ball to reach the catcher, you can calculate how fast the ball was spinning.

References

1. R. Cross, Aerodynamics of a party balloon. Phys. Teach. **45**, 334–336 (2007)
2. R. Cross, The sweet spot of a baseball bat. Am. J. Phys. **66**, 772–779 (1998)
3. R.K. Adair, Comment on "The sweet spot of a baseball bat," by Rod Cross [Am. J. Phys. **66**(9), 772–779 (1998)] Am. J. Phys. **69**, 229–230 (2001)
4. R. Cross, The bounce of a ball. Am. J. Phys. **67**, 222–227 (1999)
5. R. Cross, Impact of a ball with a bat or racket. Am. J. Phys. **67**, 692–702 (1999)

Conversion Factors

Most scientific measurements and calculations make use of the SI (metric) system of units. The English system of units is more commonly used by baseball and softball fans. Some useful conversion factors and baseball/softball measures are listed below.

Length

1 inch (in.) = 25.4 mm   12 in. = 1 foot = 304.8 mm   3 feet = 0.9144 m
1 metre (m) = 100 cm = 1,000 mm = 39.370 in. = 3.281 feet (ft)
1 km = 1,000 m = 0.6214 mile
1 mile = 5,280 ft = 1.609 km
Diameter of 12 in. softball: 3.780 in. (96.01 mm) to 3.859 in. (98.03 mm)
Diameter of baseball: 2.865 in. (72.766 mm) to 2.944 in. (74.788 mm)
Length of typical bat: 33 in. = 838.2 mm   34 in. = 863.6 mm
Distance between bases in softball = 60 ft = 18.288 m
Distance between bases in baseball = 90 ft = 27.432 m

Speed

$1\,\mathrm{m\,s^{-1}} = 3.6\,\mathrm{km\,h^{-1}} = 3.281\,\mathrm{ft\,s^{-1}} = 2.237\,\mathrm{mph}$ $50\,\mathrm{m\,s^{-1}} = 111.85\,\mathrm{mph}$
$100\,\mathrm{mph} = 160.9\,\mathrm{km\,h^{-1}} = 44.70\,\mathrm{m\,s^{-1}}$
$100\,\mathrm{km\,h^{-1}} = 62.15\,\mathrm{mph} = 27.778\,\mathrm{m\,s^{-1}}$

Acceleration

Acceleration due to gravity $= 9.80\,\mathrm{m\,s^{-2}} = 32.15\,\mathrm{ft\,s^{-2}}$

R. Cross, *Physics of Baseball & Softball*, DOI 10.1007/978-1-4419-8113-4,

Area

1 sq in. = 6.4516 sq cm 1 sq cm = 0.1550 sq in.
10,000 cm^2 = 1 m^2

Mass

1 kg = 1,000 g = 2.205 lb = 35.27 oz 1 lb = 16 oz = 0.4536 kg
1 oz = 28.35 g
Mass of 12 in. softball: 6.25 oz (177.19 g) to 7 oz (198.45 g)
Mass of baseball: 5 oz (141.75 g) to 5.25 oz (148.84 g)
Typical bat mass: 30 oz = 850.5 g 31 oz = 878.85 g 32 oz = 907.2 g

Density

Density of dry air (0% humidity) = 1.204 kg m^{-3} = 0.0752 lb ft^{-3} at 20°C (68°F) and standard atmospheric pressure (1013.2 millibars)

1 kg m^{-3} = 0.06243 lb ft^{-3}

Moment of Inertia

1 oz in.^{2} = 1.829×10^{-5} kg m^2
1 kg m^2 = 54,674 oz in.^{2}

The swing weight of a bat for rotation about an axis in the handle, 6 in. from the knob, is given in oz in.^{2} to a good approximation by $I_6=(0.76B/L - 0.2)ML^2$ where M is the bat mass in oz, L is bat length in inches, and B is the balance distance from the center of mass to the knob in inches.

Spin

1 rev s^{-1} = 60 rpm = 6.28 rad s^{-1}
1,000 rpm = 16.67 rev s^{-1} = 104.7 radian s^{-1}
3,000 rpm = 50 rev s^{-1} = 314.16 radian s^{-1}

Force

1 Newton (N) = 0.2248 lb 1 lb = 4.448 N (lb here is a unit of force)

In the USA 1 lb can be a unit of mass or force in everyday use.

Pressure

Standard air pressure at sea level = 1013.2 millibars = 101.32 kPa = 14.70 psi
1 millibar = 100 Pa   1 Pa = 1 N m^{-2} 1 psi = 6,895 N m^{-2}

Temperature

0°C = 32°F = 273.1°K

To convert a temperature, T_C in degrees Celsius to degrees Fahrenheit (T_F) use $T_F = 32 + 1.8T_C$. For example, 0°C = 32°F, 30°C = 86°F and 40°C = 104°F.

Stiffness

1 N m^{-1} = 0.00571 lb $in.^{-1}$ 1 lb $in.^{-1}$ = 175.13 N m^{-1}

The dynamic stiffness of a baseball or softball (when compressed rapidly) is typically about 1,000,000 N m^{-1} = 5,710 lb $in.^{-1}$ or more. The static stiffness, when compressed slowly, is about 1,500 lb $in.^{-1}$

Energy

1 J = 0.7376 ft–lb   1 ft–lb = 1.356 J

Index